the analysis of
linear
integral
equations

McGraw-Hill Series in Modern Applied Mathematics
Menahem M. Schiffer *Consulting Editor*

COCHRAN: The Analysis of Linear Integral Equations
FLUGGE-LOTZ: Discontinuous and Optimal Control
LEITMANN: An Introduction to Optimal Control
TRUESDELL: Rational Thermodynamics

The Analysis of Linear Integral Equations

James Alan Cochran
Bell Telephone Laboratories

McGraw-Hill Book Company   New York   St. Louis   San Francisco   Düsseldorf
Johannesburg   Kuala Lumpur   London   Mexico   Montreal   New Delhi
Panama   Rio de Janeiro   Singapore   Toronto

The Analysis of
Linear
Integral
Equations

Library of Congress Catalog Card Number 79-159301
07-011527-3

1234567890KPKP7987654321

This book was set in Times Roman by Textbook Services, Inc., and printed and bound by Kingsport Press, Inc. The editors were Howard S. Aksen, Lee W. Peterson, and Joan Stern. Sally Ellyson supervised production.

Contents

Preface

This book is designed to present a comprehensive treatment of the theory and analysis of linear integral equations. As such it is intended primarily for applied mathematicians, mathematical physicists, and others with strong mathematical interests. It is anticipated that the book will appeal to student and researcher alike. For the former it should provide a thorough grounding in fundamentals while hopefully awakening in him an appreciation of the scope and beauty of the subject matter; for the latter it should furnish a valuable resource (with an extensive bibliography) in which a significant body of extant knowledge is delineated and frontiers probed.

The format and content of the book can be traced to many sources. Courses and research at Stanford University and extended lecture series on several occasions at Bell Telephone Laboratories have been fundamental to the particular selection of material to be covered as well as to the approach taken and emphasis given in various parts of the final presentation. The advice of students and colleagues alike has also occasioned any number of significant inclusions and/or alterations. In this latter regard, it is a pleasure to acknowledge Dale Swann's special contributions to Chapters 15 and 16. In a series of unpublished manuscripts beginning in 1965, Swann suggested the foundations for much of the presentation as it appears in those chapters. It seems appropriate here especially to recognize also the authors of several earlier accounts on the general subject of integral equations, notably Riesz and Sz.-Nagy, Smithies, and Mikhlin, lest the various acknowledgments appearing in the text fail to convey sufficiently the impact that these men's work has undoubtedly had on our treatment of the subject matter.

Throughout the evolution of this book, we have been guided by a sincere belief that although a proper theoretical setting for linear integral equations may be found in general operator theory, there is much to be gained by avoiding unnecessary encumbrances and the heavy reliance on functional analysis that characterize several other works in this field. In keeping with this approach, therefore, although most of our results are equally valid for instance for completely continuous operators on abstract separable Hilbert spaces, attention is directed primarily to Fredholm integral equations of the second kind with square-integrable or Hilbert-Schmidt kernels with two independent real variables. Even so, a certain level of mathematical sophistication on the part of the reader must be assumed. This should include background in linear algebra and real and complex analysis, or what might be most succinctly stated as "a working knowledge of the foundations of applied mathematics." In order to be more specific, however, and to provide a convenient source to which interested readers may refer as needed, most of the relevant fundamental concepts and results have been collected in Appendix B.

The book divides naturally into three parts. The first six chapters deal with what is often termed *the classical theory of general linear integral equations*. With selected additions from the later portions of the text, this material makes a good one-semester advanced undergraduate-graduate course. The next seven chapters are concerned with Hermitian kernels, the existence (both theoretical and constructive) of their characteristic values and characteristic functions, and various means for the estimation and approximation of such kernels, values, and functions. Although specific procedures are not by and large examined from a numerical point of view, the important techniques are all investigated and error bounds are often given. A chapter on perturbation theory is included as well.

In the last five chapters other special kernels are considered. The goal here is to describe what can be said concerning the nature of integral equations and/or their solution in cases wherein the given kernel has some non-Hermitian, but otherwise beneficial, properties or structure. Normal kernels, symmetrizable kernels, nuclear kernels, positive kernels, and differentiable kernels, for example, are all treated. Another term of the above-mentioned course can be built easily around topics selected from these last portions of the text, and it is hoped that teachers will welcome the expanse of material which this book provides.

In a very real sense, this book grew out of lectures delivered over the past six years in the Out-of-Hours Course Program carried on at Bell Telephone Laboratories. Naturally I am greatly indebted to them for their support in this entire undertaking. Special thanks are also due Miss Lynn Esposito for her marvelous typing assistance: Mrs. Barbara Guinsler for her help with corrections and indexing; James Haidt, Robert O' Malley, Beresford Parlett, my understanding wife, Katherine for her proofreading assistance; and Morton Schwartz for critical readings of significant portions of the manuscript; and to many other students, colleagues, and friends who contributed so much in various ways toward bringing this book to print.

Finally, let me acknowledge the tremendous debt owed Menahem M. Schiffer. It was from this distinguished scholar and teacher that I learned the first rudiments of integral equations theory. Moreover, he later suggested the possibility of this book, encouraged its writing, advised on its content, invited me to spend the 1968–1969 academic year as Visiting Professor of Mathematics at Stanford University in order to complete the manuscript, and welcomed the finished product as part of the McGraw-Hill Applied Mathematics Series, for which he is consulting editor. As a token of my respectful thanks, therefore, I hereby dedicate this book to Professor Schiffer, a truly "uncommon" man.

JAMES ALAN COCHRAN

**the analysis of
linear
integral
equations**

chapter one

An Introduction to General Linear Integral Equations

An equation in which an unknown function appears in a linear fashion under one or more signs of integration is called a *linear integral equation*. Although the analysis of certain specific such equations dates back to the early 1800s, the general theory, it must be said, belongs to the twentieth century, having its origins in two papers by the Swedish geometer Fredholm (1900 and 1903) and the early work of Volterra, Hilbert, and E. Schmidt. In the intervening 60+ years the names of numerous other investigators have become associated with various aspects of the subject matter, and we shall be discussing many of these efforts in succeeding chapters. To a substantial degree much of the significant research over the years appears to have been motivated by the fundamental desire to analyze or otherwise understand various physical processes. It seems appropriate, therefore, that we begin our study of integral equations by illustrating how such equations arise rather naturally in various practical situations.

1.1 EXAMPLES OF INTEGRAL EQUATIONS IN SIMPLE PRACTICAL PROBLEMS

Consider the linear ordinary differential equation

$$\frac{d\phi}{dx} = A(x)\phi + B(x) \tag{1}$$

for real values of the independent variable x in the finite interval $[a, b]$. If this interval is equipartitioned with points $a = x_0 < x_1 < x_2 < \cdots < x_{n-1} < x_n = b$, designating the spacing $x_\nu - x_{\nu-1}$ by h and $\phi(x_\nu)$ by ϕ_ν, then (1) may be approximated by the set of difference equations

$$\frac{1}{h}(\phi_\nu - \phi_{\nu-1}) = A(x_\nu)\phi_{\nu-1} + B(x_\nu) \qquad \nu = 1, 2, \ldots, n \tag{2}$$

The problem of solving (2) can be viewed as a special case of the problem of solving a set of n linear algebraic equations in n unknowns:

$$\frac{1}{n}\sum_{\mu=1}^{n} A_{\nu\mu}\phi_\mu = B_\nu \qquad \nu = 1, 2, \ldots, n$$

As n tends to infinity ($h \to 0$), this system goes over into its continuous analogue

$$\int_a^b A(x, y)\phi(y)\, dy = B(x) \qquad a \le x \le b \tag{3}$$

a linear *Fredholm integral equation of the first kind*.

As a second example let us consider a generalized optical experiment in which light from a finite one-dimensional source passes through appropriate optical "hardware" to form an image or illuminate a receiver. Such a situation may be described typically by the integral equation

$$R(x) = \int_a^b K(x - y)S(y)\, dy \qquad a \le x \le b \tag{4}$$

in which S and R are, in essence, the transmitted and received optical intensities, respectively. For the case of a slit in an otherwise opaque screen separating source and receiver, the *kernel* K takes the form

$$K(x - y) = C\,\frac{\sin^2 \alpha(x - y)}{(x - y)^2}$$

It is of physical interest to inquire whether, for a given *transformation function K*, certain objects can essentially reproduce themselves; i.e., do solutions of the homogeneous *Fredholm integral equation of the second kind*

$$S(x) = \lambda \int_a^b K(x - y)S(y) \, dy \qquad a \le x \le b \tag{5}$$

exist?

Finally, let us examine the problem of determining the form assumed by the classical tightly stretched flexible string of length L under the influence of an external force.

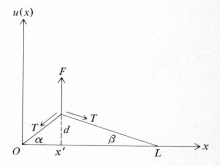

If, as indicated in the figure, a force F is applied at the point $x = x'$ causing a displacement d there, then the displacement u at any other point is given by

$$u(x) = \begin{cases} \dfrac{d}{x'} x & 0 \le x \le x' \\[3mm] \dfrac{d}{L - x'} (L - x) & x' \le x \le L \end{cases} \tag{6}$$

In terms of the tension T within the string, the equilibrium conditions for small displacements are

$$T \cos \alpha \doteq T \cos \beta$$

and

$$F = T \sin \alpha + T \sin \beta$$

$$\doteq T(\tan \alpha + \tan \beta)$$

$$= T \frac{Ld}{x'(L - x')}$$

If we replace d in (6) by its value as given by this last relation, the displacement equation becomes

$$u(x) = \frac{F}{T} \, g(x,x')$$

where
$$g(x,x') = \begin{cases} \dfrac{x(L - x')}{L} & x \leq x' \\[2mm] \dfrac{x'(L - x)}{L} & x' \leq x \end{cases}$$

is the Green's function or impulse response associated with this problem [see Courant and Hilbert (1953), Friedman (1956), Tricomi (1957), or Garabedian (1964) for background information concerning Green's functions; also see Appendix A]. More generally, when a continuous load is applied along the string with a force per unit length at the point x' equal to $F(x')$, the displacement may be expressed as

$$u(x) = \frac{1}{T} \int_0^L g(x,x')F(x') \, dx' \tag{7}$$

If a desired displacement $u(x)$ is given, this relation becomes an integral equation of the first kind for the force function F.

With a vibrating string, there is associated an inertial force per unit length given by

$$I(x,t) = -\mu(x) \frac{\partial^2 u(x,t)}{\partial t^2}$$

where $\mu(x)$ is the linear mass density at the point x. To describe equilibrium (uniform) motion we can use (7) with F replaced by I or, equivalently,

$$u(x,t) = -\frac{1}{T} \int_0^L g(x,x')\mu(x') \frac{\partial^2 u(x',t)}{\partial t^2} \, dx' \tag{8}$$

Given $\mu(x)$ this is an *integro-differential equation* for the displacement $u(x,t)$. Under the periodicity assumption associated with simple harmonic motion

$$u(x,t) = \varphi(x) \sin \omega t$$

we have
$$\varphi(x) = \frac{\omega^2}{T} \int_0^L g(x,x')\mu(x')\varphi(x') \, dx'$$

an integral equation of the second kind for $\varphi(x)$.

1.2 CLASSIFICATION OF EQUATIONS

In the foregoing examples we have encountered essentially two types of linear integral equations:

$$\psi(x) = \int_a^b K(x,y)\varphi(y)\,dy \tag{1}$$

and

$$\varphi(x) = f(x) + \lambda \int_a^b K(x,y)\varphi(y)\,dy \tag{2}$$

where $K(x,y)$, $\psi(x)$, and $f(x)$ are known functions and λ is a complex-valued parameter. Equations of the first type are termed *Fredholm integral equations of the first kind* for the unknown function $\varphi(x)$. Equations of the second type are called *Fredholm integral equations of the second kind* for $\varphi(x)$. If $f(x) \equiv 0$, (2) becomes a *homogeneous* integral equation.

In general, homogeneous Fredholm equations of the second kind have only the trivial solutions $\varphi(x) \equiv 0$. For certain critical values of λ, however, there may exist nontrivial solutions. Such values are called *characteristic values* and the corresponding φ's *characteristic functions*.[†]

The characteristic values and characteristic functions associated with a given equation depend upon the *kernel* $K(x,y)$ and the domain of integration. Although in this book we shall be concerned only with functions of real variables whose domains are essentially restricted to intervals of the real line, such need not be the case more generally. Indeed, it is worth noting that these domains may, for example, be curves or surfaces. Moreover, they may lie in spaces of any dimension and may be either bounded or unbounded. For Fredholm equations, the domain is fixed and constitutes the range of the independent variable x. The differential dy stands for the volume element or any other appropriate measure in this domain.

$K(x,y)$, $\psi(x)$, and $f(x)$ will all usually be endowed with certain specific smoothness properties, and we shall be interested in solutions of (1) and (2) which themselves have similar properties. These properties will be enumerated in each individual instance. In various applications the

---

[†] It should be noted that normally in integral-equation theory the eigenparameter λ is placed on the same side of the equation as the integral operator. This is in contrast to the practice in matrix theory where such equations generally have the form $Au = \lambda u$. We shall use the term *characteristic value* uniformly in our work and shall reserve the term *eigenvalue*, then, only for those rare instances when the integral equation has the eigenparameter placed opposite the operator and the *reciprocals* of the characteristic values are the quantities of interest.

kernel $K(x,y)$ vanishes for values of y outside a region depending upon x. If $K(x,y) \equiv 0$ for $y > x$, the equation is usually termed a *Volterra integral equation* in recognition of the role played in the study of this type of equation by the Italian mathematician Vito Volterra.

There are natural extensions to nonlinear integral equations, and the current technical literature shows considerable effort in this field [see the collection of articles in Anselone (1964), for example]. Unfortunately, however, for reasons of space, we shall by and large have to leave the consideration of such work to the interested researcher and restrict our attention to the linear case.

In this book, moreover, we shall concentrate for the most part on the theory and analysis of integral equations of the second kind.† Mathematical applications and numerical examples will be used from time to time to amplify the discussion. For a greater array of applications to physical problems, on the other hand, the reader may wish to consult Frank and von Mises (1937), Tricomi (1957), Mikhlin (1964, part 2), or Pogorzelski (1966). Similarly, for far more detailed treatment of various aspects of the numerical solution of integral equations we recommend references such as Schmeidler (1950), Bückner (1952), Fox and Goodwin (1953), Lonseth (1954), Kantorovich and Krylov (1958), Collatz (1960), P.I.C.C. (1960), Noble (1964), Mikhlin and Smolitskiy (1967), and Keller (1968).

1.3 ABEL'S INTEGRAL EQUATION

To conclude this introductory chapter, we turn to the early history of integral equations. Although Fourier in 1811 had already in essence solved the integral equation

$$F(x) = \frac{1}{\sqrt{2\pi}} \int_{-\infty}^{\infty} e^{ixy} f(y) \, dy \tag{1}$$

with the inversion formula

$$f(y) = \frac{1}{\sqrt{2\pi}} \int_{-\infty}^{\infty} e^{-ixy} F(x) \, dx \tag{2}$$

[see Fourier (1888)], probably the first significant analysis of a problem by integral-equation techniques was performed in 1823 by Abel in his research on a generalization of the *tautochrone* problem.

---

† Integral equations of the first kind are considered in Secs. 1.3, 2.3, 5.8, 18.3, and 18.4. Additional material may be found in chap. 2 of Schmeidler (1950), sec. 8.3 of Morse and Feshbach (1953), and various parts of Tricomi (1957), for example.

Suppose that a point mass m, under the influence of gravity, follows a continuously decreasing curve $y(x)$ in the (x,y) plane from an initial height $y_0 > 0$ down to the x axis ($y = 0$). Abel asked essentially the following question: Given a monotone time function $T = f(y)$, which represents the time necessary for the point mass to descend from the height y, what curve $y(x)$ could give rise to this function?

We know that $v^2 = (ds/dt)^2 = 2g(y_0 - y)$ for $y \leq y_0$ where g is the constant acceleration due to gravity and s is the arc-length parameter. Therefore,

$$T = \int_0^T dt = \int_0^s \frac{ds}{\sqrt{2g(y_0 - y)}}$$

$$= -\frac{1}{\sqrt{2g}} \int_0^{y_0} \frac{(ds/dy)\,dy}{\sqrt{y_0 - y}}$$

The problem thus reduces to: Given $f(y)$, find $\varphi(y)$ such that

$$f(y) = \int_0^y \frac{\varphi(y')}{\sqrt{y - y'}}\,dy' \tag{3}$$

[Here $\varphi(y) \equiv -(ds/dy)/\sqrt{2g}$.] Abel effected the solution of this Volterra integral equation of the first kind, which bears his name, by observing that

$$\int_0^z \frac{f(y)}{\sqrt{z - y}}\,dy = \int_0^z \frac{1}{\sqrt{z - y}} \left[\int_0^y \frac{\varphi(y')}{\sqrt{y - y'}}\,dy' \right] dy$$

$$= \int_0^z \varphi(y') \left[\int_{y'}^z \frac{dy}{\sqrt{(z - y)(y - y')}} \right] dy'$$

$$= \pi \int_0^z \varphi(y')\,dy'$$

Thence
$$\varphi(z) = \frac{1}{\pi} \frac{d}{dz} \left[\int_0^z \frac{f(y)}{\sqrt{z - y}}\,dy \right] \tag{4}\dagger$$

Of course, the question needs to be settled whether the derivative appearing in the *Abel inversion formula* (4) actually exists for the given function $f(y)$. It can be verified easily, however, that if the derivative of the integral does exist (in some appropriate sense), then $\varphi(z)$ is a solution of the *Abel integral equation* (3).

---

† See Peters (1963 and 1968) for cases in which the solution of seemingly more difficult Fredholm integral equations can be reduced to the solution of Abel's integral equation.

1.4 EXTENSIONS OF ABEL'S APPROACH

Consider the function $F_n(z)$ where the function and its first $(n-1)$ derivatives vanish at the origin while its nth derivative satisfies $F_n{}^{(n)}(z) = f(z)$. This function can be represented as the *n-tuple integral*

$$F_n(z) = \frac{1}{(n-1)!} \int_0^z (z-y)^{n-1} f(y)\, dy$$

An appropriate generalization for the case of nonintegral n is

$$F_\lambda(z) = \frac{1}{\Gamma(\lambda)} \int_0^z (z-y)^{\lambda-1} f(y)\, dy \tag{1}$$

with Re $\lambda > 0$ where $\Gamma(\lambda)$ is the familiar gamma function. In terms of this notation, the Abel inversion formula may be written as

$$\varphi(z) = \frac{1}{\sqrt{\pi}} \frac{d}{dz} F_{1/2}(z)$$

Liouville (1832), Riemann (1847), Weyl (1917), M. Riesz (1949), and others have considered *fractional integration* formulas similar to (1) in various investigations.† In the D-operator notation of ordinary differential equations we have

$$D^{-\lambda} f(z) = F_\lambda(z) \qquad\qquad \text{Re } \lambda > 0$$

and $\qquad\qquad D^\mu f(z) = D^n[D^{\mu-n} f(z)] \qquad \text{Re } (n-\mu) > 0$

Operating formally with a calculus based upon these operators, it is straightforward to show that the *generalized Abel equation*

$$f(y) = \int_0^y \frac{\varphi(y')}{(y-y')^\alpha}\, dy' \qquad 0 < \alpha < 1 \tag{2}$$

has the inversion formula

$$\varphi(z) = \frac{\sin \pi\alpha}{\pi} \frac{d}{dz} \left[\int_0^z \frac{f(y)}{(z-y)^{1-\alpha}}\, dy \right]$$

---

† See also Erdélyi and Sneddon (1962), Sneddon (1962), and Erdélyi (1968) for application of methods based upon fractional integration to the solution of dual integral equations. See Fox (1963) and Higgins (1964) for use of fractional integration in the development of integral transforms.

Tonelli (1928) has studied the conditions under which this formalism is valid [see also Liouville (1839) and Bosanquet (1930)], while Sakaljuk (1960) has analyzed still more generalized versions of (2).

PROBLEMS

1.   **a.** Classify the linear integral equation

$$\varphi(x) = x + \int_0^x (y - x)\varphi(y)\,dy$$

     b. Show that $\varphi(x) = \sin x$ satisfies this equation.
     c. Can you surmise what is the solution of

$$\varphi(x) = x - \int_0^x (y - x)\varphi(y)\,dy$$

2.   Demonstrate that $\varphi(x) = \sin \omega x$ is a solution of the vibrating string equation

$$\varphi(x) = \omega^2 \int_0^1 g(x,y)\varphi(y)\,dy \qquad 0 \le x \le 1$$

with
$$g(x,y) = \begin{cases} x(1 - y) & x \le y \\[2mm] y(1 - x) & y \le x \end{cases}$$

only if ω is an integer multiple of π.

3.   Verify Huygens' conjecture that if the time function $T = f(y)$ appearing in the generalized tautochrone problem is independent of the height, the resulting curve is a cycloid.

4.   **a.** Show that for the fractional integration formula (1.4-1)

$$D^{-\lambda}[D^{-\mu}f(z)] = D^{-(\lambda+\mu)}f(z) \qquad \text{Re } \lambda, \text{ Re } \mu > 0$$

     b. Verify that

$$D^{-\lambda}f(z) = \sum_{k=0}^{n-1} \frac{z^{\lambda+k}f^{(k)}(0)}{\Gamma(\lambda + k + 1)} + D^{-(\lambda+n)}f^{(n)}(z)$$

for all positive integers n. In this manner the fractional integration formula may be analytically continued into the region Re $\lambda \le 0$.
     c. Satisfy yourself that if $D^{-\lambda}f(z) = F_\lambda(z)$ is known, then formally

$$f(z) = D^\lambda F_\lambda(z) = \frac{1}{\Gamma(-\lambda)} \int_0^z \frac{F_\lambda(t)}{(z - t)^{\lambda+1}}\,dt$$

5. Demonstrate that

$$\varphi(z) = \frac{\sin \pi\alpha}{\pi} \left[\frac{f(0)}{z^{1-\alpha}} + \int_0^z \frac{f'(y)}{(z-y)^{1-\alpha}} \, dy \right]$$

is formally a solution of the generalized Abel equation (1.4-2).

chapter two

Fredholm Integral Equations and the Fredholm Alternative

2.1 SQUARE-INTEGRABLE FUNCTIONS AND KERNELS

Although many of the results with which we shall be concerned remain valid for functions defined on more arbitrary Lebesgue measurable sets, in the sequel we shall generally restrict our attention to complex-valued functions $f(x)$ and $K(x,y)$, the domain of whose independent variable(s) consists of either a finite interval $[a,b]$ of the real line or the corresponding square $a \leq x, y \leq b$ of the real plane. The natural context for our discussions, moreover, will be within the Hilbert space of square-integrable such functions. For convenience, we shall use the symbol $\mathfrak{L}^2$ and the term *square integrable* to denote this space and describe these functions irrespective of whether we are talking about functions of one or two independent variables. In like manner $\|\cdot\|$ will be used in both cases to designate the finite $\mathfrak{L}^2$ norms of the functions in question, that is,

$$\|f\| \equiv \left\{ \int_a^b |f(x)|^2 \, dx \right\}^{\frac{1}{2}} \qquad \|K\| \equiv \left\{ \int_a^b \int_a^b |K(x,y)|^2 \, dx \, dy \right\}^{\frac{1}{2}}$$

As discussed in Appendix B, for functions of a single variable square integrable over the same interval $[a,b]$ we can define a *complex inner product* (f,g) by the relation†

$$(f,g) \equiv \int_a^b f(x)\overline{g}(x)\ dx$$

where the bar, as usual, denotes complex conjugation. This inner product has the property that

$$(\alpha_1 f_1 + \alpha_2 f_2, g) = \alpha_1(f_1, g) + \alpha_2(f_2, g)$$
$$(f, \beta_1 g_1 + \beta_2 g_2) = \overline{\beta}_1(f, g_1) + \overline{\beta}_2(f, g_2)$$
$$(g, f) = \overline{(f, g)}$$

and is related to the norm functional by the identity

$$(f,f) = \|f\|^2$$

It can be easily established that the inner product also satisfies the following fundamental inequality, usually known as *Schwarz's inequality*:

$$|(f,g)| \leq \|f\| \cdot \|g\|$$

A result closely related to this, which we shall likewise have occasion to use often, is *Minkowski's inequality*:

$$\|f + g\| \leq \|f\| + \|g\|$$

[These two basic results are proved in any number of places. See, for example, Beckenbach and Bellman (1965), pp. 20ff ; Hardy, Littlewood, and Pôlya (1952), pp. 132 and 146; and Smithies (1962), pp. 7ff]

If a square-integrable function $K(x,y)$ of two variables serves as the kernel of an integral equation, it will be called an $\mathfrak{L}^2$ *kernel*. By Minkowski's and Schwarz's inequalities extended to functions of two variables, the set of all such square-integrable kernels forms a complex vector space that is closed under *composition*; that is, if $K_1(x,y)$ and $K_2(x,y)$ are $\mathfrak{L}^2$ kernels, so also is

$$L(x,y) \equiv K_1 K_2 \equiv \int_a^b K_1(x,z) K_2(z,y)\ dz$$

---

† If $(f,g) = 0$, we say that f and g are *orthogonal* to one another. Within $\mathfrak{L}^2$ there are an infinite number of functions orthogonal to any one (or finite collection of) function(s).

Moreover, the norm of this composite kernel satisfies the inequality

$$\|L\| = \|K_1 K_2\| \leq \|K_1\| \cdot \|K_2\|$$

in terms of the norms of the two original kernels. As a conseqence of Schwarz's inequality we also have the fact that if $K(x,y)$ is an $\mathfrak{L}^2$ kernel and $\varphi(y)$ is an $\mathfrak{L}^2$ function, the formula†

$$\psi(x) = \int_a^b K(x,y)\varphi(y)\, dy \tag{1}$$

defines an $\mathfrak{L}^2$ function $\psi(x)$ for almost all x, that is, almost everywhere in $[a,b]$. The linear transformation (1), which we shall often write symbolically as $\psi = K\varphi$, thus takes $\mathfrak{L}^2$ functions into $\mathfrak{L}^2$ functions, with

$$\|\psi\| \leq \|K\| \cdot \|\varphi\|$$

In succeeding chapters we shall be concerned for the most part with integral equations having square-integrable or $\mathfrak{L}^2$ kernels. When we speak of solutions of such equations or characteristic functions associated with such kernels, therefore, we shall generally mean the $\mathfrak{L}^2$ solutions thereof or the $\mathfrak{L}^2$ characteristic functions belonging thereto. In the few cases where the kernel has additional smoothness (e.g., continuity), however, we shall be interested in the effect of this supplementary behavior on the characteristic functions and/or similar quantities. In this regard, it is important to appreciate that kernels which are *equivalent*, i.e., differ at most by an $\mathfrak{L}^2$ function of vanishing norm, have equivalent characteristic functions and lead to equivalent solutions of integral equations in which they appear (see Prob. 1).

2.2 GENERAL REMARKS ABOUT KERNELS OF FINITE RANK

As we shall see shortly, Fredholm integral equations with kernels that can be expressed as *finite* sums of products of functions of x alone by functions of y alone can be solved simply by reduction to systems of linear algebraic equations. Such kernels are called *degenerate* kernels or kernels of *finite rank* [Stakgold (1967), among others, terms these *separable* kernels], and have the form

$$K(x,y) = \sum_{\nu=1}^{n} A_\nu(x)\overline{B}_\nu(y) \qquad a \leq x,y \leq b \tag{1}$$

---

† Throughout this book, equalities and inequalities between functions should generally be understood as holding in the "almost everywhere" sense.

where each of the sets $\{A_\nu\}$ and $\{B_\nu\}$ is *linearly independent* (with respect to equivalence) over the interval $[a,b]$. (If the $\{A_\nu\}$ or $\{B_\nu\}$ are not linearly independent, they can be reduced to this case trivially.) The unique integer n, which designates the number of terms appearing in the expression (1) for the kernel, is called the *rank* of K.†

Given an arbitrary finite set of $\mathfrak{L}^2$ functions $\{A_\nu\}$, it is important, therefore, to be able to test for linear dependence. One classical method involves inspection of the norm of

$$A(x) = \sum_{\nu=1}^{n} \alpha_\nu A_\nu(x)$$

where the α_ν are arbitrary complex constants. [We recall that the A_ν are linearly independent if and only if (iff) $A(x) \equiv 0$ implies $\alpha_\nu = 0$, $\nu = 1, 2, \ldots, n$.] In this case, taking the inner product of A with itself, we see that the relation

$$Q[\alpha] \equiv \|A\|^2 = \sum_{\mu=1}^{n} \sum_{\nu=1}^{n} (A_\mu, A_\nu)\alpha_\mu \overline{\alpha_\nu}$$

defines a *nonnegative definite* quadratic form in the α_ν with a Hermitian coefficient matrix $((A_\mu, A_\nu))$. We know then that there exists an appropriate linear transformation of coordinates under which Q can be rewritten as

$$Q[\alpha] = \sum_{\nu=1}^{n} \gamma_\nu |\beta_\nu(\alpha)|^2$$

where the γ_ν are the nonnegative eigenvalues of this coefficient matrix. It follows readily that if all of these eigenvalues are strictly positive, the A_ν must be linearly independent. On the other hand, if at least one of the γ's is zero, the A_ν's are linearly dependent.

The determinant of the coefficient matrix is the so-called *Gram determinant* and satisfies ‡

$$\det((A_\mu, A_\nu)) = \prod_{\nu=1}^{n} \gamma_\nu \geq 0$$

---

† Another appropriate expression for a kernel of finite rank (at most n) is

$$K(x,y) = \sum_{\mu,\nu=1}^{n} (K\psi_\nu, \psi_\mu)\psi_\mu(x)\overline{\psi_\nu}(y)$$

where $(\psi_\mu, \psi_\nu) = \delta_{\mu\nu}$, the Kronecker delta. In subsequent chapters, we shall have occasion to use extensively this form of representation.

‡ The concept of a nonnegative Gram determinant may be viewed as a generalization of the Schwarz inequality since the latter follows in the special case $n = 2$.

The linear dependence or independence of the set $\{A_\nu\}$ may also be decided, therefore, by evaluating the Gram determinant. If it is (non)vanishing, the A_ν are linearly (in)dependent.

It should be noted that kernels of finite rank for which the $A_\nu(x)$ and $B_\nu(y)$ are $\mathfrak{L}^2$ functions are necessarily $\mathfrak{L}^2$ kernels. We have, in fact, from (1) and Schwarz's inequality that

$$\|K\|^2 = \sum_{\mu=1}^{n} \sum_{\nu=1}^{n} (A_\mu, A_\nu)(\overline{B_\mu, B_\nu})$$

$$\leq \sum_{\mu=1}^{n} \sum_{\nu=1}^{n} \|A_\mu\| \cdot \|A_\nu\| \cdot \|B_\mu\| \cdot \|B_\nu\|$$

$$= \left(\sum_{\nu=1}^{n} \|A_\nu\| \cdot \|B_\nu\| \right)^2$$

2.3 EQUATIONS OF THE FIRST KIND WITH DEGENERATE KERNELS

Let us now consider the solution of equations of the form

$$\psi(x) = \int_a^b K(x,y)\varphi(y)\, dy \qquad a \leq x, y \leq b \tag{1}$$

where $\psi(x)$ is a given $\mathfrak{L}^2$ function and $K(x,y)$ is a degenerate kernel of rank n, with square-integrable factors. Using the representation (2.2-1), we see that

$$\psi(x) = \sum_{\nu=1}^{n} A_\nu(x) \int_a^b \overline{B_\nu(y)}\varphi(y)\, dy$$

from which it is clear that a *necessary* condition for a solution of (1) to exist is that $\psi(x)$ be a linear function of the $\{A_\nu\}$. We assume, therefore, that

$$\psi(x) = \sum_{\nu=1}^{n} \alpha_\nu A_\nu(x) \tag{2}$$

It follows then, since the $A_\nu(x)$ are linearly independent, that

$$\alpha_\nu = \int_a^b \overline{B_\nu(y)}\varphi(y)\, dy = (\varphi, B_\nu) \qquad \nu = 1, 2, \ldots, n \tag{3}$$

Thus our actual problem, in the case of a degenerate kernel and with the assumption (2), has been reduced to the solution of the subsidiary system of n integral equations (3) in which the α_ν are known complex constants and the $B_\nu(y)$ are known functions.

We first note that *at least one* solution of (3) exists. To see this let φ be of the form

$$\varphi(y) = \sum_{\mu=1}^{n} \beta_\mu B_\mu(y)$$

where the β_μ are yet to be determined. Substitution in (3) yields

$$\alpha_\nu = \sum_{\mu=1}^{n} \beta_\mu (B_\mu, B_\nu) \qquad \nu = 1, 2, \ldots, n$$

or in matrix notation $\alpha = B\beta$ where $B = \left(\overline{(B_\nu, B_\mu)}\right)$. In view of the linear independence of the $B_\mu(y)$, B is a positive-definite Hermitian matrix, and this set of equations for β therefore has a solution.

Actually there is an *infinity* of solutions of the original system (3). For let

$$\tilde{\varphi}(y) = \varphi(y) + \omega(y)$$

where φ is the known solution and $\omega(y)$ has been chosen orthogonal to all of the $B_\nu(y)$. Then

$$\begin{aligned}
\alpha_\nu &= (\varphi, B_\nu) \\
&= (\tilde{\varphi}, B_\nu) - (\omega, B_\nu) \qquad \nu = 1, 2, \ldots, n \\
&= (\tilde{\varphi}, B_\nu)
\end{aligned}$$

and thus $\tilde{\varphi}$ is also a solution to (3) for any such ω. The original conclusion follows immediately since it is known that there exists an infinite number of nontrivial $\mathfrak{L}^2$ functions ω satisfying $(\omega, B_\nu) = 0$, $\nu = 1, 2, \ldots, n$.

Before we summarize the above results, let us inspect for a moment the homogeneous equation which is *adjoint* to (1), namely†

$$0 = \int_a^b \overline{K(y,x)}\,\Omega(y)\,dy$$

---

† The *adjoint* (or conjugate transpose) of an $\mathfrak{L}^2$ kernel $K(x,y)$ is given by the relation

$$K^*(x,y) = \overline{K(y,x)}$$

It follows from this that $(\phi, K\psi) = (K^*\phi, \psi)$ for every pair of $\mathfrak{L}^2$ functions ϕ and ψ. In more general settings, this latter relation is often used as the definition of the adjoint transformation associated with a given linear operator on an abstract Hilbert space.

For degenerate K of the form (2.2-1) we obtain

$$\sum_{\nu=1}^{n} B_{\nu}(x)(\Omega, A_{\nu}) = 0$$

It is clear that there exists an infinite number of $\mathcal{L}^2$ solutions of this equation, i.e., those $\mathcal{L}^2$ functions $\Omega(y)$ that are orthogonal to all of the A_{ν}. (Since the B_{ν} are linearly independent, these must be the only $\mathcal{L}^2$ solutions.) Moreover, in view of the assumed form of $\psi(x)$ given originally by (2), it follows that

$$(\Omega, \psi) = 0$$

for all such Ω.

In summary, then, for the equations

$$\psi(x) = \int_{a}^{b} K(x,y)\varphi(y)\,dy \tag{4}$$

$$0 = \int_{a}^{b} K(x,y)\omega(y)\,dy \tag{5}$$

$$0 = \int_{a}^{b} K^*(x,y)\Omega(y)\,dy \tag{6}$$

where K is an $\mathcal{L}^2$ kernel of finite rank, we can draw the following conclusions:

1. There exist infinitely many nontrivial $\mathcal{L}^2$ solutions of both (5) and (6).
2. A necessary and sufficient condition (nasc) for (4) to have an $\mathcal{L}^2$ solution is that the $\mathcal{L}^2$ function ψ be orthogonal to all the $\mathcal{L}^2$ solutions of the homogeneous adjoint equation (6).
3. If there exists a solution φ to (4), it is obviously *not* unique since $\varphi + \omega$ would also be a solution, where ω is any nontrivial solution of (5).

2.4 EQUATIONS OF THE SECOND KIND WITH DEGENERATE KERNELS

We move now to a consideration of equations of the form

$$\varphi(x) = f(x) + \lambda \int_{a}^{b} K(x,y)\varphi(y)\,dy \qquad a \le x, y \le b \tag{1}$$

where the *free term* $f(x)$ is a prescribed $\mathfrak{L}^2$ function, λ is a given complex constant, and again $K(x,y)$ is a degenerate kernel of rank n given by

$$K(x,y) = \sum_{\nu=1}^{n} A_\nu(x)\overline{B}_\nu(y) \tag{2}$$

with the A_ν and B_ν square integrable and linearly independent. Substituting the series expansion (2) into (1), we have

$$\varphi(x) = f(x) + \lambda \sum_{\nu=1}^{n} (\varphi, B_\nu) A_\nu(x) \tag{3}$$

$$= f(x) + \sum_{\nu=1}^{n} \alpha_\nu A_\nu(x) \tag{3'}$$

Solution of (1) and hence (3), therefore, is completely equivalent to determining the appropriate α_ν in (3'). With this in mind we formally replace φ where it occurs in (3) by its representation (3'). Thus

$$\sum_{\nu=1}^{n} \alpha_\nu A_\nu(x) = \lambda \sum_{\nu=1}^{n} \left[(f, B_\nu) + \sum_{\mu=1}^{n} \alpha_\mu (A_\mu, B_\nu) \right] A_\nu(x)$$

which, in view of the linear independence of the A_ν, implies that

$$\alpha_\nu = \lambda(f, B_\nu) + \lambda \sum_{\mu=1}^{n} \alpha_\mu (A_\mu, B_\nu) \qquad \nu = 1, 2, \ldots, n \tag{4}$$

Writing β as the column vector $(\beta_\nu) \equiv \left((f, B_\nu)\right)$ and C as the matrix

$$(C_{\nu\mu}) \equiv (\delta_{\nu\mu} - \lambda c_{\nu\mu}) \equiv \left(\delta_{\nu\mu} - \lambda(\overline{B_\nu, A_\mu}) \right)$$

where $\delta_{\nu\mu}$ is the *Kronecker* delta, we can express compactly in matrix notation the system of algebraic equations (4) as

$$C\alpha = \lambda\beta \tag{5}$$

If the matrix C is nonsingular, then this equation has a unique solution α given in terms of the inverse matrix C^{-1} by

$$\alpha = \lambda C^{-1}\beta$$

or, equivalently, in terms of the cofactors $C^{\mu\nu}$ of $C_{\mu\nu}$ in C by Cramer's rule

$$\alpha_\nu = \frac{\lambda}{\det C} \sum_{\mu=1}^{n} C^{\mu\nu} \beta_\mu \qquad \nu = 1, 2, \ldots, n$$

This, in turn, implies that the $\mathfrak{L}^2$ function

$$\varphi(x) = f(x) + \frac{\lambda}{\det C} \sum_{\mu=1}^{n} \sum_{\nu=1}^{n} C^{\mu\nu} \beta_\mu A_\nu(x) \tag{6}$$

uniquely satisfies the original Fredholm equation (1).

The existence of the above solution depends upon determining the α_ν, that is, inverting the matrix C. This can be done if

$$D_n(\lambda) \equiv \det C = \det (\delta_{\nu\mu} - \lambda c_{\nu\mu}) \neq 0$$

It is important to note in this regard that

1. $D_n(\lambda)$ is a polynomial in λ of at most the nth degree.
2. $D_n(0) = \det (\delta_{\nu\mu}) = 1$, so that $D_n(\lambda) \not\equiv 0$.

Thus there exist *at most*† n distinct values of λ for which $D_n(\lambda) = 0$. It follows in these cases that the homogeneous versions of (5) and hence of (1) have nontrivial solutions, and therefore such values of λ may be rightly termed *characteristic values* (cv's). The corresponding nontrivial $\mathfrak{L}^2$ solutions of (1) are designated the *characteristic functions* (cf's) of K. The number of linearly independent such cf's associated with each characteristic value is equal to the *nullity* of the matrix C, that is, the dimension of its null space for that value of λ. Together with the zero function, they form a finite-dimensional vector subspace of $\mathfrak{L}^2$ called the *characteristic subspace* belonging to λ. The dimension of this subspace is often termed the *rank* of the cv λ.

All values of λ for which $D_n(\lambda)$ is nonvanishing are termed *regular values*. For each such λ the algebraic equations (4) and (5) can be solved for the α_ν and the unique $\mathfrak{L}^2$ solution of the integral equation (1) can be expressed as in (6).

Results completely analogous to those given above are also valid for the adjoint equation with kernel $K^*(x,y) = \overline{K(y,x)}$ since $D_n^*(\overline{\lambda}) =$

---

† There need not be any, as is the case with the kernel $K(x,y) = \sin x \cos y$ for $0 \leq x, y \leq \pi$.

$\overline{D_n(\lambda)}$. It follows, too, that if Ω is a nontrivial $\mathfrak{L}^2$ function satisfying the homogeneous adjoint equation

$$\Omega(x) = \overline{\lambda} \int_a^b K^*(x,y)\Omega(y) \, dy \tag{7}$$

then taking its inner product with any $\mathfrak{L}^2$ solution φ of (1) yields

$$\begin{aligned}
(\Omega,\varphi) &= (\Omega,f) + (\Omega,\lambda K\varphi) \\
&= (\Omega,f) + (\overline{\lambda}K^*\Omega,\varphi) \\
&= (\Omega,f) + (\Omega,\varphi)
\end{aligned}$$

For an $\mathfrak{L}^2$ solution of the general inhomogeneous equation (1) to exist when λ is a characteristic value, therefore, it is *necessary* that the inner product (Ω,f) vanish for every $\mathfrak{L}^2$ solution of the homogeneous adjoint equation. That this condition is also *sufficient* follows directly from a known result in matrix theory.† (As is to be expected, there is an analogous necessary and sufficient condition relating to the existence of solutions of the *in*homogeneous adjoint equation.)

2.5   THE FREDHOLM ALTERNATIVE

This brings us to the classical restatement of the main results obtained above, first provided by Fredholm (1903) and subsequently named in his honor:

THEOREM 2.5 (THE FREDHOLM ALTERNATIVE FOR KERNELS OF FINITE RANK)

Either λ is a regular value, in which case the inhomogeneous integral equations

$$\varphi = f + \lambda K\varphi \qquad \psi = g + \overline{\lambda}K^*\psi \tag{1}$$

with degenerate $\mathfrak{L}^2$ kernels $K(x,y)$ and $K^*(x,y) \equiv \overline{K(y,x)}$ have unique $\mathfrak{L}^2$ solutions φ, ψ for any given $\mathfrak{L}^2$ functions f, g; ... *or* λ

---

† For a concise statement of the fundamental theorems of the theory of linear algebraic equations, application of which leads directly to the so-called Fredholm alternative for linear integral equations of the second kind, see Courant and Hilbert (1953, sec. 1.1-3), for example.

is a characteristic value, in which case the homogeneous equations

$$\omega = \lambda K \omega \qquad \Omega = \overline{\lambda} K^* \Omega \tag{2}$$

have nontrival $\mathfrak{L}^2$ solutions, with the number of linearly independent such solutions being the same finite number for both equations.

If the latter situation prevails, the inhomogeneous equations (1) have (nonunique) $\mathfrak{L}^2$ solutions if and only if f is orthogonal to every $\mathfrak{L}^2$ solution Ω and g is orthogonal to every $\mathfrak{L}^2$ solution ω of the respective homogeneous equations (2).

We shall see in the next chapter that the alternative expressed by this theorem actually is more generally valid and holds even for equations with arbitrary $\mathfrak{L}^2$ kernels.

PROBLEMS

1. Assume that the two $\mathfrak{L}^2$ kernels K and L are equivalent, i.e.,

$$K = L + N$$

where $\|N\| = 0$.
   a. Show that every characteristic function of K is a characteristic function of L and conversely.
   b. In similar fashion, verify that an $\mathfrak{L}^2$ function ϕ is a solution of the equation

$$\phi = f + \lambda K \phi$$

   iff it is also a solution of the equation

$$\phi = f + \lambda L \phi$$

2. Show that if either of the sets $\{A_\nu\}, \{B_\nu\}$ appearing in (2.2-1) is not linearly independent, it can be reduced easily to this case.
3. a. Rewrite the kernel $K(x,y) = 1 + y + 3xy$ for $-1 \le x, y \le 1$ in the form appropriate for a kernel of finite rank.
   b. Verify that the functions $\{A_\nu(x)\}$ and $\{B_\nu(y)\}$ are linearly independent.
   c. Solve

$$x - 1 = \int_{-1}^{1} (1 + y + 3xy)\phi(y)\, dy$$

4. Given a finite set $\{B_\nu\}$ of $\mathfrak{L}^2$ functions, satisfy yourself that there does indeed exist an infinite number of nontrivial $\mathfrak{L}^2$ functions ω for which $(\omega, B_\nu) = 0$, $\nu = 1, 2, \ldots, n$.

5. Prove that when $D_n(\lambda) = 0$, $(\Omega, f) = 0$ for all solutions Ω of the homogeneous adjoint equation (2.4-7) implies the existence of at least one solution to the general inhomogeneous equation (2.4-1) with degenerate kernel.

6. Let an $\mathfrak{L}^2$ kernel K have the representation (see the footnote on page 14)

$$K(x,y) = \sum_{\mu,\nu=1}^{2} k_{\mu\nu} \psi_\mu(x) \overline{\psi}_\nu(y)$$

where the square-integrable functions ψ_μ satisfy $(\psi_\mu, \psi_\nu) = \delta_{\mu\nu}$ and the coefficient matrix has the form

$$(k_{\mu\nu}) = \begin{pmatrix} \alpha & \beta \\ 0 & 0 \end{pmatrix}$$

Discuss the specific implications of the Fredholm alternative (Theorem 2.5) for this degenerate kernel.

7. a. Calculate $D_n(\lambda)$ when $K(x,y) = x + y$ for $0 \le x,y \le 1$.
   b. Solve

$$\phi(x) = x + \lambda \int_0^1 (x + y)\phi(y)\, dy$$

for arbitrary values of λ.

chapter three

Schmidt Theory and the Resolvent Kernel

3.1 THE NEUMANN SERIES

One important method of finding the general $\mathfrak{L}^2$ solution of an inhomogeneous Fredholm integral equation of the second kind involves a combination of the approach taken for degenerate kernels and Picard's method of successive approximations.[†] Let us look, then, at an application of the latter technique to the equation

$$\varphi(x) = f(x) + \lambda \int_a^b K(x,y)\varphi(y)\, dy \qquad (1)$$

We first introduce the *iterated kernels*

$$K^1(x,y) \equiv K(x,y)$$

$$K^2(x,y) \equiv KK \equiv \int_a^b K(x,z)K(z,y)\, dz$$

$$K^3(x,y) \equiv KK^2 \equiv \int_a^b K(x,z)K^2(z,y)\, dz \qquad \text{etc.}$$

[†]See Wouk (1964) for interesting historical remarks concerning this time-honored method.

defined by the recurrence formula

$$K^\nu(x,y) \equiv KK^{\nu-1} \equiv \int_a^b K(x,z)K^{\nu-1}(z,y)\,dz \qquad \nu \geq 2$$

[It is a simple matter to show that also

$$K^{\nu+\mu}(x,y) = K^\nu K^\mu = \int_a^b K^\nu(x,z)K^\mu(z,y)\,dz$$

for arbitrary positive integers ν and μ.] Successive approximations to the solution of (1) would then take the form

$$\varphi_0 = f,$$
$$\varphi_1 = f + \lambda K\varphi_0,$$
$$\vdots$$
$$\varphi_n = f + \lambda K\varphi_{n-1}$$

or, substituting successively,

$$\varphi_n = f + \sum_{\nu=1}^n \lambda^\nu (K^\nu f) \tag{2}$$

$$n \geq 1$$

$$= f + \lambda \left(\sum_{\nu=1}^n \lambda^{\nu-1} K^\nu \right) f \tag{2'}$$

It should be noted particularly that, with the terms grouped as in (2), at each new step in the substitution process we reobtain the result of the previous step plus one additional term. Clearly, this situation is often of considerable value for computational purposes.

If we let n tend to infinity in (2'), we obtain formally

$$\varphi_\infty(x) = f(x) + \lambda \int_a^b \left[\sum_{\nu=1}^\infty \lambda^{\nu-1} K^\nu(x,y) \right] f(y)\,dy$$

$$= f(x) + \lambda \int_a^b R_K(x,y;\lambda)f(y)\,dy \tag{3}$$

where, by definition,

$$R_K(x,y;\lambda) \equiv \sum_{\nu=1}^\infty \lambda^{\nu-1} K^\nu(x,y) \tag{4}$$

R_K is termed the *resolvent kernel*. Although Liouville (1837 and 1838) had used a development analogous to the series in (2), Carl Neumann (1877) in his studies in potential theory appears to have been the first to rigorously establish the convergence of these successive approximations. The infinite version of the series (2) is known as the *Liouville-Neumann* or *Neumann series*, and the expansion (4) is designated the *Neumann series for the resolvent*.

The question exists, of course, whether the series (4) converges and thence whether the representation (3) defines a solution to the equation (1). We note, however, that by Schwarz's inequality

$$\|K^2\| \leq \|K\|^2$$

and, in general,

$$\|K^\nu\| \leq \|K\|^\nu \qquad \nu \geq 2$$

For $\nu > 2$ we thus have

$$
\begin{aligned}
|\lambda^{\nu-1} K^\nu(x,y)| &= |\lambda|^{\nu-1} |KK^{\nu-2}K(x,y)| \\
&\leq |\lambda|^{\nu-1} k_1(x) \|K^{\nu-2}\| k_2(y) \\
&\leq |\lambda|^{\nu-1} \|K\|^{\nu-2} k_1(x) k_2(y)
\end{aligned}
$$

where

$$k_1(x) \equiv \left\{ \int_a^b |K(x,y)|^2 \, dy \right\}^{\frac{1}{2}} \qquad k_2(y) \equiv \left\{ \int_a^b |K(x,y)|^2 \, dx \right\}^{\frac{1}{2}}$$

This result is easily seen to be valid for $\nu = 2$ also, so that, neglecting the first term, the Neumann series for the resolvent (4) must be majorized by the series

$$|\lambda| k_1(x) k_2(y) \sum_{\nu=2}^\infty \left(|\lambda| \cdot \|K\| \right)^{\nu-2} \tag{5}$$

If $K(x,y)$ is an $\mathcal{L}^2$ kernel, the summation in (5) converges for

$$|\lambda| < \frac{1}{\|K\|} \tag{6}$$

Moreover, for square-integrable K, it follows from Fubini's theorem† that $k_1(x)$ and $k_2(y)$ exist almost everywhere and are $\mathcal{L}^2$ functions. In this case, owing to the form of the majorant (5), the Neumann series (4)

---

†Specific mention of Fubini's theorem will often be omitted and its application merely implied by the operations performed.

converges *almost* or *relatively uniformly*† and absolutely for $|\lambda| \cdot \|K\| < 1$ and defines the square-integrable resolvent kernel $R_K(x, y; \lambda)$. It is a simple matter, then, to show that, neglecting zero functions, the $\mathfrak{L}^2$ solution to the original equation (1) is given uniquely by the representation (3), whatever be the square-integrable free term $f(x)$.

In the case of kernels $K(x, y)$ continuous for $a \leq x, y \leq b$, the sufficient condition (6) is often replaced by

$$|\lambda| M(b - a) < 1 \tag{7}$$

where M is the least upper bound of $|K(x, y)|$ in the fundamental domain [see Riesz and Sz.-Nagy (1955, p. 146), for example]. This inequality, however, not only gives a result weaker than (6), but it also suffers from the defect that it has no meaning as the finite interval $[a, b]$ becomes infinitely large. ‡

It should be noted that the condition (6) is by no means a necessary one for the convergence of the Neumann series for the resolvent. For example, the degenerate kernel $K(x, y) = x$ on $0 \leq x, y \leq 1$ has a norm equal to $1/\sqrt{3}$ but a Neumann series that easily can be shown to converge for $|\lambda| < 2$. For the kernel cited in the footnote on page 19, moreover, the Neumann series reduces to one term and hence is perforce convergent for *all* finite λ. In the general case, various summability methods can be used to sum the formal Neumann series even when the eigenparameter λ is larger in absolute value than the radius of convergence of the expansion (4) [see Bückner (1948, 1949, and 1952), Bellman (1950), and Scott and Burgmeier (1969), for example]. In Kantorovich and Krylov (1958, sec. 2.2-2) several direct methods of analytic continuation of (4) are also considered.

---

† A sequence $\{F_n(x, y)\}$ of functions square integrable for $a \leq x, y \leq b$ is said to be relatively uniformly convergent to the ($\mathfrak{L}^2$) limiting function $F(x, y)$ if there exists a nonnegative $\mathfrak{L}^2$ function $P(x, y)$ such that, given $\epsilon > 0$, there is a positive integer $n_0(\epsilon)$ for which

$$|F_n(x, y) - F(x, y)| \leq \epsilon P(x, y) \qquad n \geq n_0 \qquad a \leq x, y \leq b$$

There is an analogous definition in the simpler case of functions of one independent variable [See Smithies (1962, pp. 23ff) or Moore (1910, pp. 30ff), who first introduced this notion.] For our purposes, it is important to note that a sequence $\{F_n\}$ is relatively uniformly convergent if it forms a Cauchy sequence in the same relatively uniform sense.

‡ Using Hölder's inequality, Parodi (1965) has been able to improve somewhat the inequality (7) for continuous kernels.

3.2 THE FREDHOLM IDENTITIES

In the preceding section, we defined the resolvent kernel for small λ in terms of the kernel $K(x,y)$ by the Neumann series (3.1-4). Viewed as a function of λ (for fixed x and y), this power series represents an analytic function that is *regular*, at least within the circle given by (3.1-6). As we have observed, however, its domain of regularity could be much larger. In fact, we shall see later that the resolvent kernel is, in general, a *meromorphic* function of λ throughout the entire complex λ plane.

If we just concern ourselves for the moment, though, with sufficiently small values of λ, the solution of

$$\varphi = f + \lambda K \varphi \tag{1}$$

can be given symbolically in terms of the resolvent kernel as in (3.1-3) by

$$\varphi = f + \lambda R_K(\lambda) f \tag{2}$$

It follows, then, that

$$\lambda R_K(\lambda) f = \varphi - f = \lambda K \varphi = \lambda K f + \lambda^2 K R_K(\lambda) f$$

which, in view of the arbitrariness of f, gives rise to the relation

$$R_K(\lambda) = K + \lambda K R_K(\lambda) \tag{3}$$

Similarly, since for sufficiently small λ the solution (2) is unique, then

$$\lambda K \varphi = \varphi - f = \lambda R_K(\lambda) f = \lambda R_K(\lambda) \varphi - \lambda^2 R_K(\lambda) K \varphi$$

from which we obtain

$$R_K(\lambda) = K + \lambda R_K(\lambda) K \tag{4}$$

These two equations (3) and (4)

$$R_K(x,y;\lambda) = K(x,y) + \begin{cases} \lambda \displaystyle\int_a^b K(x,z) R_K(z,y;\lambda)\, dz \\[2em] \lambda \displaystyle\int_a^b R_K(x,z;\lambda) K(z,y)\, dz \end{cases} \tag{5}$$

are called the *Fredholm identities* and are sometimes used to define the resolvent kernel [see Smithies (1962, pp. 16ff), for example]. In such cases, the resolvent kernel is taken to be the (unique) $\mathfrak{L}^2$ solution of the integral equations (5) with the given $\mathfrak{L}^2$ kernel $K(x,y)$ if such a solution exists for the particular value of λ.

It is interesting to note that the Fredholm identities also can be "derived" in a rather formal manner, using the symbolic *identity operator* I for which $Ig = gI = g$ for every function g. Equations (1) and (2) then suggest that

$$[I + \lambda R_K(\lambda)]f = \varphi = [I - \lambda K]^{-1}f$$

or
$$[I + \lambda R_K(\lambda)] = [I - \lambda K]^{-1}$$

Hence $[I - \lambda K][I + \lambda R_K(\lambda)] = I = [I + \lambda R_K(\lambda)][I - \lambda K]$

from which the desired relations (3) and (4) follow.

The Fredholm identities are fundamental to the study of integral equations. In fact, it might be said that the answers to questions of existence and uniqueness are "imbedded" in the relations (5). They are certainly valid (almost everywhere as a function of x and y) within the circle $|\lambda| \cdot \|K\| < 1$, as can be verified by direct substitution of the Neumann series representation (3.1-4). Moreover, since all of the terms in (5) are analytic in λ, these identities must actually hold throughout the whole domain of existence of the resolvent kernel in the complex λ plane. In particular, this implies that, given λ, if $R_K(x,y;\lambda)$ exists as an $\mathfrak{L}^2$ function, then by forming

$$\varphi(x) = f(x) + \lambda \int_a^b R_K(x,y;\lambda)f(y) \, dy \tag{6}$$

and using (5) it follows that $\varphi(x)$ is in $\mathfrak{L}^2$ and satisfies the original integral equation

$$\varphi(x) = f(x) + \lambda \int_a^b K(x,y)\varphi(y) \, dy \tag{7}$$

for arbitrary $\mathfrak{L}^2$ functions $f(x)$. Conversely, if there exists an $\mathfrak{L}^2$ function $\varphi(x)$ solving (7), then $f = \varphi - \lambda K\varphi$ and

$$f + \lambda R_K(\lambda)f = \varphi - \lambda K\varphi + \lambda R_K(\lambda)\varphi - \lambda^2 R_K(\lambda)K\varphi$$
$$= \varphi - \lambda[K - R_K(\lambda) + \lambda R_K(\lambda)K]\varphi$$
$$= \varphi$$

The solution $\varphi(x)$, therefore, must be given uniquely by (6).†

In view of these results and in keeping with our earlier terminology, the values of λ for which $R_K(x,y;\lambda)$ exists as an $\mathfrak{L}^2$ kernel may be called *regular values* of $K(x,y)$.‡ Note especially that $\lambda = 0$ is always a regular value since $R_K(x,y;0) = K(x,y)$.

3.3 THE NATURE OF SCHMIDT'S METHOD

In the preceding sections we have seen that, for a completely general $\mathfrak{L}^2$ kernel, the unique $\mathfrak{L}^2$ solution of the Fredholm integral equation of the second kind

$$\varphi(x) = f(x) + \lambda \int_a^b K(x,y)\varphi(y)\,dy \tag{1}$$

with square-integrable f, can be expressed as

$$\varphi(x) = f(x) + \lambda \int_a^b R_K(x,y;\lambda)f(y)\,dy \tag{2}$$

whenever the parameter λ is a regular value of K. In particular, these results are valid for all sufficiently small values of λ, in which case the resolvent kernel has the convergent power series representation

$$R_K(x,y;\lambda) = \sum_{\nu=1}^{\infty} \lambda^{\nu-1} K^\nu(x,y) \tag{3}$$

Earlier, for the degenerate kernel of rank n

$$K(x,y) = \sum_{\nu=1}^{n} A_\nu(x)\overline{B}_\nu(y)$$

we had expressed the solution of (1) as [see (2.4-6)]

$$\varphi(x) = f(x) + \frac{\lambda}{D_n(\lambda)} \sum_{\mu=1}^{n} \sum_{\nu=1}^{n} C^{\mu\nu} A_\nu(x)(f,B_\mu) \tag{4}$$

---

†For some applications, including various problems in optimal filtering and radiative-transfer theory, an alternative representation may have certain computational advantages [see Kagiwada, Kalaba, and Schumitzky (1969) and the references listed therein].

‡It is straightforward to show that the resolvent of K^* for $\overline{\lambda}$ is $[R_K(\lambda)]^*$ and that λ is a regular value of the $\mathfrak{L}^2$ kernel K if and only if $\overline{\lambda}$ is a regular value of K^*.

In view of (2), we can now make the association

$$R_K(x,y;\lambda) = \frac{1}{D_n(\lambda)} \sum_{\mu=1}^{n} \sum_{\nu=1}^{n} C^{\mu\nu} A_\nu(x)\, \overline{B}_\mu(y) \qquad (5)$$

for kernels of finite rank.

Formulas (3) and (5) have distinctly different characteristics. One is valid for all regular values of the parameter λ but is only applicable for degenerate kernels; the other is generally applicable but is only meaningful for sufficiently small λ. In the remainder of this chapter, however, we shall show that knowledge of these two different situations actually is all that one needs in order to handle the general case. In other words, the resolvent kernel and hence the general $\mathfrak{L}^2$ solution of the integral equation (1) with $\mathfrak{L}^2$ kernel K and $\mathfrak{L}^2$ function f can be found by a systematic interplay of already studied techniques. This approach, due to E. Schmidt (1907), was subsequently developed by Radon (1919).

Schmidt's method is based upon the knowledge that the class of degenerate kernels is dense in the space of square-integrable kernels. [A proof of this fundamental result appears many places; for example, see Riesz and Sz.-Nagy (1955, pp. 158ff) or Smithies (1962, pp. 40–41).] It follows then that an arbitrary $\mathfrak{L}^2$ kernel $K(x,y)$ can be approximated by kernels $K_n(x,y)$ of finite rank so that the norm of the difference kernel

$$L_n(x,y) \equiv K(x,y) - K_n(x,y) \qquad (6)$$

is arbitrarily small. To be precise, given any positive number ω_n, there exists a degenerate $\mathfrak{L}^2$ kernel†

$$K_n(x,y) = \sum_{\nu=1}^{n} A_\nu(x)\, \overline{B}_\nu(y) \qquad (7)$$

such that $\|L_n\| < 1/\omega_n$, where the $\mathfrak{L}^2$ functions $\{A_\nu\}$ and $\{B_\nu\}$ may be assumed to be linearly independent. It follows then that the resolvent kernel associated with the "small" kernel L_n is a regular analytic function of λ, at least for $|\lambda| < \omega_n$. In view of the arbitrariness of ω_n, we are thus able to obtain results valid for any finite value of the parameter λ by using an appropriate decomposition or *dissection* of the form (6).

3.4 APPLICATION OF SCHMIDT'S APPROACH

If the decomposition $K = K_n + L_n$ of (3.3-6) is employed, the original equation (3.3-1) becomes

---

† Special forms for this degenerate kernel approximant are extensively discussed in Chap. 14.

$$\varphi(x) = g_n(x;\lambda) + \lambda \int_a^b L_n(x,y)\varphi(y)\,dy \tag{1}$$

where
$$g_n(x;\lambda) \equiv f(x) + \lambda \int_a^b K_n(x,y)\varphi(y)\,dy$$

Assuming, for the moment, that $g_n(x;\lambda)$ is known, we can formally express the solution of (1) in terms of the resolvent R_L of L_n, namely,

$$\varphi(x) = g_n(x;\lambda) + \lambda \int_a^b R_L(x,y;\lambda)g_n(y;\lambda)\,dy$$

Rewriting this relation using the definition of $g_n(x;\lambda)$, we obtain

$$\varphi(x) = F_n(x;\lambda) + \lambda \int_a^b \tilde{K}_n(x,y;\lambda)\varphi(y)\,dy \tag{2}$$

with
$$F_n(x;\lambda) \equiv f(x) + \lambda \int_a^b R_L(x,y;\lambda)f(y)\,dy$$

and
$$\tilde{K}_n(x,y;\lambda) \equiv K_n(x,y) + \lambda \int_a^b R_L(x,z;\lambda)K_n(z,y)\,dz$$

Now the $\mathcal{L}^2$ function F_n and the $\mathcal{L}^2$ kernel $\tilde{K}_n$ that appear in (2) may be easily determined, at least for $|\lambda| < \omega_n$, since f is given, $K_n(x,y)$ is known, and R_L may be calculated from its Neumann series representation. Furthermore, and most importantly, $\tilde{K}_n$ is a degenerate kernel since K_n is of finite rank. Thus, for $|\lambda| < \omega_n$ and hence for arbitrary finite λ, the solution of the original equation (3.3-1) with a general $\mathcal{L}^2$ kernel may be reduced to the solution of the subsidiary equation (2) with a degenerate kernel, and this latter equation can be solved by the familiar algebraic procedures of Chap. 2. It should be recognized, of course, that as $\omega_n \to \infty$, the rank n of the degenerate kernel $\tilde{K}_n$ generally becomes infinitely large also and these algebraic procedures may become more cumbersome to apply.

To investigate this reduction more closely, we note that with the definition of $K_n(x,y)$ as in (3.3-7) we have for $\tilde{K}_n(x,y;\lambda)$

$$\tilde{K}_n(x,y;\lambda) = \sum_{\nu=1}^{n} \left[A_\nu(x) + \lambda \int_a^b R_L(x,z;\lambda)A_\nu(z)\,dz \right] \overline{B}_\nu(y)$$

$$\equiv \sum_{\nu=1}^{n} A_\nu(x;\lambda)\overline{B}_\nu(y) \tag{3}$$

If, in analogy with the notation of the preceding chapter, we define $\beta_\nu(\lambda) \equiv (F_n, B_\nu)$, $c_{\nu\mu}(\lambda) \equiv (\overline{B_\nu, A_\mu})$, $C_{\nu\mu}(\lambda) \equiv \delta_{\nu\mu} - \lambda c_{\nu\mu}(\lambda)$, and $C^{\nu\mu}$ as the cofactor of $C_{\nu\mu}$, then the algebraic methods of Chap. 2 show that (2) has a unique $\mathcal{L}^2$ solution, at least for $|\lambda| < \omega_n$, given by

$$\varphi(x) = F_n(x; \lambda) + \frac{\lambda}{D_n(\lambda)} \sum_{\mu=1}^{n} \sum_{\nu=1}^{n} C^{\mu\nu}(\lambda) A_\nu(x; \lambda) \beta_\mu(\lambda) \tag{4}$$

whenever†

$$D_n(\lambda) \equiv \det\left(\delta_{\nu\mu} - \lambda c_{\nu\mu}(\lambda)\right) \tag{5}$$

is different from zero. Moreover, the points of vanishing $D_n(\lambda)$ are characteristic values of $\widetilde{K}_n(x, y; \lambda)$.

As before, the number of such characteristic values is limited. In view of the definitions given above, it is easy to see that the $c_{\nu\mu}(\lambda)$, which appear in the definition of $D_n(\lambda)$, are regular analytic as functions of λ wherever R_L is and, in particular, when $|\lambda| < \omega_n$. $D_n(\lambda)$, therefore, while not in general being a polynomial in λ, nevertheless is analytic in λ (and regular for $|\lambda| < \omega_n$), since it consists of a finite combination (by addition and multiplication) of analytic functions. Since $D_n(0) = 1$, this determinant $D_n(\lambda)$ does not vanish identically, which ensures, then, that there is at most a *finite* number of characteristic values of $\widetilde{K}_n$ with modulus less than ω_n.

If we employ the notation suggested by (3.3-5) and call

$$r_n(x, y; \lambda) = \frac{1}{D_n(\lambda)} \sum_{\mu=1}^{n} \sum_{\nu=1}^{n} C^{\mu\nu}(\lambda) A_\nu(x; \lambda) \overline{B}_\mu(y)$$

the resolvent kernel associated with $\widetilde{K}_n$, then the above solution (4) can be expressed compactly as

$$\varphi(x) = F_n(x; \lambda) + \lambda \int_a^b r_n(x, y; \lambda) F_n(y; \lambda) \, dy$$

The solution of the original equation (3.3-1) may now be conveniently determined using the definition of F_n. Thus,

$$\varphi(x) = f(x) + \lambda \int_a^b R_K(x, y; \lambda) f(y) \, dy \tag{6}$$

---

†In this equation we should really write $\widetilde{D}_n(\lambda)$ to express the fact that the determinant is derived from the components of $\widetilde{K}_n$; here and elsewhere we have merely suppressed the tilde for convenience.

where †

$$R_K(x,y;\lambda) \equiv R_L(x,y;\lambda) + r_n(x,y;\lambda) + \lambda \int_a^b r_n(x,z;\lambda) R_L(z,y;\lambda)\, dz \qquad (7)$$

As might be expected, R_K is the resolvent kernel associated with the given $\mathfrak{L}^2$ kernel K. For every λ of modulus less than ω_n such that $D_n(\lambda) \neq 0$, it can easily be shown to satisfy the Fredholm identities (3.2-5). Moreover, for these regular values of λ, the solution of the original integral equation is expressed uniquely in terms of this square-integrable kernel by (6).

It should also be noted that, in view of the form of r_n, the resolvent R_K given by (7) may be rewritten as the *quotient* of two functions, each regular analytic, at least for $|\lambda| < \omega_n$. Since ω_n is completely arbitrary, this immediately implies that R_K is a *meromorphic* function of λ. In fact, given *any* large circle of radius ω in the complex λ plane, we can find a degenerate kernel K_n that approximates (in the mean) the given $\mathfrak{L}^2$ kernel K well enough to ensure the existence of R_L and hence of R_K in this circle. *The only singularities of R_K are poles at the zeros of the regular analytic function $D_n(\lambda)$ (associated with the kernel $\widetilde{K}_n$) in this circle, and these are only finite in number.* It follows, then, that these poles are at most denumerable and have no accumulation points, except perhaps the point at infinity.

3.5 THEORETICAL INVESTIGATION OF THE RESOLVENT KERNEL

In order to continue our study of the nature of the resolvent kernel as given by (3.4-7) it will be helpful to introduce the notion of the *trace* of a general $\mathfrak{L}^2$ kernel $K(x,y)$. Whenever $K(x,x)$ is measurable and integrable on the interval $a \leq x \leq b$, we define

$$\text{Trace } K = \text{Tr } (K) \equiv \int_a^b K(x,x)\, dx$$

It is clear that if K is either degenerate or the composite of two $\mathfrak{L}^2$ kernels, then $\text{Tr } (K)$ exists. In fact, it is not difficult to show that if $K(x,y)$

---

† If we let

$$\overline{B}_\mu(y;\lambda) \equiv \overline{B}_\mu(y) + \lambda \int_a^b R_L(x,y;\lambda) \overline{B}_\mu(x)\, dx$$

(7) can be rewritten in the more compact, yet less instructive, form

$$R_K(x,y;\lambda) = R_L(x,y;\lambda) + \frac{1}{D_n(\lambda)} \sum_{\mu=1}^n \sum_{\nu=1}^n C^{\mu\nu}(\lambda) A_\nu(x;\lambda) \overline{B}_\mu(y;\lambda)$$

and $L(x,y)$ are two arbitrary $\mathfrak{L}^2$ kernels, then the linear trace operator is such that

    1. $|\operatorname{Tr}(KL)| \le \|K\| \cdot \|L\|$.

    2. $\operatorname{Tr}(K^{\nu})$ exists for $\nu \ge 2$ and $|\operatorname{Tr}(K^{\nu})| \le \|K\|^{\nu}$.   (1)

    3.   If $\|K_n - K\| \to 0$ and $\|L_n - L\| \to 0$, then $\operatorname{Tr}(K_n L_n) \to \operatorname{Tr}(KL)$.

    4.   If $\|K_n - K\| \to 0$, then $\operatorname{Tr}(K_n{}^{\nu}) \to \operatorname{Tr}(K^{\nu})$ for $\nu \ge 2$.

Consider now

$$\operatorname{Tr}\,[R_K(\lambda) - K] \tag{2}$$

where we assume that λ is a regular value of the given kernel K so that a unique resolvent R_K exists in $\mathfrak{L}^2$ for this value of λ. On the one hand, by the Fredholm identities we have

$$\lambda \operatorname{Tr}\,[KR_K(\lambda)] = \lambda \operatorname{Tr}\,[R_K(\lambda)K] = \operatorname{Tr}\,[R_K(\lambda) - K]$$

from which it follows that the trace (2) exists. On the other hand, under the Schmidt "dissection" of K into $K_n + L_n$ we have from (3.4-7) for $|\lambda| < \omega_n$ that

$$\operatorname{Tr}\,[R_K(\lambda) - K] = \operatorname{Tr}\,[R_L(\lambda) + r_n(\lambda) + \lambda r_n(\lambda)R_L(\lambda) - K_n - L_n]$$

$$= -\operatorname{Tr}\,(K_n) + \operatorname{Tr}\,[R_L(\lambda) - L_n]$$

$$+ \operatorname{Tr}\,[r_n(\lambda) + \lambda r_n(\lambda)R_L(\lambda)] \tag{3}$$

Let us examine each of the terms on the right-hand side of this equation.

    The first term is well defined since $K_n(x,y)$ is degenerate. In fact, with K_n given by (3.3-7), we obtain

$$\operatorname{Tr}\,(K_n) = \sum_{\nu=1}^{n} (A_\nu, B_\nu)$$

The second term is of the same form as (2), and hence it not only exists but can be rewritten, say, as

$$\lambda \operatorname{Tr}\,[L_n R_L(\lambda)] \tag{4}$$

For the third term we have (see the footnote, page 33)

$$\operatorname{Tr}\,[r_n(\lambda) + \lambda r_n(\lambda)R_L(\lambda)] = \frac{1}{D_n(\lambda)} \sum_{\mu=1}^{n} \sum_{\nu=1}^{n} C^{\mu\nu}(\lambda) \int_a^b A_\nu(x;\lambda)\overline{B}_\mu(x;\lambda)\,dx$$

We leave it as an exercise (Prob. 9) for the reader to show that this relation can be reexpressed as

$$\mathrm{Tr}\,[r_n(\lambda) + \lambda r_n(\lambda)R_L(\lambda)] = -\frac{1}{D_n(\lambda)} \sum_{\mu=1}^{n} \sum_{\nu=1}^{n} C^{\mu\nu}(\lambda)\,\frac{\partial C_{\mu\nu}(\lambda)}{\partial \lambda}$$

It follows, then, from the formula for the derivative of a determinant [due originally to Jacobi (1841); see also Muir (1906, pp. 261–262), and Pipes (1963, p. 15)] that

$$\mathrm{Tr}\,[r_n(\lambda) + \lambda r_n(\lambda)R_L(\lambda)] = -\frac{D'_n(\lambda)}{D_n(\lambda)} \tag{5}$$

The substitution of (4) and (5) into (3) yields

$$\mathrm{Tr}\,[R_K(\lambda) - K] = -\mathrm{Tr}\,(K_n) + \lambda\,\mathrm{Tr}\,[L_n R_L(\lambda)] - \frac{D'_n(\lambda)}{D_n(\lambda)} \tag{6}$$

Using this equation we may make the following significant observations:

1. Since the left-hand side of (6) is completely determined by the given kernel $K(x,y)$ and its unique resolvent, the apparent dependence upon n of the right-hand side is really extrinsic.

2. On the other hand, (6) suggests again quite clearly that the only points within the circle $|\lambda| < \omega_n$ where R_K is not defined are the zeros of $D_n(\lambda)$. Since only a finite number of zeros of $D_n(\lambda)$ can occur in any such bounded region, these poles of R_K form either a finite sequence (perhaps void) or an infinite sequence tending in modulus to ∞.

3. In evaluating $D_n(\lambda)$ and determining its zeros, it may well occur that $D_n(\lambda) \neq 0$ for all $|\lambda| < \omega_n$. If, however, there are zeros of $D_n(\lambda)$ within this region, they are automatically poles of $R_K(x,y;\lambda)$. As ω_n and hence n tend to ∞, the position and multiplicity of these zeros remain unchanged.

4. The poles of $\mathrm{Tr}\,[R_K - K]$ are all *simple* and have residues that are negative *integers* (Prob. 9).

In view of these properties and the above results we may uniquely form

$$\tilde{D}(\lambda) \equiv \exp\left[-\int_0^\lambda \mathrm{Tr}\,[R_K(\mu) - K]\,d\mu\right]$$

$$= \exp\left[-\int_0^\lambda \mu\mathrm{Tr}\,[KR_K(\mu)]\,d\mu \right]$$

$$= \exp\left[-\int_0^\lambda \mu\mathrm{Tr}\,[R_K(\mu)K]\,d\mu \right]$$

$$= D_n(\lambda)\exp\left[\lambda\mathrm{Tr}\,(K_n) - \int_0^\lambda \mu\mathrm{Tr}\,[L_nR_L(\mu)]\,d\mu \right] \qquad (7)$$

which we call the *modified Fredholm determinant* of $K(x,y)$. Owing to properties 1 and 4 above, $\tilde{D}(\lambda)$ is not only independent of n but also is uninfluenced by the particular path chosen in the evaluation of the integrals appearing in its definition (7). Moreover, it is an especially well-behaved function of λ, being expressible, for arbitrary n, as the product of a regular analytic function $D_n(\lambda)$ with an exponential function. It follows, then, that $\tilde{D}(\lambda)$ is an *entire* (or *integral*) function with zeros (within $|\lambda| < \omega_n$ for arbitrary ω_n) coinciding in location and multiplicity with the zeros of $D_n(\lambda)$.

   This last observation has an important implication, for consider the function $\tilde{D}(x,y;\lambda)$ defined by the expression

$$\tilde{D}(x,y;\lambda) \equiv \tilde{D}(\lambda)R_K(x,y;\lambda)$$

It is evident that $\tilde{D}(x,y;\lambda)$, which is termed the *modified first Fredholm minor*, is also an entire function of λ since the factor $D_n(\lambda)$ in $\tilde{D}(\lambda)$ exactly cancels all the singularities of R_K, which are contributed by r_n, for arbitrary ω_n. As a consequence, we thus have the final relation

$$R_K(x,y;\lambda) = \frac{\tilde{D}(x,y;\lambda)}{\tilde{D}(\lambda)} \qquad (8)$$

which exhibits the resolvent as the quotient of two functions that are regular analytic throughout the entire complex λ plane.

3.6   THE FREDHOLM ALTERNATIVE FOR ARBITRARY $\mathfrak{L}^2$ KERNELS

The analysis of the preceding section shows that the resolvent R_K associated with a general square-integrable kernel K is independent of the particular dissection $K = K_n + L_n$ since it is an analytic function of λ for all finite λ and is unique where it exists.† This result was not to be ex-

---

†It can quite naturally occur, however, that for a particular dissection which leads to results valid for $|\lambda| < \omega_n$, more singularities of the resolvent are noted than for another dissection with $\omega_m < \omega_n$. In such a situation the additional poles determined under the first dissection will lie in the annulus $\omega_m \leq |\lambda| < \omega_n$, the singularities within the lesser circle $|\lambda| < \omega_m$ agreeing in both cases.

pected at first glance from the construction, although it is inherent therein. We have seen, moreover, that R_K is regular analytic in λ and exists as an $\mathfrak{L}^2$ function of x and y wherever the modified Fredholm determinant $\tilde{D}(\lambda)$ is nonvanishing. For these regular values of λ an $\mathfrak{L}^2$ solution to the original equation

$$\varphi(x) = f(x) + \lambda \int_a^b K(x,y)\varphi(y)\,dy \tag{1}$$

exists and is given uniquely by

$$\varphi(x) = f(x) + \lambda \int_a^b R_K(x,y;\lambda)f(y)\,dy$$

In particular, if $f = 0$, then $\varphi = 0$.

The situation is understandably different when $\tilde{D}(\lambda)$ vanishes for some particular value of λ. There then exists an (infinite number of) appropriate dissection(s) with $\omega_n > |\lambda|$ and for which $D_n(\lambda) = 0$. It follows that the degenerate kernel $\tilde{K}_n$ associated with this dissection has characteristic functions. Since vanishing F_n is equivalent to vanishing f [see (3.4-2)], this, in turn, implies that the homogeneous equation

$$\omega(x) = \lambda \int_a^b K(x,y)\omega(y)\,dy \tag{2}$$

also has nontrivial $\mathfrak{L}^2$ solutions, and hence the values of λ for which $\tilde{D}(\lambda) = 0$ are truly characteristic values of the given kernel K. In fact, if

$$\omega_0 = \lambda \tilde{K}_n \omega_0$$

then, using the Fredholm identities (3.2-5) in conjunction with our earlier definitions, we find

$$\begin{aligned}
\omega_0 &= \lambda K_n \omega_0 + \lambda^2 R_L K_n \omega_0 \\
&= \lambda K \omega_0 - \lambda L_n \omega_0 - \lambda^2 R_L L_n \omega_0 + \lambda^2 R_L K \omega_0 \\
&= \lambda K \omega_0 - \lambda R_L (\omega_0 - \lambda K \omega_0)
\end{aligned}$$

or

$$(\omega_0 - \lambda K \omega_0) + \lambda R_L (\omega_0 - \lambda K \omega_0) = 0$$

Since λ is a regular value of L_n, it must be that $(\omega_0 - \lambda K \omega_0)$ vanishes and thus the characteristic functions of $\tilde{K}_n$ are perforce characteristic functions of the original kernel K also.

Although we shall have more to say regarding this correspondence in a subsequent chapter, the reader should clearly understand that the duality between the cv's and cf's of the given kernel K and those of the degenerate kernel $\tilde{K}_n$ is primarily of theoretical significance. More important from an applications point of view is the relationship between the cv's and cf's of K and those of the original unmodified approximating kernels K_n from which the $\tilde{K}_n$ are formed. In this regard, it is known that enumerations of the cv's and allied cf's of K_n can be selected so that as $n \to \infty$ these quantities, in a manner that can be made precise, tend toward corresponding quantities associated with K [see Dunford and Schwartz (1963, sec. 11.9), Yamamoto (1968), or Gohberg and Krein (1969, sec. 1.5), for example]. The "proximity" of corresponding quantities, moreover, can be "measured" in certain special cases, and these matters will be essential to our later discussions.

To complete our analysis relative to the validity of the Fredholm alternative in this general setting, we need to study the equations

$$\psi(x) = g(x) + \bar{\lambda} \int_a^b K^*(x,y)\psi(y)\, dy \tag{1'}$$

and

$$\Omega(x) = \bar{\lambda} \int_a^b K^*(x,y)\Omega(y)\, dy \tag{2'}$$

adjoint to (1) and (2). Here some care must be exercised since a straightforward application of the earlier dissection and reduction process to the adjoint kernel K^* does not automatically lead to a subsidiary degenerate kernel which is the adjoint of $\tilde{K}_n$ in (3.4-2). We begin, however, in the usual manner.

Using the decomposition $K^* = K_n^* + L_n^*$, (1') becomes symbolically

$$\psi = g_n^* + \bar{\lambda} L_n^* \psi \tag{3}$$

with

$$g_n^* \equiv g + \bar{\lambda} K_n^* \psi$$

Regarding (3) as an integral equation with "known" g_n^*, we can write its solution in terms of the resolvent $R_{L^*} = R_L^*$ of L_n^* as

$$\psi = g_n^* + \bar{\lambda} R_L^* g_n^* \tag{4}$$

In contrast to the approach of Sec. 3.4, we now substitute this expression for ψ into the definition of g_n^* rather than vice versa. This yields the integral equation

$$g_n^* = g + \bar{\lambda} \tilde{K}_n^* g_n^* \tag{5}$$

with the desired degenerate kernel

$$\widetilde{K}_n^* = K_n^* + \bar{\lambda} K_n^* R_L^*$$

adjoint to $\widetilde{K}_n$.

The above construction for the conjugate case allows us to reason in a manner completely analogous to that employed previously. The Fredholm alternative is certainly valid for (5) since the kernel K_n^* is of finite rank, but by means of (4) we can relate results regarding g_n^* to similar results regarding ψ. In particular, if $D_n^*(\bar{\lambda}) \neq 0$, (5) and hence (1') have unique solutions for arbitrary square-integrable g. When D_n^* vanishes, then the homogeneous version of (5) admits nontrivial $\mathfrak{L}^2$ solutions and these give rise, via (4), to nontrivial $\mathfrak{L}^2$ solutions of (2'). Owing to the conjugate relationship of $\widetilde{K}_n$ and $\widetilde{K}_n^*$, and of K and K^*, the number of linearly independent such characteristic solutions must be the same finite number in this case as for the original equation.

It only remains to be observed that, with F_n given as in (3.4-2), we have from (4) and (5),

$$
\begin{aligned}
(F_n, g_n^*) &= (f + \lambda R_L f, g_n^*) \\
&= (f, g_n^* + \bar{\lambda} R_L^* g_n^*) \\
&= (f, \psi)
\end{aligned}
$$

This relation (and its adjoint counterpart) show that in the event λ is a characteristic value, the fundamental inner product of free term and solution of homogeneous adjoint equation for both the original equation and its reduced degenerate analogue under the dissection process are entirely equivalent. With this detail the necessary correspondence has been completed, and we see that Schmidt's approach with its attendant decomposition into degenerate kernels and arbitrarily small remainders has permitted the Fredholm alternative of Chap. 2 to be carried over entirely to the general case and its validity to be established for all square-integrable kernels.†

---

† Various forms of the Fredholm alternative as well as analogues of many other fundamental results in the theory of linear integral equations are valid for considerably wider classes of kernels. In particular, such is the case for kernels, some one of whose higher-order iterates is square integrable, a class which includes kernels with so-called "weak singularities" (see Probs. 12-14). The interested reader may find further information in Zaanen (1953, sec. 11.2), Tricomi (1957, sec. 3.17), and Mikhlin (1964, sec. 1.10). The theory of integral equations with certain classes of even more singular kernels is discussed in Muskhelishvili (1953), chap. 4 of Tricomi (1957), and chap. 3 of Mikhlin (1964); see also Chap. 18 of this text.

PROBLEMS

1. Show that for the iterated kernels,

$$K^{\nu+\mu} = K^\nu K^\mu$$

   for arbitrary positive integers ν and μ.

2. Assume $K(x,y)$ is a continuous kernel for $a \le x,y \le b$ with $|K(x,y)| < M$.
   a. Show the resolvent kernel is defined and jointly continuous in all three variables x, y, and λ for $|\lambda| M(b-a) < 1$, $a \le x,y \le b$.
   b. Verify that for fixed x and y, R_K is an analytic function of λ, regular at least in the same circle $|\lambda| M(b-a) < 1$.
   c. Satisfy yourself that the sufficient condition (3.1-7) is indeed weaker in general than the condition (3.1-6).

3. The system of n simultaneous equations

$$\varphi_\mu(x) = f_\mu(x) + \lambda \int_a^b \sum_{\nu=1}^n K_{\mu\nu}(x,y)\varphi_\nu(y)\, dy \qquad \mu = 1, 2, \ldots, n$$

   can be written in the form of a single equation

$$\varphi = f + \lambda \int K\varphi$$

   in which φ and f are column vectors and K is a matrix. Discuss the convergence of the Neumann series (3.1-4) for this latter equation.

4. Given an $\mathfrak{L}^2$ kernel $K(x,y)$, show that the resolvent of K^* for $\overline{\lambda}$ is $(R_K)^*$ and that λ is a regular value of K if and only if $\overline{\lambda}$ is a regular value of K^*.

5. a. For the degenerate kernel $K(x,y) = a(x)\,\overline{b}(y)$, calculate the resolvent kernel from the Neumann series for small λ.
   b. Verify that the formula (3.3-5), which gives the resolvent according to the theory applicable to kernels of finite rank, leads to the analytic continuation of this Neumann series.

6. Given the linearly independent set of $\mathfrak{L}^2$ functions $A_\nu(x)$, $\nu = 1, 2, \ldots, n$, verify that the set formed by

$$A_\nu(x) + \lambda \int_a^b R_K(x,y;\lambda) A_\nu(y)\, dy$$

   is linearly independent also for any regular value λ of R_K, the resolvent kernel associated with a given $\mathfrak{L}^2$ kernel K.

7. Demonstrate that for every λ such that $D_n(\lambda) \ne 0$ with $|\lambda| < \omega_n$, R_K as given by (3.4-7) satisfies the Fredholm identities (3.2-5).

8. Study the dissection of the kernel $K(x,y) = x + y$ for $0 \le x,y \le 1$ when $K_n(x,y) = y$. In particular, observe that the smallest characteristic value of $\widetilde{K}_n(x,y)$ agrees, as expected, with the smallest characteristic value of $K(x,y)$. How does it compare with the smallest characteristic value of $K_n(x,y)$?

9. Let $A_\nu(x;\lambda)$ and $\overline{B}_\mu(y;\lambda)$ be as defined in (3.4-3) and the footnote on page 33, respectively. Using the fact that the resolvent kernel satisfies the relationship

$$\frac{\partial R_L}{\partial \lambda} = R_L{}^2$$

which will be proved later, show that

$$\frac{\partial C_\nu(x;\lambda)}{\partial \lambda} = \int_a^b R_L(x,y;\lambda) A_\nu(y;\lambda)\, dy$$

and hence

$$\frac{\partial C_{\mu\nu}(\lambda)}{\partial \lambda} \equiv \frac{\partial \left[\delta_{\mu\nu} - \lambda \int_a^b A_\nu(x;\lambda)\overline{B}_\mu(x)\, dx \right]}{\partial \lambda}$$

$$= -\int_a^b A_\nu(x;\lambda)\overline{B}_\mu(x;\lambda)\, dx$$

10. **a.** Verify that both degenerate and composite kernels have well-defined traces.
    b. Show that the linear trace operator satisfies the several properties listed on page 34.

11. Satisfy yourself that (3.5-6) implies that the only singularities of $\text{Tr}\,[R_K(\lambda) - K]$ in the complex λ plane are simple poles having negative integer residues.

12. Let the kernel K have a "weak singularity," i.e., be of the form

$$K(x,y) = \frac{H(x,y)}{|x-y|^\alpha}$$

where $H(x,y)$ is bounded and $0 < \alpha < 1$. Satisfy yourself that for $0 < \alpha < \frac{1}{2}$, K is square integrable, while for $\frac{1}{2} \le \alpha < 1$, there exists an n (which depends upon α) such that the iterates K^ν are square integrable for all $\nu \ge n$.

13. **a.** Verify that if ϕ is a characteristic function of a given kernel K associated with the characteristic value λ, then for each positive integer ν it is also a cf of the iterated kernel K^ν belonging to the cv λ^ν.
    b. Conversely, show that if ϕ is a cf of K^n associated with the cv λ^n, then at least one of the functions

$$\Phi_\nu = \phi + \sum_{\mu=1}^{n-1} \varepsilon^{\mu\nu}\,\lambda^\mu K^\mu \phi \qquad \nu = 1, 2, \ldots, n$$

with $\varepsilon = e^{2\pi i/n}$ constitutes a cf of K belonging to the cv $\varepsilon^\nu \lambda$. (Note particularly that $\sum_{\nu=1}^{n} \Phi_\nu = n\phi$.)

14. Use the results of the preceding problem to establish in formal fashion that the Fredholm alternative is valid for the kernel K iff it holds for the iterated kernel K^n. (Try first the case $n = 2$.)

chapter four

Fredholm Formulas and the Resolvent Equation

4.1 FREDHOLM-CARLEMAN THEORY

Proceeding from a point of view rather different from that of the previous chapter, Fredholm (1900 and 1903) in his original theory (applicable to continuous, and hence bounded kernels) showed that the solution of the integral equation

$$\varphi(x) = f(x) + \lambda \int_a^b K(x,y)\varphi(y) \, dy$$

whenever it exists, is given likewise by a unique relation of the form

$$\varphi(x) = f(x) + \lambda \int_a^b R_K(x,y;\lambda)f(y) \, dy$$

The resolvent of the given kernel K is represented as the quotient of two entire functions of the complex variable λ (the unmodified first Fredholm minor and the Fredholm determinant), the zeros of the denominator being the characteristic values of the kernel K and conversely.   It follows, of course, that R_K is unique and satisfies the Fredholm identities (3.2-5).†

As a principal result Fredholm obtained series expansions for the determinant and minors, which bear his name.   Hilbert (1904) subsequently developed modified expressions valid for certain discontinuous kernels.‡   Somewhat later Carleman (1921) [see also Smithies (1941) and Mikhlin (1944)] extended Hilbert's work and showed that similar results are actually applicable for arbitrary $\mathcal{L}^2$ kernels.   In this general situation, the unique resolvent kernel has the form (3.5-8) encountered earlier, namely,

$$R_K(x,y;\lambda) = \frac{\widetilde{D}(x,y;\lambda)}{\widetilde{D}(\lambda)} \tag{1}$$

where $\widetilde{D}(x,y;\lambda)$ and $\widetilde{D}(\lambda)$ are the modified first Fredholm minor and the modified Fredholm determinant, respectively.   We now give various representations for these Fredholm functions.

Since $\widetilde{D}(\lambda)$ is an entire function, it has an everywhere convergent power series of the form

$$\widetilde{D}(\lambda) = \sum_{\nu=0}^{\infty} \widetilde{d}_\nu \lambda^\nu \tag{2}$$

Carleman's expression for the coefficients $\widetilde{d}_\nu$ was

$$\widetilde{d}_0 = 1$$

$$\widetilde{d} = \frac{(-1)^\nu}{\nu!} \int\int \cdots \int \begin{vmatrix} 0 & K(s_1,s_2) & \cdots & K(s_1,s_\nu) \\ K(s_2,s_1) & 0 & \cdots & K(s_2,s_\nu) \\ \cdots & \cdots & \cdots & \cdots \\ K(s_\nu,s_1) & K(s_\nu,s_2) & \cdots & 0 \end{vmatrix} \tag{3}$$

$$\cdot\, ds_1 \cdots ds_\nu \qquad \nu \geq 1$$

We shall establish here the recursion relations

$$\widetilde{d}_0 = 1 \qquad \widetilde{d}_1 = 0$$

$$\nu \widetilde{d}_\nu = -\sum_{\mu=2}^{\nu} k_\mu \widetilde{d}_{\nu-\mu} \qquad \nu \geq 2 \tag{4}$$

---

† A good account of the approach of Fredholm can be found in Riesz and Sz. -Nagy (1955, pp. 172ff) or Smithies (1962, pp. 65ff).

‡ A survey account (by E. Hellinger) of Hilbert's work in integral equations' theory is to be found in Hilbert (1935).

where $k_\mu = \text{Tr}\,(K^\mu)$ (the trace of K^μ), which we know exists for $\mu \geq 2$. [Smithies (1941 and 1962) has developed other alternative formulas.]

For λ small enough we have from (3.5-7) and (3.1-4) that

$$\tilde{D}(\lambda) \equiv \exp\left[-\int_0^\lambda \text{Tr}\,(R_K - K)d\lambda \right]$$

$$= \exp\left(-\sum_{\mu=2}^\infty \frac{\lambda^\mu}{\mu}\,k_\mu \right)$$

and hence $\displaystyle\sum_{\nu=0}^\infty \tilde{d}_\nu \lambda^\nu = \exp\left(-\sum_{\mu=2}^\infty \frac{\lambda^\mu}{\mu}\,k_\mu \right)$ (small λ)

Setting $\lambda = 0$, it is clear that $\tilde{d}_0 = 1$. For the general case, we first take derivatives with respect to λ, obtaining thereby

$$\sum_{\nu=1}^\infty \nu \tilde{d}_\nu \lambda^{\nu-1} = -\sum_{\mu=2}^\infty \lambda^{\mu-1} k_\mu \exp\left(-\sum_{\mu=2}^\infty \frac{\lambda^\mu}{\mu}\,k_\mu \right)$$

$$= -\sum_{\mu=2}^\infty \lambda^{\mu-1} k_\mu \sum_{\gamma=0}^\infty \tilde{d}_\gamma \lambda^\gamma$$

$$= -\sum_{\mu=2}^\infty k_\mu \sum_{\nu=\mu}^\infty \tilde{d}_{\nu-\mu} \lambda^{\nu-1}$$

$$= -\sum_{\nu=2}^\infty \left(\sum_{\mu=2}^\nu k_\mu \tilde{d}_{\nu-\mu} \right) \lambda^{\nu-1}$$

Equating coefficients of like powers then yields the desired result valid for small λ. We know, however, that the same relations must be applicable for arbitrary finite λ since $\tilde{D}(\lambda)$ is an entire function.

Corresponding formulas can easily be established for the modified first Fredholm minor. Consider

$$\tilde{D}(x,y;\lambda) = \sum_{\nu=0}^\infty \tilde{D}_\nu(x,y)\lambda^\nu \qquad (5)$$

and use the Fredholm identity

$$R_K(x,y;\lambda) = K(x,y) + \lambda \int_a^b R_K(x,z;\lambda)K(z,y)\,dz$$

Then $\quad \tilde{D}(x,y;\lambda) = K(x,y)\tilde{D}(\lambda) + \lambda \int_a^b \tilde{D}(x,z;\lambda)K(z,y)\,dz$

or $\quad \sum_{\nu=0}^{\infty} \tilde{D}_\nu(x,y)\lambda^\nu = K(x,y)\sum_{\nu=0}^{\infty} \tilde{d}_\nu \lambda^\nu + \sum_{\nu=0}^{\infty} \lambda^{\nu+1} \int_a^b \tilde{D}_\nu(x,z)K(z,y)\,dz$

We obtain the result, therefore, that

$$\tilde{D}_0(x,y) = K(x,y)$$

$$\tilde{D}_\nu(x,y) = \tilde{d}_\nu K(x,y) + \int_a^b \tilde{D}_{\nu-1}(x,z)K(z,y)\,dz$$

$$\quad \nu \geq 1 \qquad (6)$$

$$= \tilde{d}_\nu K(x,y) + \int_a^b K(x,z)\tilde{D}_{\nu-1}(z,y)\,dz$$

Equivalently, by successive substitution, we derive

$$\tilde{D}_\nu(x,y) = \sum_{\mu=0}^{\nu} \tilde{d}_{\nu-\mu} K^{\mu+1}(x,y) \qquad \nu \geq 0 \qquad (7)$$

There are a number of related expressions that are similar to (7) in that they show the interplay between the $\tilde{d}_\nu$ and the $\tilde{D}_\nu$. For instance, from (7) we have

$$\text{Tr}\,(\tilde{D}_\nu - \tilde{d}_\nu K) = \text{Tr}\left(\sum_{\mu=1}^{\nu} \tilde{d}_{\nu-\mu} K^{\mu+1}\right)$$

$$= \sum_{\mu=1}^{\nu} \tilde{d}_{\nu-\mu} k_{\mu+1}$$

$$= -(\nu + 1)\,\tilde{d}_{\nu+1}$$

Hence, using (6),

$$\nu\tilde{d}_\nu = -\text{Tr}\,(\tilde{D}_{\nu-1} - \tilde{d}_{\nu-1}K)$$

$$= -\text{Tr}\,(\tilde{D}_{\nu-2}K) \qquad\qquad \nu \geq 2$$

$$= -\int_a^b \int_a^b \tilde{D}_{\nu-2}(x,y)K(y,x)\,dy\,dx$$

These last relations are often an aid to computation of the modified Fredholm determinant and first Fredholm minor.

If $k_1 = \text{Tr}\,(K)$ is meaningful, the unmodified functions originally suggested by Fredholm are known to exist. When such is the case, it turns out that the identities

$$D(\lambda) = e^{-k_1\lambda}\,\tilde{D}(\lambda)$$

$$D(x,y;\lambda) = e^{-k_1\lambda}\,\tilde{D}(x,y;\lambda)$$

(8)

provide entirely consistent definitions of these unmodified Fredholm functions. Expressions completely analogous to those given above for the general case may then be easily derived. In particular, we have

$$D(\lambda) = \sum_{\nu=0}^{\infty} d_\nu\,\lambda^\nu$$

(9)

and

$$D(x,y;\lambda) = \sum_{\nu=0}^{\infty} D_\nu(x,y)\lambda^\nu$$

where $\qquad d_0 = 1 \qquad D_0(x,y) = K(x,y)$

$$\nu d_\nu = -\sum_{\mu=1}^{\nu} k_\mu\,d_{\nu-\mu} \qquad\qquad \nu \geq 1$$

(10)

and $\qquad D_\nu(x,y) = d_\nu K(x,y) + \displaystyle\int_a^b K(x,z)\,D_{\nu-1}(z,y)\,dz$

$\qquad\qquad\qquad\qquad\qquad\qquad\qquad\qquad \nu \geq 1$

$$= \sum_{\mu=0}^{\nu} d_{\nu-\mu}\,K^{\mu+1}(x,y)$$

Alternatively, $\qquad\qquad \nu d_\nu = -\mathrm{Tr}\,(D_{\nu-1})$

It is of interest to observe that for the unmodified Fredholm functions, if D_ν vanishes for any value of ν, then d_μ and D_μ vanish for *all* $\mu > \nu$. It follows, then, in this case that $D(\lambda)$ and $D(x,y;\lambda)$ reduce to polynomials in λ.

Finally, we note that the classical formula for d_ν due to Fredholm (1903) differs from (3) by the appearance of terms $K(s_i,s_i)$ in place of the zeros on the main diagonal, with a somewhat similar result existing for the $D_\nu(x,y)$ (see Prob. 2). Other relations for the Fredholm functions in terms of iterates of the kernel and their traces were first given by Plemelj (1904) and considered subsequently by Smithies (1941 and 1962).

4.2   THE FACTORIZATION OF THE MODIFIED FREDHOLM DETERMINANT

Since for general $\mathfrak{L}^2$ kernels the characteristic values appear as the zeros of the modified Fredholm determinant, it is valuable to have a representation of the entire function $\tilde{D}(\lambda)$ that exhibits this relationship explicitly. In order to derive such an expression we shall need certain definitions and results from the theory of entire functions. These are to be found in standard texts such as Copson (1935) or Boas (1954) and are repeated here for convenience.

1. An entire (integral) function $f(z)$ is said to be of finite order if there exists a positive real number k, independent of r, such that the maximum modulus $M(r)$ of $f(z)$ on the circle $|z| = r$ satisfies

$$\log M(r) < r^k \qquad \text{as} \qquad r \to \infty$$

The *infimum* (or greatest lower bound) of such k is called the *order ρ* of $f(z)$. In other words,

$$\rho = \limsup_{r \to \infty} \frac{\log \log M(r)}{\log r}$$

Moreover, if the power series expansion for $f(z)$ has the form

$$f(z) = \sum_{\nu=0}^{\infty} a_\nu z^\nu$$

it can be shown that

$$\rho = \limsup_{\nu \to \infty} \frac{\nu \log \nu}{\log (1/|a_\nu|)}$$

This last relation is often used as an alternative definition of the order ρ of $f(z)$.

2. The *convergence exponent ρ_1* of the zeros z_ν of $f(z)$ is the infimum of positive numbers γ for which

$$\sum_{\nu=1}^{\infty} |z_\nu|^{-\gamma} < \infty \qquad (1)$$

If the z_ν are arranged in order of increasing modulus, then

$$\rho_1 = \limsup_{\nu \to \infty} \frac{\log \nu}{\log |z_\nu|}$$

[The series (1) need not converge for $\gamma = \rho_1$.] It can be easily established [Copson (1935, p. 170)] that $\rho_1 \le \rho$ for entire functions of finite order.

3. The smallest *positive integer* γ for which the series (1) converges is denoted by $(p + 1)$, and p is termed the *genus of the set of zeros of* $f(z)$.

4. If $f(z)$ is an entire function of finite order ρ, which has a set of zeros $z_\nu \neq 0$ with a convergence exponent ρ_1, the following factorization (due to Hadamard) is valid:

$$f(z) = e^{Q(z)} \prod_{\nu=1}^{\infty} \left(1 - \frac{z}{z_\nu}\right) \exp\left[\frac{z}{z_\nu} + \frac{1}{2}\left(\frac{z}{z_\nu}\right)^2 + \cdots + \frac{1}{p}\left(\frac{z}{z_\nu}\right)^p\right]$$

Here p is the genus of the set of zeros of $f(z)$ and $Q(z)$ is a polynomial of degree $q \leq [\rho]$. We have, in fact, $\max (q, \rho_1) = \rho$. The greater of the integers p and q is called the *genus* of the function $f(z)$, and clearly it is a number less than the order ρ.

In applying the above results to the modified Fredholm determinant

$$\tilde{D}(\lambda) = \sum_{\nu=0}^{\infty} \tilde{d}_\nu \lambda^\nu$$

we first note that for $\mathfrak{L}^2$ kernels in general the inequality

$$|\tilde{d}_\nu| < \frac{C^\nu}{\nu^{\nu/2}}$$

where C is a constant independent of ν, can be established [see Smithies (1962, pp. 94ff), for example]. It then follows that the order of $\tilde{D}(\lambda)$ and hence its genus is at most 2. Carleman (1918) has exhibited *continuous* kernels for which the order of $\tilde{D}(\lambda)$ precisely equals 2, so that, a fortiori, this limit cannot be lowered for the class $\mathfrak{L}^2$.

Regarding the zeros of $\tilde{D}(\lambda)$, a result of Schur (1909) as extended to $\mathfrak{L}^2$ kernels by Carleman (1921) shows that

$$\sum_{\nu=1}^{\infty} |\lambda_\nu|^{-2} \leq \|K\|^2 \tag{2}$$

where the λ_ν are the characteristic values of $K(x, y)$ each repeated according to its multiplicity as a zero of $\tilde{D}(\lambda)$. Unless contrarily stated, we shall always assume that these cv's are ordered in the natural manner according to increasing modulus; that is, $0 < |\lambda_1| \leq |\lambda_2| \leq \cdots$. Although we shall have more to say about the relation (2) in subsequent chapters, we note at least now the implication that the convergence exponent of the zeros satisfies $\rho_1 \leq 2$ and the genus of the zeros is thus either *zero* or *one*.

In view of the above results, the Hadamard factorization for $\tilde{D}(\lambda)$ becomes

$$\tilde{D}(\lambda) = e^{a\lambda^2 + b\lambda + c} \prod_{\nu=1}^{\infty} \left(1 - \frac{\lambda}{\lambda_\nu}\right) e^{\lambda/\lambda_\nu}$$

This particular expansion, which is uniformly convergent on compacta, however, may be considerably simplified. Since $\tilde{D}(0) = 1$ and $\tilde{D}'(0) = 0$, both of the constants b and c must vanish. Furthermore, an extension of an early proof of Carleman (1917) shows that a also equals zero. Thus,

$$\tilde{D}(\lambda) = \prod_{\nu=1}^{\infty} \left(1 - \frac{\lambda}{\lambda_\nu}\right) e^{\lambda/\lambda_\nu} \tag{3}$$

Observe, then, that from definition 4 above, the order of $\tilde{D}(\lambda)$ turns out to be precisely equal to ρ_1, the convergence exponent of its zeros.

As we have discussed previously, when the $\mathfrak{L}^2$ kernel $K(x, y)$ is such that $k_1 = \text{Tr}(K)$ exists, the unmodified Fredholm determinant also has meaning. For this case, it follows from (4.1-8) that the expansion for $D(\lambda)$ analogous to (3) has the form

$$D(\lambda) = e^{-k_1\lambda} \prod_{\nu=1}^{\infty} \left(1 - \frac{\lambda}{\lambda_\nu}\right) e^{\lambda/\lambda_\nu} \tag{4}$$

an expression that is clearly the $\mathfrak{L}^2$ analogue of the typical product expansion for the Fredholm determinant associated with a degenerate kernel.

4.3 VARIOUS DEDUCTIONS FROM THE FACTORIZATION FORMULA

It follows immediately from (4.2-3) that

$$|\tilde{D}(\lambda)|^2 = \prod_{\nu=1}^{\infty} \left(1 - \frac{\lambda}{\lambda_\nu}\right)\left(1 - \frac{\bar{\lambda}}{\bar{\lambda}_\nu}\right) e^{\lambda/\lambda_\nu} e^{\bar{\lambda}/\bar{\lambda}_\nu}$$

$$= \prod_{\nu=1}^{\infty} \left[1 - 2\,\text{Re}\left(\frac{\lambda}{\lambda_\nu}\right) + \left|\frac{\lambda}{\lambda_\nu}\right|^2\right] e^{2\,\text{Re}\,(\lambda/\lambda_\nu)}$$

$$\leq \prod_{\nu=1}^{\infty} e^{-2\,\text{Re}\,(\lambda/\lambda_\nu) + |\lambda/\lambda_\nu|^2} e^{2\,\text{Re}\,(\lambda/\lambda_\nu)}$$

$$= \prod_{\nu=1}^{\infty} e^{|\lambda/\lambda_\nu|^2}$$

In view of the Schur-Carleman result (4.2-2), we then obtain the interesting and important inequality

$$|\tilde{D}(\lambda)| \leq e^{\frac{1}{2}|\lambda|^2 \|K\|^2} \tag{1}$$

for the modified Fredholm determinant. Examples can be easily constructed which demonstrate that this bound is sharp for square-integrable kernels K (see Prob. 5).

There is a corresponding inequality established by Carleman (1921) for the modified first Fredholm minor, which takes the form

$$|\tilde{D}(x,y;\lambda)| \leq e^{\frac{1}{2}|\lambda|^2 \|K\|^2} \left\{ |K(x,y)| \right.$$
$$\left. + |\lambda| \sqrt{e} \left[\int_a^b |K(x,z)|^2 \, dz \int_a^b |K(z,y)|^2 \, dz \right]^{\frac{1}{2}} \right\}$$

for almost all $a \leq x, y \leq b$. Using Schwarz's inequality we can easily show, then, that

$$\|\tilde{D}(x,y;\lambda)\| \leq \|K\| e^{\frac{1}{2}|\lambda|^2 \|K\|^2} \left(1 + |\lambda| \sqrt{e} \|K\| \right) \tag{2}$$

As we shall see later in Chap. 14, however, this bound can be substantially improved.

One last observation is worth making at this time. As we have seen in Sec. 4.1, the expression

$$\tilde{D}(\lambda) = \exp\left(- \sum_{n=2}^{\infty} \frac{\lambda^n}{n} k_n \right)$$

or

$$\log \tilde{D}(\lambda) = - \sum_{n=2}^{\infty} \frac{\lambda^n}{n} k_n$$

is valid for small λ. On the other hand, the expansion (4.2-3) implies that for small λ

$$\log \tilde{D}(\lambda) = \sum_{\nu=1}^{\infty} \left[\log \left(1 - \frac{\lambda}{\lambda_\nu} \right) + \frac{\lambda}{\lambda_\nu} \right]$$
$$= - \sum_{\nu=1}^{\infty} \sum_{n=2}^{\infty} \frac{\lambda^n}{n} \left(\frac{1}{\lambda_\nu} \right)^n$$

Comparing these two results, we see that

$$\text{Tr}(K^n) \equiv k_n = \sum_{\nu=1}^{\infty} \left(\frac{1}{\lambda_\nu} \right)^n \tag{3}$$

for all $n \geq 2$. It is natural to surmise that this relationship holds also when $n = 1$. *Such is not the case, however, even if the left-hand side of (3) is meaningful for the kernel under consideration.* In fact, it may be inferred from results of Salem (1954) that there are continuous symmetric kernels with denumerably many characteristic values for which $\Sigma(1/\lambda_\nu)$ does not even exist.

4.4 THE RESOLVENT EQUATION

We want now to derive some interesting and important relationships for the resolvent kernel, which we have seen to be, in general, a meromorphic function of λ. To do this we shall use the Neumann series representation for R_K. Thus, for small enough λ and μ

$$R_K(\lambda)R_K(\mu) \equiv \int_a^b R_K(x,z;\lambda)R_K(z,y;\mu)\, dz$$

$$= \sum_{\alpha=1}^{\infty} \sum_{\beta=1}^{\infty} \lambda^{\alpha-1}\mu^{\beta-1}K^{\alpha+\beta}(x,y)$$

$$= \sum_{\alpha=1}^{\infty} \lambda^{\alpha-1} \sum_{\nu=\alpha+1}^{\infty} \mu^{\nu-\alpha-1}K^{\nu}(x,y)$$

$$= \sum_{\nu=2}^{\infty} \mu^{\nu-2}K^{\nu}(x,y) \sum_{\alpha=1}^{\nu-1} \left(\frac{\lambda}{\mu}\right)^{\alpha-1}$$

For $\nu \geq 2$, however, the inner summation gives rise to

$$\sum_{\alpha=1}^{\nu-1} \left(\frac{\lambda}{\mu}\right)^{\alpha-1} = \sum_{\alpha=0}^{\nu-2} \left(\frac{\lambda}{\mu}\right)^{\alpha}$$

$$= \frac{(\lambda/\mu)^{\nu-1} - 1}{(\lambda/\mu) - 1}$$

$$= \mu^{2-\nu}\frac{\lambda^{\nu-1} - \mu^{\nu-1}}{\lambda - \mu}$$

It follows, then, that for $\lambda \neq \mu$

$$\int_a^b R_K(x,z;\lambda)R_K(z,y;\mu)\, dz = \frac{1}{\lambda-\mu}\sum_{\nu=2}^{\infty}\left(\lambda^{\nu-1} - \mu^{\nu-1}\right)K^{\nu}(x,y)$$

$$= \frac{1}{\lambda-\mu}\left[R_K(x,y;\lambda) - R_K(x,y;\mu)\right]$$

or in symbolic notation

$$R_K(\lambda)R_K(\mu) = \frac{R_K(\lambda) - R_K(\mu)}{\lambda - \mu} \tag{1}$$

Although derived under the assumption of small λ and μ, this expression, called the *resolvent equation*, is clearly valid for all complex values of λ and μ ($\lambda \neq \mu$) that are regular values of the $\mathfrak{L}^2$ kernel K. It contains the familiar Fredholm identities (3.2-5) as special cases when either $\mu = 0$ or $\lambda = 0$.†

Several other results follow from the resolvent equation. For instance, letting $\mu = \sigma + \lambda$, we can rewrite (1) as

$$R_K(\sigma + \lambda) = R_K(\lambda) + \sigma R_K(\lambda)R_K(\sigma + \lambda)$$

Symmetry dictates that

$$R_K(\sigma + \lambda) = R_K(\lambda) + \sigma R_K(\sigma + \lambda)R_K(\lambda)$$

is a valid relationship also. These two expressions, by analogy with the Fredholm identities, show that *the resolvent kernel, with respect to σ, of $R_K(\lambda)$ is $R_K(\sigma + \lambda)$* and thus

$$R_K(x,y; \sigma + \lambda) = \sum_{\nu=1}^{\infty} \sigma^{\nu-1} R_K{}^{\nu}(x,y; \lambda) \tag{2}$$

at least for small enough σ. By means of this result we see how the resolvent kernel, originally defined in terms of iterates of K by the Neumann series (3.1-4), may be analytically continued throughout the open set formed by the regular values of the given kernel K.

---

† An alternate derivation of (1), based upon the Fredholm identities, takes the following form [see Riesz and Sz.-Nagy (1955, p. 153)]:

$$\lambda R_K(\lambda)[R_K(\mu) - K] = \lambda R_K(\lambda)[\mu K R_K(\mu)]$$

$$= [\lambda R_K(\lambda)K]\mu R_K(\mu)$$

$$= [R_K(\lambda) - K]\mu R_K(\mu)$$

Thus, $$(\lambda - \mu)R_K(\lambda)R_K(\mu) = \lambda R_K(\lambda)K - \mu K R_K(\mu)$$

$$= R_K(\lambda) - K + K - R_K(\mu)$$

$$= R_K(\lambda) - R_K(\mu)$$

Also see Plemelj (1904) and Radon (1919).

Proceeding one step further, since $R_K(\sigma + \lambda)$ is a meromorphic function of both λ and σ, it can be expanded around any regular value λ as

$$R_K(x,y;\sigma + \lambda) = \sum_{\nu=0}^{\infty} \frac{\sigma^\nu}{\nu!} \frac{\partial^\nu}{\partial\lambda^\nu} R_K(x,y;\sigma + \lambda) \bigg|_{\sigma=0}$$

$$= \sum_{\nu=0}^{\infty} \frac{\sigma^\nu}{\nu!} \frac{\partial^\nu}{\partial\lambda^\nu} R_K(x,y;\lambda)$$

Comparison with (2) then yields the following relationship for the various iterates of R_K:

$$R_K{}^{\nu+1}(x,y;\lambda) = \frac{1}{\nu!} \frac{\partial^\nu}{\partial\lambda^\nu} R_K(x,y;\lambda) \tag{3}$$

valid for integer ν and λ a regular value of K. As a special case we have for the second iterate

$$R_K{}^2(x,y;\lambda) = \frac{\partial}{\partial\lambda} R_K(x,y;\lambda)$$

a result that could have been obtained directly from (1) by merely letting $\mu \to \lambda$.

It is of interest also to apply the trace operator to the above expressions. We first observe that for small enough λ

$$\mathrm{Tr}(R_K - K) = \sum_{n=2}^{\infty} \lambda^{n-1} k_n$$

which, using (4.3-3) and rearranging, gives rise to

$$\mathrm{Tr}(R_K - K) = \sum_{n=2}^{\infty} \lambda^{n-1} \sum_{\nu=1}^{\infty} \left(\frac{1}{\lambda_\nu}\right)^n$$

$$= \sum_{\nu=1}^{\infty} \frac{\lambda}{\lambda_\nu{}^2} \sum_{n=2}^{\infty} \left(\frac{\lambda}{\lambda_\nu}\right)^{n-2}$$

$$= \sum_{\nu=1}^{\infty} \frac{\lambda}{\lambda_\nu(\lambda_\nu - \lambda)} \tag{4}$$

This last result, valid for all $\lambda \neq \lambda_\nu$ by analytic continuation, clearly shows the simple poles, with negative integer residues, of $\mathrm{Tr}(R_K - K)$

(see Chap. 3, Prob. 11). If we now combine (1) and (4), we find for all λ and μ regular values of K that

$$\text{Tr}\left[R_K(\lambda)R_K(\mu)\right] = \sum_{\nu=1}^{\infty} \frac{1}{(\lambda_\nu - \lambda)(\lambda_\nu - \mu)} \tag{5}$$

Similarly, from (3) and (4) we obtain

$$\text{Tr}\left[R_K{}^n(\lambda)\right] = \sum_{\nu=1}^{\infty} \left(\frac{1}{\lambda_\nu - \lambda}\right)^n \qquad n \geq 2 \tag{6}$$

4.5 A SIMPLE APPLICATION

In order to suggest the type of calculations that may be performed in evaluating the Fredholm functions associated with a given kernel, let us consider the square-integrable degenerate kernel of rank 1:

$$K(x,y) = A(x)\overline{B}(y) \qquad a \leq x,y \leq b$$

It follows, then, from the recursion relations (4.1-10) that $d_0 = 1$, $D_0(x,y) = A(x)\overline{B}(y)$, $d_1 = -k_1$, and $D_1(x,y) = 0$. Thus $D_\nu(x,y) = 0 = d_\nu$ for $\nu \geq 2$, and

$$D(\lambda) = 1 - \lambda k_1$$
$$D(x,y;\lambda) = A(x)\overline{B}(y)$$

We then have

$$R_K(x,y;\lambda) = \frac{A(x)\overline{B}(y)}{1 - \lambda k_1}$$

or, in general, for each integer $\nu \geq 1$

$$R_K{}^\nu(x,y;\lambda) = k_1{}^{\nu-1}\frac{A(x)\overline{B}(y)}{(1 - \lambda k_1)^\nu}$$

The resolvent equation (4.4-1) and the relationship (4.4-3) may now be easily verified for this particular kernel.

PROBLEMS

1. Establish the validity of the expressions (4.1-9) and (4.1-10) for the unmodified Fredholm functions.

2. [Smithies (1962, pp. 70ff).] If we employ the definition

$$
K\begin{pmatrix} x_1, \ldots, x_\nu \\ y_1, \ldots, y_\nu \end{pmatrix} \equiv \begin{vmatrix} K(x_1,y_1) & K(x_1,y_2) & \cdots & K(x_1,y_\nu) \\ K(x_2,y_1) & K(x_2,y_2) & \cdots & K(x_2,y_\nu) \\ \cdots\cdots\cdots\cdots\cdots\cdots\cdots \\ K(x_\nu,y_1) & K(x_\nu,y_2) & \cdots & K(x_\nu,y_\nu) \end{vmatrix}
$$

the classical formulas for the higher-order coefficients d_ν and $D_\nu(x,y)$ appearing in the power series expansions of the Fredholm determinant and the first Fredholm minor (4.1-9), respectively, become

$$
d_\nu = \frac{(-1)^\nu}{\nu!} \int_a^b \cdots \int_a^b K\begin{pmatrix} x_1, \ldots, x_\nu \\ x_1, \ldots, x_\nu \end{pmatrix} dx_1 \cdots dx_\nu
$$

$$
\nu \geq 1
$$

$$
D_\nu(x,y) = \frac{(-1)^\nu}{\nu!} \int_a^b \cdots \int_a^b K\begin{pmatrix} x, x_1, \ldots, x_\nu \\ y, x_1, \ldots, x_\nu \end{pmatrix} dx_1 \cdots dx_\nu
$$

Verify that these expressions satisfy the recurrence relations (4.1-10).

3. Using the formulas (4.1-9) and (4.1-10), calculate $D(\lambda)$ and $D(x,y;\lambda)$, and hence form $R_K(x,y;\lambda)$ for the kernel

$$
K(x,y) = x + y \qquad 0 \leq x,y \leq 1
$$

4. Calculate the modified Fredholm functions $\tilde{D}(\lambda)$ and $\tilde{D}(x,y;\lambda)$ for the kernel

$$
K(x,y) = A(x)\overline{B}(y) \qquad a \leq x,y \leq b
$$

5. Given arbitrary complex λ, construct a degenerate kernel of rank 1 for which the equality sign holds in (4.3-1). [Carleman (1921, p. 199) studied the degenerate kernel K_n with squared norm Ω given in terms of the real orthonormalized functions ϕ_ν by

$$
K_n(x,y) = e^{i\theta} \sqrt{\frac{\Omega}{2n}} \left[\sum_{\nu=1}^{n} \phi_\nu(x)\phi_\nu(y) - \sum_{\nu=n+1}^{2n} \phi_\nu(x)\phi_\nu(y) \right]
$$

For this kernel we have

$$
\tilde{D}(\lambda) = \left(1 - \lambda^2 \frac{\Omega}{2n} e^{2i\theta} \right)^n
$$

from which it follows that

$$
\lim_{n \to \infty} \tilde{D}(\lambda) = \exp\left[-\tfrac{1}{2} \lambda^2 \Omega e^{2i\theta} \right]
$$

If $\theta + \arg \lambda = \tfrac{1}{2}\pi$, therefore, the bound in (4.3-1) may be approached arbitrarily closely.]

6. Show that $k_\nu = 0$ for all ν greater than some positive integer $n \ (\geq 2)$ is a necessary and sufficient condition for the absence of characteristic values of a given $\mathfrak{L}^2$ kernel $K(x,y)$.

7. Verify the resolvent equation (4.4-1) for the kernel of Prob. 3.

8. Let

$$K(x,y) = \sum_{\nu=0}^{\infty} \alpha_\nu \cos \nu x \cos \nu y \qquad 0 \leq x,y \leq \pi$$

where $\Sigma \, |\alpha_\nu| < \infty$.

a. Use the Neumann series representation to show that the resolvent associated with this kernel is given by

$$R_K(x,y;\lambda) = \frac{\alpha_0}{1 - \pi\alpha_0\lambda} + \sum_{\nu=1}^{\infty} \frac{\alpha_\nu \cos \nu x \cos \nu y}{1 - (\pi/2)\alpha_\nu \lambda}$$

b. Verify the resolvent equation for this kernel. [N.B.: the $\cos \nu x$ are the characteristic functions of this kernel associated with the characteristic values $2/(\pi\alpha_\nu)$ for $\nu \neq 0$ and $1/(\pi\alpha_0)$ for $\nu = 0$.]

9. Let R_n be the resolvent associated with the iterated kernel K^n.

a. Show that

$$n\lambda^{n-1} R_n(\lambda^n) = \sum_{\mu=0}^{n-1} \epsilon^\mu R_K(\epsilon^\mu \lambda)$$

where $\epsilon \equiv e^{2\pi i/n}$

b. If

$$H_n(\lambda) \equiv \sum_{\nu=1}^{n-1} \lambda^{\nu-1} K^\nu$$

satisfy yourself that

$$R_K(\lambda) = H_n(\lambda) + \lambda^{n-1} R_n(\lambda^n) + \lambda^n H_n(\lambda) R_n(\lambda^n)$$

10. Explore the ramifications of the results of the preceding problem and thereby develop an alternative approach to Prob. 14, Chap. 3.

chapter five

The Nature of Volterra Equations

5.1 VOLTERRA KERNELS

In Appendix A we show that integral equations of the Volterra type occur quite naturally in physical applications that lead to initial-value problems. In this and succeeding sections of this chapter we shall investigate the behavior that is characteristic of such equations and sample a variety of results that arise with Volterra kernels.

Consider a Volterra integral equation

$$\varphi(x) = f(x) + \lambda \int_a^x K(x,y)\varphi(y)\, dy \qquad a \le x, y \le b \qquad (1)$$

where we assume that $K(x,y)$, square integrable for $y \le x$, is defined to be *zero* for $y > x$. It follows that

$$K^\nu(x,y) = \begin{cases} \displaystyle\int_y^x K(x,z)K^{\nu-1}(z,y)\, dz & y \le x \\[2ex] 0 & y > x \end{cases}$$

for $\nu \geq 2$, and thus each of the iterates, as well as the resolvent R_K, of a Volterra kernel is also of the Volterra type (and so is the composite of any two Volterra kernels). The Fredholm identities (3.2-5) take the special form

$$R_K(x,y;\lambda) = K(x,y) + \begin{cases} \lambda \int_y^x K(x,z)R_K(z,y;\lambda)\,dz \\ \\ \lambda \int_y^x R_K(x,z;\lambda)K(z,y)\,dz \end{cases} \tag{2}$$

Moreover,
$$k_2 = k_3 = \cdot\cdot\cdot = 0$$

so that the modified Fredholm determinant reduces to

$$\tilde{D}(\lambda) \equiv \exp\left(-\sum_{\nu=2}^{\infty} \frac{\lambda^\nu}{\nu}\,k_\nu\right)$$
$$= 1$$

We may conclude, therefore, that $\mathfrak{L}^2$ Volterra kernels possess no characteristic values† and have resolvent kernels that are regular analytic throughout the entire complex λ plane. In other words, the Neumann series (3.1-4) converges for arbitrary λ.‡

5.2   VOLTERRA EQUATIONS WITH DIFFERENCE KERNELS

Many problems of interest arise from physical phenomena that exhibit an aftereffect behavior. In such cases an appropriate description often takes the form of an integral equation of convolution type in which $K(x,y)$ is a *difference kernel*. A canonical example is given by§

$$\varphi(x) = f(x) + \lambda \int_0^x k(x-y)\varphi(y)\,dy$$

where $k(\sigma) = 0$ for $\sigma < 0$. For suitable k, this particular type of equa-

---

† Smithies (1962, p. 35) discusses the celebrated example of an $\mathfrak{L}^2$ Volterra kernel with a non-$\mathfrak{L}^2$ characteristic function due to Bôcher (1909).

‡ These same conclusions, of course, can be derived (and often are) from "first principles"; see Sec. 5.6.

§ For a study of an interesting generalization of this equation, see Friedman and Shinbrot (1967) and Friedman (1969). For a discussion of difference kernels on infinite intervals, see Chap. 18.

tion can always be solved by a very elegant approach [Paley and Wiener (1933, pp. 785–791), (1934, sec. 18)].

Given any appropriate function $f(x)$, we recall the definition of its Laplace transform, viz.,

$$\mathscr{L}[f] = \int_0^\infty e^{-xs} f(x)\, dx$$

Applying this transformation to the above equation yields

$$\mathscr{L}[\varphi] = \mathscr{L}[f] + \lambda\, \mathscr{L}[k] \cdot \mathscr{L}[\varphi]$$

or

$$\mathscr{L}[\varphi] = \frac{\mathscr{L}[f]}{1 - \lambda \mathscr{L}[k]}$$

(see Prob. 2). It should be noted that we have gained significantly by this technique. An integral equation with an arbitrary kernel $k(x-y)$ has been transformed into an integral equation with a fixed kernel e^{-xs}; that is,

$$\frac{F(s)}{1 - \lambda K(s)} = \Phi(s) = \int_0^\infty e^{-xs}\varphi(x)\, dx$$

and for this last equation we have the inversion formula

$$\varphi(x) = \frac{1}{2\pi i} \int_{\sigma - i\infty}^{\sigma + i\infty} e^{xs}\Phi(s)\, ds$$

where σ is larger than the real part of any of the singularities of $\Phi(s)$. [See Titchmarsh (1948), Widder (1941), or Erdélyi (1954), for example.] A related technique will be useful in dealing with Fredholm kernels of the Wiener-Hopf type, as we shall see in a subsequent chapter.

As Tricomi (1957) observes, for difference kernels of the Volterra type, all of the iterates as well as the resolvent kernel are also of this type; e.g.,

$$\int_y^x k(x-z)k(z-y)\, dz = \int_0^{x-y} k(x-y-\tau)k(\tau)\, d\tau$$

See Tricomi (1940 and 1957), Davis (1930), Fock (1924), and Doetsch (1923 and 1925) for various other results and applications.

The approach outlined above, naturally, is useful also for convolution equations of the first kind [see Titchmarsh (1948, pp. 322ff), for example]. In certain fortuitous situations, notably those in which the kernel is simply related to an orthogonal polynomial, the solution takes the form of a convolution transform of essentially the same type as the original equation [Widder (1963), Higgins (1964)]. Reudink (1967) has recently shown that kernel functions whose Laplace transforms are polynomials to the minus one-half power allow such inversions.

5.3 DEGENERATE VOLTERRA KERNELS OF RANK 1

If $K(x,y)$ is such that

$$K(x,y) = \begin{cases} A(x)B(y) & y \leq x \\ 0 & y > x \end{cases}$$

then

$$K^2(x,y) = A(x)B(y) \int_y^x A(z)B(z) \, dz$$

Similarly,

$$K^3(x,y) = \frac{1}{2}\left[A(x)B(y) \int_y^x A(z)B(z) \int_z^x A(t)B(t) \, dt \, dz \right.$$

$$\left. + A(x)B(y) \int_y^x A(z)B(z) \int_y^z A(t)B(t) \, dt \, dz \right]$$

$$= \frac{A(x)B(y)}{2} \left[\int_y^x A(z)B(z) \, dz \right]^2$$

and, in general,

$$K^\nu(x,y) = \frac{A(x)B(y)}{(\nu - 1)!} \left[\int_y^x A(z)B(z) \, dz \right]^{\nu - 1} \tag{1}$$

This last expression may be obtained by induction.

It follows from (1) that

$$R_K(x,y;\lambda) = A(x)B(y) \exp\left[\lambda \int_y^x A(z)B(z) \, dz \right] \tag{2}$$

a result which should be contrasted with that obtained in the case of a degenerate Fredholm kernel of the same form, namely,

$$R_K(x,y;\lambda) = \frac{A(x)B(y)}{1 - \lambda \int_a^b A(z)B(z) \, dz}$$

Unfortunately, the compact expression (2) does not generalize to degenerate Volterra kernels of higher rank. Goursat (1933) has shown, however, that the determination of the resolvents of such kernels is completely equivalent to the solution of certain linear ordinary differential equations.

5.4 ELEMENTARY PERTURBATION THEORY FOR VOLTERRA KERNELS

Although the analysis of perturbations of general integral operators is often a complex subject, as we shall see in Chap. 12 [see also Kato (1966)], the situation is somewhat simplified for Volterra kernels. We have, for example, the following theorem [see also Porath (1968)].

THEOREM 5.4

If the Volterra kernel $K(x,y;\epsilon)$, $a \leq x,y \leq b$, depending on a smallness parameter ϵ satisfies

$$K(x,y;\epsilon) = K_0(x,y) + \epsilon K_1(x,y;\epsilon)$$

where K_0, K_1 are bounded (in norm) independently of ϵ, then the resolvent $R_K(x,y;\lambda,\epsilon)$ may be expressed as

$$R_K(x,y;\lambda,\epsilon) = R_0(x,y;\lambda) + \epsilon R_1(x,y;\lambda,\epsilon)$$

where R_0 is the resolvent of the Volterra kernel K_0 and R_1 is likewise bounded independently of ϵ.

Proof   Substituting in the first of the Fredholm identities (5.1-2), we obtain

$$R_0 + \epsilon R_1 = K_0 + \epsilon K_1 + \lambda \int_y^x (K_0 + \epsilon K_1)(R_0 + \epsilon R_1) \, dz$$

or $$R_1 = K_1 + \lambda \int_y^x K_1 R_0 \, dz + \lambda \int_y^x K R_1 \, dz$$

since R_0 is the resolvent of K_0. For fixed y, we recognize this as a Volterra integral equation, with kernel K, for the function R_1. It follows then that

$$R_1 = K_1 + \lambda \int_y^x K_1 R_0 \, dz + \lambda \int_y^x R_K \left(K_1 + \lambda \int_y^z K_1 R_0 \, dt \right) dz$$

Since K_1 is bounded (in norm) independently of ϵ, so also is K and thus so also is R_K. This, coupled with the boundedness of K_0 and hence of R_0, ensures the desired behavior for R_1.

5.5   POSITIVE KERNELS AND COMPARISON THEOREMS

For nonnegative kernels, each of the iterates K^ν and hence the resolvent R_K as well is also nonnegative. It follows trivially in such cases that

$$R_K(x,y;\lambda) \geq K(x,y) \geq 0 \qquad (1)$$

Moreover, if f is nonnegative on the interval $[a,b]$, then the solution of the integral equation (5.1-1) for arbitrary positive λ must satisfy a related inequality, viz.,

$$\varphi(x) \geq f(x) \geq 0 \qquad (2)$$

These elementary conclusions are all that is needed to establish for Volterra integral equations the following analogues of classical comparison theorems for differential equations [see Cole (1968, chap. 9), for example].

THEOREM 5.5-1

Let the square-integrable functions f_i and Volterra kernels $K_i (i = 1, 2)$ satisfy

$$|f_1(x)| \leq f_2(x) \qquad |K_1(x,y)| \leq K_2(x,y) \qquad (3)$$

The inequality

$$|\varphi_1(x)| \leq \varphi_2(x) + |f_1(x)| - f_2(x)$$

then holds between the unique $\mathfrak{L}^2$ solutions of the integral equations

$$\varphi_i(x) = f_i(x) + \int_a^x K_i(x,y)\varphi_i(y)\, dy \qquad i = 1, 2 \qquad (5.1\text{-}1')$$

Proof   Symbolically we have

$$|\varphi_1| \leq |f_1| + |K_1||\varphi_1|$$
$$\leq |f_1| + K_2|\varphi_1|$$

so that $\varphi_2 - |\varphi_1| \geq f_2 - |f_1| + K_2(\varphi_2 - |\varphi_1|)$

If g denotes the (positive) difference between the left- and right-hand sides of this inequality, it follows from (2) and the first of (3) that

$$\varphi_2 - |\varphi_1| \geq f_2 - |f_1| + g$$

which leads to the desired result.

For real-valued square-integrable Volterra kernels we have the following theorem.

THEOREM 5.5-2

Under the assumptions that

$$0 \leq f_2(x) \qquad f_1(x) \leq f_2(x)$$
$$0 \leq K_1(x,y) \leq K_2(x,y) \tag{4}$$

the unique $\mathfrak{L}^2$ solutions of (5.1-1') satisfy the inequality

$$\varphi_1(x) \leq \varphi_2(x) + f_1(x) - f_2(x) \tag{5}$$

Proof Denoting the resolvent kernels associated with K_1, K_2 evaluated at $\lambda = 1$ by R_1 and R_2, respectively, we know that the unique $\mathfrak{L}^2$ solutions of (5.1-1') may be expressed as

$$\varphi_i = f_i + R_i f_i \qquad i = 1, 2$$

It follows, then, that

$$\varphi_2 - \varphi_1 = f_2 - f_1 + R_2 f_2 - R_1 f_1$$
$$= f_2 - f_1 + (R_2 - R_1)f_2 + R_1(f_2 - f_1) \tag{6}$$

Since $K_1 \leq K_2$, we have $K_1{}^\nu \leq K_2{}^\nu$ (for all ν) and thus $R_1 \leq R_2$. The relations (1), (4), and (6) now lead to

$$\varphi_2 - \varphi_1 \geq f_2 - f_1 + K_1(f_2 - f_1)$$

which is an even stronger inequality than (5).

COROLLARY

In the case wherein K_1 and K_2 are *equivalent* kernels, the inequality (5) is valid for arbitrary $f_2(x) \geq f_1(x)$.

The above results appear to have been presented in published form first by Beesack (1969a). The corollary of Theorem 5.5-2 contains an extension of an earlier result of Chu and Metcalf (1967), which itself generalized the classical inequality of Gronwall (1919) [see Prob. 5; see also Beckenbach and Bellman (1965, pp. 134ff)].

5.6 NONLINEAR VOLTERRA INTEGRAL EQUATIONS

The important results for nonlinear Volterra integral equations of the second kind follow basically from the classical method of successive approximations, or the *Picard process*.† Recall its application to the integral equation

$$y(x) = y_0 + \int_a^x F(t, y(t)) \, dt$$

where, for instance, $F(x,y)$ is continuous in the rectangle $R = \{(x,y): |x-a| \le c, |y-y_0| \le d\}$, with $c > 0$ and $d > 0$, and satisfies there a uniform Lipschitz condition in the second variable of the form

$$|F(x,y) - F(x,z)| \le A |y - z|$$

with A a positiv constant [see Coddington and Levinson (1955, chap. 1) and Cole (19 8, chap. 2); the original reference here is Lipschitz (1880)]. The procedure involves the recursive algorithm

$$y_0(x) = y_0$$
$$y_{\nu+1}(x) = y_0 + \int_a^x F(t, y_\nu(t)) \, dt \qquad \nu \ge 0 \tag{1}$$

Note particularly, then, that

$$|y_{\nu+1}(x) - y_0| \le \pm \int_a^x |F(t, y_\nu(t))| \, dt \le \text{const} \, |x - a|$$

so that during this process we never leave a fundamental domain of the type R.

---

† Discussion of the nonlinear situation under appropriate restrictions is hardly more complicated than a careful treatment of the linear case from this point of view. When applied to the linear situation, this classical approach shows, as expected, that the Neumann series for the resolvent converges for arbitrary $\mathfrak{L}^2$ Volterra kernels.

From (1) we have (by induction)

$$| y_{\nu+1}(x) - y_\nu(x) | \le \left| \int_a^x [F(t,y_\nu) - F(t,y_{\nu-1})] \, dt \right|$$

$$\le A \int_a^x | y_\nu(t) - y_{\nu-1}(t) | \, dt$$

$$\le \text{const} \, \frac{A^\nu |x - a|^{\nu + 1}}{(\nu + 1)!} \tag{2}$$

where the constant is independent of ν.

This line of reasoning can be extended to show that the y_ν form a Cauchy sequence in the space of continuous functions. It then follows from the Arzela-Ascoli theorem† [since the $y_\nu(x)$ are actually equiuniformly continuous] that there exists a uniformly convergent subsequence and hence a continuous limit function $Y(x)$. Alternatively we could argue that since the continuous functions are dense in $\mathfrak{L}^2$ (which is a complete space), $Y(x)$ is at least square integrable and satisfies

$$Y(x) = y_0 + \int_a^x F(t, Y(t)) \, dt$$

Since the right-hand side is continuously differentiable with respect to x, it then results that $Y(x)$ actually has that property also. A third possibility is to use the fact that

$$Y(x) = \lim_{\nu \to \infty} y_\nu(x) = y_0(x) + \left[y_1(x) - y_0(x) \right] + \left[y_2(x) - y_1(x) \right] + \cdots$$

a series which, in view of (2), is majorized by

$$\frac{\text{const}}{A} \, (e^{A|x-a|} - 1)$$

Thus, $Y(x)$ is the sum of an absolutely and uniformly convergent series of continuous functions from which the desired properties follow.

Considerations analogous to those presented above are readily

---

† Although many authors attribute this result to one or the other alone, both Ascoli (1883–1884, pp. 545–549) and Arzela (1889 and 1895, pp. 56–60) made contributions toward the theorem that bears their names. See Hobson (1957, vol. 2, pp. 168–169) or Garabedian (1964, p. 380) for more modern versions of the proof.

applicable to general nonlinear Volterra equations [see Tricomi (1957, pp. 42ff); also see Wouk (1964) and Nohel (1964)]. For instance, let

$$\varphi(x) = f(x) + \int_a^x K\Big(x, y; \varphi(y)\Big)\, dy \qquad a \leq x, y \leq b \qquad (3)$$

where f is in $\mathfrak{L}^2$ and

(a) $\qquad |K(x, y; \varphi) - K(x, y; \psi)| \leq A(x, y)\,|\varphi - \psi|$

(b) $\qquad \int_a^b \int_a^x A^2(x, y)\, dy\, dx \leq B^2$

(c) $\qquad \int_a^b \left| \int_a^x K\Big(x, y; \varphi(y)\Big)\, dy \right|^2 dx \leq C^2$

with B and C positive constants.† The recurrence formula

$$\varphi_0(x) = f(x)$$

$$\varphi_{\nu+1}(x) = f(x) + \int_a^x K\Big(x, y; \varphi_\nu(y)\Big)\, dy$$

gives rise to

$$|\varphi_{\nu+1}(x) - \varphi_\nu(x)|^2 \leq \left[\int_a^x \left| K\Big(x, y; \varphi_\nu(y)\Big) - K\Big(x, y; \varphi_{\nu-1}(y)\Big) \right| dy \right]^2$$

$$\leq \left[\int_a^x A(x, y)\,|\varphi_\nu(y) - \varphi_{\nu-1}(y)|\, dy \right]^2$$

$$\leq \int_a^x A^2(x, y)\, dy \int_a^x |\varphi_\nu(y) - \varphi_{\nu-1}(y)|^2\, dy$$

Substituting successively in this inequality and making the same observation as in Sec. 5.3 then yields

$$|\varphi_{\nu+2}(x) - \varphi_{\nu+1}(x)|^2 \leq C^2 \int_a^x A^2(x, y)\, dy\, \frac{B^{2\nu}}{\nu!} \qquad \nu \geq 0$$

But this just shows, as before, that the series $\varphi_\nu(x)$ converges *relatively uniformly* to a square-integrable limit function $\Phi(x)$.

---

† If K is linear in φ and square integrable in x, y, conditions (a) and (b) are automatically fulfilled. Condition (c) then will be satisfied if we restrict attention, as usual, to the space of $\mathfrak{L}^2$ functions.

It is a simple matter to show that $\Phi(x)$ is an almost everywhere solution of (3) (see Prob. 6 at the end of this chapter). The uniqueness question can also be answered affirmatively, for let $\Phi_1(x)$ and $\Phi_2(x)$ both be $\mathfrak{L}^2$ functions satisfying the given integral equation. We then have

$$\Phi_2(x) - \Phi_1(x) = \int_a^x \left[K\Big(x,y;\Phi_2(y)\Big) - K\Big(x,y;\Phi_1(y)\Big) \right] dy$$

from which it follows as above that

$$|\Phi_2(x) - \Phi_1(x)|^2 \le \int_a^x A^2(x,y)\,dy \int_a^x |\Phi_2(y) - \Phi_1(y)|^2 dy$$

Successive substitution then leads to

$$\int_a^x |\Phi_2(x) - \Phi_1(x)|^2\,dx \le \|\Phi_2 - \Phi_1\|^2 \frac{B^{2\nu}}{\nu!}$$

for arbitrary integer ν, which implies the desired result that $\Phi_2(x) = \Phi_1(x)$ almost everywhere. Erdélyi (1962) has shown how certain useful elementary growth characteristics of the solution can be obtained by strengthening essentially condition (b) above.

The existence of solutions to (3) actually follows under restrictions weaker than those given previously. For example, if the kernel K merely satisfies a generalized Lipschitz or Hölder condition

$$|K(x,y;\varphi) - K(x,y;\psi)| \le A(x,y)\,|\varphi - \psi|^\alpha \tag{4}$$

with $0 < \alpha < 1$, an argument similar to that carried out above leads to the appropriate conclusion. With regard to the uniqueness of solutions of the given integral equation, however, the situation is not as clear-cut. Although condition (a) is not a necessary condition, (4) unfortunately is not generally sufficient for uniqueness (see Probs. 7 and 11).

5.7   THE UNIQUENESS OF SOLUTIONS OF BOUNDARY-VALUE PROBLEMS

If initial-value problems are reformulated as Volterra integral equations, the considerations of the previous section show that the existence and uniqueness questions can both be easily settled under fairly broad assumptions. In the case of boundary-value problems, however, wherein the situation is represented by equations of the Fredholm type (see Appendix A), proof of the uniqueness of solutions is of a more complex nature and requires careful analysis regarding the possible existence of characteristic functions for the related homogeneous problem. These difficulties can be obviated, at least formally, by exploiting the possibility of representing boundary-value problems as *pseudo* initial-value problems and hence as equivalent *pseudo* Volterra integral equations. This approach, which has similarities with the *shooting method* of Henrici (1962) [see also Keller (1968, chap. 2)], was applied to a particular class of problems arising in singular perturbation theory by Cochran (1962). Subsequently in Cochran (1968), the approach was used to derive necessary and sufficient conditions (nasc) for the uniqueness of solutions of two-point boundary-value problems in terms of the resolvent kernels associated with the "related" Volterra integral equations. (For specific sets of boundary conditions the nasc become particularly simple and practical for application.) Somewhat related techniques based upon integral-equation methods and useful in general situations are discussed in Erdélyi (1964), Willett (1964), and Aalto (1966a and b).

For simplicity of expression let us consider the second-order equation

$$[p(x)u'(x)]' + q(x)u(x) = f(x) \qquad 0 \le x \le 1 \tag{1}$$

where all coefficients are assumed to be real valued and sufficiently smooth and $p(x)$ is taken, say, to be positive throughout the unit interval. The related integral equation may then be put in the form

$$u(x) = u(0) + u'(0)p(0) \int_0^x [p(z)]^{-1}\, dz$$
$$+ \int_0^x (f - qu) \left\{ \int_y^x [p(z)]^{-1}\, dz \right\} dy$$

If we assume for the moment that $u(0)$ and $u'(0)$ are known, this is a Volterra equation with kernel

$$K(x,y) = -q(y) \int_y^x [p(z)]^{-1}\, dz$$

and it has a solution expressible (after some manipulation) in terms of the resolvent kernel $R(x,y) \equiv R_K(x,y;1)$ as

$$u(x) = u(0) \left[1 + \int_0^x R(x,y)\, dy \right] - \frac{u'(0)p(0)R(x,0)}{q(0)} - \int_0^x f(y) \frac{R(x,y)}{q(y)}\, dy$$

We now return to a consideration of (1) as a two-point boundary-value problem and take the boundary conditions to be *regular*, or *self-adjoint* [see Ince (1956), for example], i.e., of the form

$$\begin{pmatrix} m_{11} & m_{12} \\ m_{21} & m_{22} \end{pmatrix} \begin{pmatrix} u(0) \\ u'(0) \end{pmatrix} + \begin{pmatrix} n_{11} & n_{12} \\ n_{21} & n_{22} \end{pmatrix} \begin{pmatrix} u(1) \\ u'(1) \end{pmatrix} = \begin{pmatrix} \alpha \\ \beta \end{pmatrix} \qquad (2)$$

where the combined matrix $(m_{ij} : n_{ij})$ is assumed to be of rank 2 and

$$(m_{11} m_{22} - m_{12} m_{21}) p(1) = (n_{11} n_{22} - n_{12} n_{21}) p(0)$$

The following theorem then can be shown to be valid.

THEOREM 5.7-1

The boundary-value problem (1) with the self-adjoint boundary conditions (2) has a *unique* solution *if and only if* (iff) the determinant

$$\begin{vmatrix} m_{11} + A n_{11} + C n_{12} & m_{12} + B n_{11} + D n_{12} \\ m_{21} + A n_{21} + C n_{22} & m_{22} + B n_{21} + D n_{22} \end{vmatrix}$$

is *nonvanishing*. Here

$$A \equiv 1 + \int_0^1 R(1, y) \, dy$$

$$B \equiv -\frac{p(0) R(1, 0)}{q(0)}$$

$$C \equiv -\left[p(1) \right]^{-1} \int_0^1 q(x) \left[1 + \int_0^x R(x, y) \, dy \right] dx$$

$$D \equiv p(0) \left[p(1) \right]^{-1} \left[1 + \int_0^1 q(x) \frac{R(x, 0)}{q(0)} \, dx \right]$$

As immediate consequences we have the following corollaries.

COROLLARY 1

The Dirichlet problem associated with (1), wherein

$$m_{12} = m_{22} = n_{12} = n_{22} = 0$$

has a unique solution iff $B \neq 0$.

COROLLARY 2

The Neumann problem associated with (1), wherein

$$m_{11} = m_{21} = n_{11} = n_{21} = 0$$

has a unique solution iff $C \neq 0$.

5.8 VOLTERRA EQUATIONS OF THE FIRST KIND

We close this chapter with a brief discussion of Volterra integral equations typified by

$$f(x) = \int_a^x K(x,y)\varphi(y)\, dy \qquad a \leq x \leq b \tag{1}$$

Such equations are of the first kind and need not be amenable to solution. Nevertheless let us proceed formally.

Differentiation of (1) gives rise to

$$f'(x) = K(x,x)\varphi(x) + \int_a^x \frac{\partial K(x,y)}{\partial x}\,\varphi(y)\, dy$$

If $K(x,x)$ does not vanish in the fundamental interval $[a,b]$, then we can divide through by it, thus obtaining a Volterra integral equation of the second kind. This latter equation can be solved if $f'(x)/K(x,x)$ and $[\partial K(x,y)/\partial x]/K(x,x)$ are in $\mathfrak{L}^2$, for example.

The most critical condition in the above procedure, obviously, is the nonvanishing of $K(x,x)$. [If $K(x,x) \equiv 0$, a further reduction of the problem may be effected by additional differentiations.] A zero of $K(x,x)$ for $a \leq x \leq b$ is a complication analogous to the presence of singular points in linear differential equations, and such a situation necessitates special care [see Tricomi (1957, p. 16) and Lalesco (1912), for example].

PROBLEMS

1. Show that for a *bounded* Volterra kernel $K(x,y)$, the Neumann series (3.1-4) is majorized by a uniformly and absolutely convergent series for all values of λ.

2. Satisfy yourself that for the convolution

$$\mathscr{L}\left[\int_0^x f(x-y)g(y)\, dy \right] = \mathscr{L}[f] \cdot \mathscr{L}[g]$$

where $\mathscr{L}$ denotes the Laplace transform introduced in Sec. 5.2.

3. Solve the integral equation

$$\phi(x) = 1 + \lambda \int_0^x (x - y)\phi(y)\, dy \qquad x \geq 0$$

by the following alternative methods:

(a) Reduction to the equivalent initial-value problem
(b) Use of the Neumann series and the resolvent kernel
(c) Laplace-transform techniques

4. Find a closed–form expression for the resolvent kernel associated with the equation

$$\phi(x,y) = f(x,y) - \tfrac{1}{2}\lambda^2 \int_0^x \int_0^y A(s)B(t)\phi(s,t)\, ds\, dt$$

in the region $x \geq 0$, $y \geq 0$.

5. a. Verify that the corollary of Theorem 5.5-2 implies that if

$$\phi(x) \leq f(x) + \int_a^x K(x,y)\phi(y)\, dy \qquad a \leq x \leq b$$

where f and K are square integrable and K is also nonnegative, then

$$\phi(x) \leq f(x) + \int_a^x R_K(x,y; 1)f(y)\, dy \qquad a \leq x \leq b$$

b. Using (5.3-2), show that in the special case wherein $f(x)$ is a constant c and $K(x,y) = B(y)$, the above result reduces to Gronwall's inequality

$$\phi(x) \leq c \exp\left[\int_a^x B(y)\, dy\right] \qquad a \leq x \leq b$$

6. Let $\Phi(x) = \lim_{\nu \to \infty} \phi_\nu(x) = \phi_\nu(x) + R_\nu(x)$ be as defined in Sec. 5.6. Show that $\Phi(x)$ is an almost everywhere solution of the integral equation (5.6-3) by verifying that

$$\int_a^b \left| \Phi(x) - f(x) - \int_a^x K\big(x,y; \Phi(y)\big)\, dy \right|^2 dx = 0$$

wherever $\int_a^x A^2(x,y)\, dy < \infty$ [Tricomi (1957, p. 45)].

7. Let the conditions (a), (b), and (c) of Sec. 5.6 be replaced by

$$(a') \quad |K(x,y;\phi) - K(x,y;\psi)| \leq A(x,y)|\phi - \psi|^\alpha$$

(b') $\displaystyle\int_a^b \left[\int_a^x A^2(x,y)\, dy\right]^\alpha dx \leq B^2$

(c') $\displaystyle\int_a^b \left|\int_a^x K\big(x,y;\phi(y)\big)\, dy\right|^{2\alpha} dx \leq C^2$

with $0 < \alpha < 1$ and B and C positive constants greater than unity. Show that even under these weaker restrictions the method of successive substitutions still gives rise to a solution of the nonlinear Volterra integral equation (5.6-3). *Hint*:

$$|\phi_{\nu+2}(x) - \phi_{\nu+1}(x)|^2 \leq C^2 \left[\int_a^x A^2(x,y)\, dy\right] \frac{B^{2\nu}}{F(\alpha,\nu)}$$

where $F(\alpha,\nu) = (1+\alpha)^{\alpha^{\nu-2}} (1+\alpha+\alpha^2)^{\alpha^{\nu-3}} \cdots (1+\alpha+\cdots+\alpha^{\nu-1})$

8.   For any point P in the (x,y) plane, let $S(P)$ denote the rectangle with one vertex at P and the diametrically opposite vertex at the origin. Discuss the method of successive approximations as applied to the Volterra equation

$$\phi(P) = f(P) + \lambda \iint_{S\,(P)} K(P,Q)\phi(Q)\, dS$$

9.   Discuss the form that the considerations of Sec. 5.6 assume in the case wherein the kernel K of (5.6-3) is an analytic function of its several variables.

10.   Determine the Volterra equation equivalent to the initial-value problem

$$\frac{d^2u}{dx^2} + q(x)u = 0 \qquad 0 \leq x \leq 1$$

$$u(0) = \alpha$$

$$u'(0) = \beta$$

and discuss its solution.

11.   Solve the homogeneous nonlinear Volterra equation

$$\phi(x) = \int_a^x \phi^\alpha(y)\, dy \qquad a \leq x \leq b$$

for $0 < \alpha < 1$, and hence show that solutions of nonhomogeneous equations with such kernels cannot be unique.

chapter six

Characteristic Behavior of Integral Equations

6.1 EXPLICIT CONSTRUCTION OF CHARACTERISTIC FUNCTIONS

In Chap. 3 we showed that for a given nonhomogeneous integral equation with square-integrable kernel, the solution was readily obtainable wherever the resolvent kernel existed as a function of the eigenparameter λ. We now turn our attention to the singular points of these resolvent kernels and study anew the behavior that is representative of integral equations at characteristic values of their kernels.

The concept of *orthogonal kernels*, first introduced by Goursat (1907) and Heywood (1907), will be useful in our investigations. Two square-integrable kernels $K_1(x,y)$ and $K_2(x,y)$ on $a \le x,y \le b$ are defined to be orthogonal if

$$\int_a^b K_1(x,z)K_2(z,y)\,dz = 0 = \int_a^b K_2(x,z)K_1(z,y)\,dz$$

for almost all x,y in the fundamental rectangle. It follows immediately, then, for such kernels that $K_1{}^n$ and $K_2{}^m$ are also orthogonal, where m and n are arbitrary positive integers. Hence, if $K(x,y) = K_1(x,y) + K_2(x,y)$ with K_1 and K_2 orthogonal, then

$$K^n = K_1{}^n + K_2{}^n$$

and
$$R_K(\lambda) = R_{K_1}(\lambda) + R_{K_2}(\lambda)$$

this last result being a consequence of the Neumann series representation for the resolvent, analytically continued throughout the complex λ plane. Since the modified Fredholm determinant is given by

$$\tilde{D}(\lambda) \equiv \exp\left[-\int_0^\lambda \mathrm{Tr}\,[R_K(\mu) - K]\,d\mu \right]$$

we also obtain for this situation

$$\tilde{D}(\lambda) = \tilde{D}_1(\lambda) \cdot \tilde{D}_2(\lambda)$$

From our earlier analysis (Sec. 3.5) we know that each distinct zero of $\tilde{D}(\lambda)$ appears as a pole of the resolvent kernel

$$R_K(x,y;\lambda) = \frac{\tilde{D}(x,y;\lambda)}{\tilde{D}(\lambda)}$$

and these poles [zeros of $\tilde{D}(\lambda)$], which are perforce isolated and have no finite point of accumulation, are also truly characteristic values of the given $\mathfrak{L}^2$ kernel K. Using the first Fredholm identity (3.2-5), we shall now construct a representation for the expected solutions to the homogeneous equation associated with these λ. A similar technique employing the other identity gives rise to solutions of the homogeneous transposed equation.

If $R_K(x,y;\lambda)$ has a pole of order n† at $\lambda = \lambda_0$, we may expand it in a Laurent series about this singular point, viz.,

$$R_K(x,y;\lambda) = \sum_{\nu=1}^{n} \frac{r_\nu(x,y)}{(\lambda - \lambda_0)^\nu} + r(x,y;\lambda) \tag{1}$$

Here r and the r_ν are square integrable as functions of x,y and r is regular analytic in λ within at least a sufficiently small neighborhood of the point $\lambda = \lambda_0$. Substituting (1) into the first of (3.2-5), we obtain

---

† The order n of the pole of R_K is not necessarily the same as the order N of the zero in $\tilde{D}(\lambda)$ from which it results. In other words, $\tilde{D}(x,y;\lambda)$ may well vanish at a zero of $\tilde{D}(\lambda)$, but the order of any such zero, of course, is strictly less than N. As we shall see later in this chapter, the special case $n = 1$ occurs often in practice.

$$\sum_{\nu=1}^{n} \frac{r_\nu(x,y)}{(\lambda - \lambda_0)^\nu} + r(x,y;\lambda) = K(x,y) + (\lambda - \lambda_0 + \lambda_0)$$
$$\cdot \int_a^b K(x,z)\left[\sum_{\nu=1}^{n} \frac{r_\nu(z,y)}{(\lambda - \lambda_0)^\nu} + r(z,y;\lambda)\right] dz$$

Rearranging the right-hand side and equating like negative powers in $(\lambda - \lambda_0)$ yields

$$r_\nu(x,y) = \int_a^b K(x,z)r_{\nu+1}(z,y)\,dz + \lambda_0 \int_a^b K(x,z)r_\nu(z,y)\,dz \qquad (2)$$

for $\nu = 1, 2, \ldots, n$, where $r_{n+1}(x,y)$ vanishes by definition. Since $r_n(x,y)$ must be a nontrivial $\mathfrak{L}^2$ function for the pole λ_0 of R_K to be of order n, there is at least one fixed value y_0 of y for which $r_n(x,y_0)$ is a nontrivial square-integrable function of the single variable x. Equation (2) then shows that for this value y_0, $r_n(x,y_0)$ properly normalized, is a characteristic function of the kernel K associated with the characteristic value λ_0 (see Prob. 1).

6.2 DECOMPOSITION OF A KERNEL IN TERMS OF CHARACTERISTIC FUNCTIONS

We have just seen how to construct a solution of the homogeneous equation when $\tilde{D}(\lambda) = 0$. By analogy with the *algebraic eigenvalue problem* in finite-dimensional spaces, however, we should not expect to be able always to construct as many linearly independent solutions as the order of the zero of $\tilde{D}(\lambda)$ (see Prob. 2 at the end of this chapter for an illustrative example; also see Prob. 4). On the other hand, given solutions $\varphi(x)$ and $\psi(x)$ to the homogeneous and transposed homogeneous equations, respectively, under certain conditions we can separate the contributions of these functions from the original kernel in a unique manner. The decomposition so obtained displays the characteristic functions in explicit fashion and often permits recursive application.

To see this, let nontrivial $\mathfrak{L}^2$ functions φ_0 and ψ_0 satisfy

$$\varphi_0(x) = \lambda_0 \int_a^b K(x,y)\varphi_0(y)\,dy$$
$$\psi_0(x) = \bar{\lambda}_0 \int_a^b K^*(x,y)\psi_0(y)\,dy \qquad (1)$$

with $\lambda_0 \neq 0$, and assume $(\varphi_0, \psi_0) = 1$. For the given square-integrable kernel $K(x,y)$ we write

$$K(x,y) = K_0(x,y) + K_1(x,y)$$

$$\equiv \frac{\varphi_0(x)\overline{\psi}_0(y)}{\lambda_0} + K_1(x,y)$$

Since φ_0 and ψ_0 have unit inner product, it is then a simple matter to establish the *orthogonality* of the two kernels K_0 and K_1. It follows that

$$\tilde{D}(\lambda) = \tilde{D}_0(\lambda) \cdot \tilde{D}_1(\lambda)$$

$$= \left(1 - \frac{\lambda}{\lambda_0}\right) e^{\lambda/\lambda_0} \tilde{D}_1(\lambda)$$

If $\tilde{D}(\lambda)$ originally had an Nth-order zero at $\lambda = \lambda_0$, $\tilde{D}_1(\lambda)$ must now have only an $(N-1)$st-order zero there.

If $N - 1 \neq 0$, this process can perhaps be repeated with the kernel $K_1(x,y)$. Note, however, that owing to the orthogonality of K_0 and K_1, new characteristic functions φ_1 and ψ_1 (which we know exist) associated with K_1 and corresponding to the characteristic value λ_0 must perforce be characteristic functions associated with the original kernel K also. If $(\varphi_1,\psi_1) = 1$, therefore, we can construct

$$K(x,y) = \frac{\varphi_0(x)\overline{\psi}_0(y)}{\lambda_0} + \frac{\varphi_1(x)\overline{\psi}_1(y)}{\lambda_0} + K_2(x,y) \tag{2}$$

and duplicate the above analysis.

This procedure can be continued until the zeros of $\tilde{D}(\lambda)$ are exhausted provided that at each step $(\varphi_i,\psi_i) = 1$ [actually, of course, all that is needed is $(\varphi_i,\psi_i) \neq 0$]. It should be noted that Hermitian kernels, for which $K^*(x,y) = K(x,y)$, always yield to such analysis. For if $\varphi(x)$ is a solution of the homogeneous equation with Hermitian kernel, then it is also a solution of the adjoint homogeneous equation and certainly nontrivial φ can be chosen of unit norm.

6.3   SINGULAR DECOMPOSITION OF THE RESOLVENT KERNEL

In Sec. 6.1 we employed the Laurent series of the resolvent kernel R_K about a singular point $\lambda = \lambda_0$ in order to explicitly exhibit characteristic functions of the given square-integrable kernel K. In the sequel we shall show that this same decomposition according to singular value leads to a corresponding decomposition of K, which, although not generally of the form (6.2-2), does separate off completely that portion of the original kernel which is responsible for the characteristic value λ_0. This allows us to make some important observations regarding the characteristic behavior of general $\mathfrak{L}^2$ kernels.

As might be anticipated, we begin with the resolvent equation (4.4-1)

$$\int_a^b R_K(x,z;\lambda)R_K(z,y;\mu)\,dz = \frac{R_K(x,y;\lambda) - R_K(x,y;\mu)}{\lambda - \mu}$$

and let
$$R_K(x,y;\lambda) \equiv \frac{\rho(x,y;\lambda)}{(\lambda - \lambda_0)^n} + r(x,y;\lambda) \tag{1}$$

By reference to (6.1-1), we see that in splitting off the singular part associated with the nth-order pole of the resolvent R_K as in (1), $\rho(x,y;\lambda)$ is a polynominal in λ of at most degree $n-1$. Substituting the decomposition (1) into the resolvent equation and introducing the notation $L = \lambda - \lambda_0$ and $M = \mu - \lambda_0$, we then obtain

$$\int_a^b [\rho(x,z;\lambda) + L^n r(x,z;\lambda)]\,[\rho(z,y;\mu) + M^n r(z,y;\mu)]\,dz$$
$$= \frac{M^n\rho(x,y;\lambda) - L^n\rho(x,y;\mu)}{L-M} + L^n M^n \frac{r(x,y;\lambda) - r(x,y;\mu)}{\lambda - \mu} \tag{2}$$

Note particularly that the first term on the right-hand side is merely a polynomial of degree $n-1$ in either L or M while the second term only contains powers of L and M greater than or equal to n. Using this information we may equate like powers, say, of L, respectively, on both sides of (2) and then in the relations so obtained equate like powers of M. This procedure gives rise to

$$\int_a^b \rho(x,z;\lambda)\rho(z,y;\mu)\,dz = \frac{M^n\rho(x,y;\lambda) - L^n\rho(x,y;\mu)}{L-M} \tag{3}$$

$$\int_a^b r(x,z;\lambda)r(z,y;\mu)\,dz = \frac{r(x,y;\lambda) - r(x,y;\mu)}{\lambda - \mu} \tag{4}$$

$$\int_a^b \rho(x,z;\lambda)r(z,y;\mu)\,dz = 0 = \int_a^b r(x,z;\lambda)\rho(z,y;\mu)\,dz \tag{5}$$

Equations (3) and (4) show that both $\rho(x,y;\lambda)/(\lambda - \lambda_0)^n$ and $r(x,y;\lambda)$ satisfy the most general Fredholm identity, namely, the resolvent equation. Moreover, as (5) indicates, they are also *orthogonal* kernels for *all* values of λ and/or μ. In particular, then, the decomposition suggested by setting $\lambda = 0$ in (1), that is,

$$K(x,y) = K_0(x,y) + K_1(x,y)$$

$$\equiv \frac{\rho(x,y;0)}{(-\lambda_0)^n} + r(x,y;0) \tag{6}$$

exhibits explicitly and completely that portion of the original kernel which is associated with the characteristic value λ_0. For, indeed, the representation $K = K_0 + K_1$ gives rise to $R_K = R_{K_0} + R_{K_1}$ and $\tilde{D}(\lambda) = \tilde{D}_0(\lambda) \cdot \tilde{D}_1(\lambda)$. By association, we have†

$$R_{K_0} = \frac{\rho(x,y;\lambda)}{(\lambda - \lambda_0)^n} \qquad \text{and} \qquad R_{K_1} = r(x,y;\lambda)$$

and thus K_0, and K_0 alone, is responsible for the appearance of $\lambda = \lambda_0$ as a characteristic value.

The above analysis permits the following conclusions to be drawn:

1. The characteristic value λ_0 of $K(x,y)$ is also a characteristic value of $K_0(x,y)$ and conversely.

2. Every characteristic function of $K(x,y)$ associated with the characteristic value λ_0 is also a characteristic function of $K_0(x,y)$ and conversely.

3. $\lambda_1 \neq \lambda_0$ is a characteristic value of $K(x,y)$ if and only if it is also a characteristic value of $K_1(x,y)$.

4. A function $\varphi(x)$ is a characteristic function of $K(x,y)$ associated with $\lambda_1 \neq \lambda_0$ if and only if it is also a characteristic function of $K_1(x,y)$.

The specific proofs of conclusions 1 and 3 are trivial verifications using the product decomposition of the modified Fredholm determinant. The remaining results are all consequences of the orthogonality of K_0 and K_1. For instance, if $\varphi(x)$ is known to be a characteristic function of K associated with $\lambda_1 \neq \lambda_0$, then symbolically

$$\Phi = \varphi - \lambda_1 K_1 \varphi$$

$$= \varphi - \lambda_1 K \varphi + \lambda_1 K_0 \varphi$$

$$= \lambda_1 K_0 \varphi$$

---

†It should be especially noted that the resolvent kernel corresponding to that portion of the original kernel associated with the characteristic value λ_0 is just the principal part of the Laurent expansion of the full resolvent around $\lambda = \lambda_0$.

so that

$$\lambda_1 K_0 \Phi = \lambda_1 K_0 \varphi = \Phi$$

Since K_0 has only the single characteristic value λ_0, it follows that $\Phi = 0$ from which the desired conclusion ensues. The other proofs proceed along similar lines.

6.4   THE RANK OF CHARACTERISTIC VALUES

We shall now show that the kernel $K_0(x,y)$ derived in the previous section is *degenerate*. This is an important result, for it gives us another method of showing that each and every characteristic value of an $\mathfrak{L}^2$ kernel is of finite rank. We begin by recalling that

$$\frac{\rho(x,y;\lambda)}{(\lambda - \lambda_0)^n} \equiv \sum_{\nu=1}^{n} \frac{r_\nu(x,y)}{(\lambda - \lambda_0)^\nu}$$

Equation (6.3-3) can then be rewritten with $\mu = 0$ as

$$(\lambda - \lambda_0 + \lambda_0) \int_a^b \sum_{\nu=1}^{n} \frac{r_\nu(x,z)}{(\lambda - \lambda_0)^\nu} K_0(z,y)\, dz = \sum_{\nu=1}^{n} \frac{r_\nu(x,y)}{(\lambda - \lambda_0)^\nu} - K_0(x,y)$$

from which we obtain

$$K_0(x,y) = - \int_a^b r_1(x,z) K_0(z,y)\, dz$$

by equating terms containing only the *zeroth* power of $(\lambda - \lambda_0)$. This last relation shows that, for fixed y, $K_0(x,y)$ satisfies a homogeneous integral equation having the square-integrable kernel $r_1(x,y)$ and the eigenparameter -1.

For a fixed finite number N, let us assume, therefore, that there are at least N linearly independent characteristic functions φ_ν of this kernel r_1 associated with the characteristic value -1, and, in fact, we choose them to be orthonormalized. It then follows that

$$0 \le \int_a^b \left| r_1(x,y) + \sum_{\nu=1}^{N} \varphi_\nu(x)\overline{\varphi_\nu}(y) \right|^2 dy$$

$$= \int_a^b |r_1(x,y)|^2\, dy - \sum_{\nu=1}^{N} \varphi_\nu(x)\overline{\varphi_\nu}(x)$$

and hence

$$N = \int_a^b \sum_{\nu=1}^N \varphi_\nu(x)\overline{\varphi}_\nu(x)\, dx \le \int_a^b \int_a^b |r_1(x,y)|^2\, dy\, dx = \|r_1\|^2 < \infty$$

We conclude from this that there can only be *finitely* many linearly independent characteristic functions of the kernel r_1 associated with $\lambda = -1$. Since $K_0(x,y)$ is known to be such a function for fixed y, it must be representable, say, as a linear combination of the φ_ν in the form

$$K_0(x,y) = \sum_{\nu=1}^N \varphi_\nu(x)\overline{\psi}_\nu(y) \tag{1}$$

and thus we arrive at the desired conclusion that K_0 is degenerate.

The particular representation (1), of course, does not explicitly exhibit the characteristic functions of K_0 corresponding to the characteristic value $\lambda = \lambda_0$. There are, however, alternate representations that may be derived from the natural decomposition of the function space $\mathfrak{L}^2$ which is induced by the kernel K_0 [see Riesz and Sz.-Nagy (1955, pp. 188ff) or Schmeidler (1965, pp. 67ff), for example]. One of these takes the form

$$K_0(x,y) = \sum_{\nu=1}^p \sum_{\mu=1}^{m_\nu} \frac{\phi_{\nu\mu}(x)\overline{\psi}_{\nu\mu}(y)}{\lambda_0} \tag{2}$$

where

$$\phi_{\nu 1} = \lambda_0 K_0 \phi_{\nu 1}$$

$$\phi_{\nu\mu} = \lambda_0 K_0 \phi_{\nu\mu} - \lambda_0 \phi_{\nu,\mu-1} \qquad \mu = 2, 3, \ldots, m_\nu \tag{3}$$

and

$$\psi_{\nu\mu} = \overline{\lambda}_0 K_0^* \psi_{\nu\mu} - \overline{\lambda}_0 \psi_{\nu,\mu+1} \qquad \mu = 1, 2, \ldots, m_\nu - 1$$

$$\psi_{\nu m_\nu} = \overline{\lambda}_0 K_0^* \psi_{\nu m_\nu}$$

for all $1 \le \nu \le p$. Here the $\phi_{\nu\mu}$ and $\psi_{\nu\mu}$ are linearly independent (and perhaps generalized) characteristic functions of the kernels K_0 and K_0^*, respectively; moreover, they are assumed to have been *biorthonormalized* so that[†]

$$(\phi_{ij}, \psi_{\nu\mu}) = \delta_{\nu i}(\delta_{\mu j} + \lambda_0 \delta_{\mu, j-1}) \tag{4}$$

[†] One could equally well have chosen the characteristic functions to satisfy $(\phi_{ij}, \psi_{\nu\mu}) = \delta_{\nu i}\delta_{\mu j}$, that is, to be biorthonormal in the classical sense. The representation (2) would then have a slightly different form, of course.

With K_0 as in (2), it is then a simple matter to show that the corresponding resolvent kernel is given, in general, by

$$R_{K_0}(x,y;\lambda) = \sum_{\nu=1}^{p} \sum_{\mu=1}^{m_\nu} \sum_{\sigma=0}^{m_\nu-\mu} \left(\frac{\lambda\lambda_0}{\lambda_0-\lambda}\right)^\sigma \frac{\phi_{\nu,\mu}(x)\overline{\psi}_{\nu,\mu+\sigma}(y)}{\lambda_0-\lambda} \tag{5}$$

This relation, in which λ_0 appears as a pole of order $n = \max [m_\nu]$, simplifies considerably if we know a priori that the index of the characteristic value is unity, that is, $n = 1$.[†] In this case, which actually occurs quite often in practical situations, we obtain

$$R_{K_0}(x,y;\lambda) = \sum_{\nu=1}^{p} \frac{\phi_\nu(x)\overline{\psi}_\nu(y)}{\lambda_0-\lambda}$$

(here we have suppressed the second subscript since there are no generalized cf's). Similarly (2) becomes

$$K_0(x,y) = \sum_{\nu=1}^{p} \frac{\phi_\nu(x)\overline{\psi}_\nu(y)}{\lambda_0} \qquad \text{with } (\phi_\nu,\psi_\mu) = \delta_{\nu\mu} \tag{6}$$

a relation essentially of the same form as (1) but with considerably different meaning. We therefore have the interesting result that a pole λ_0 of the resolvent of a given $\mathfrak{L}^2$ kernel is *simple* if that portion of the kernel which is responsible for the appearance of λ_0 as a characteristic value may be represented as in (6).

This result can be restated in perhaps a more instructive fashion by noting that K_0 above is essentially *fully* (left) *symmetrizable*; i.e., it is a kernel for which there exists a *Hermitian* kernel H with the property that the composite kernel HK_0 differs at most from a Hermitian kernel by the (perhaps complex) multiplicative constant λ_0.[‡] In addition, H is such that

$$(H\phi,\phi) > 0 \qquad iff \quad K_0\phi \neq 0$$

in other words, the inner product $(H\phi,\phi)$ is positive for all functions

---

[†] The term *semisimple* is often used to describe such cv's (and the portions of the operators or kernels that give rise to them); see, for example, Kato (1966, p. 41). Transformations having this property play a central role in analytic perturbation theory (see Chap. 12).

[‡] In analogy with matrix theory, a *normal* kernel K is one for which $KK^* = K^*K$. If, in addition, $K^* = K$, the kernel is termed *Hermitian*. A Hermitian H, which accomplishes the "symmetrization" in the case in question, is $\sum_{\nu=1}^{p} \psi_\nu(x)\overline{\psi}_\nu(y)$. The class of symmetrizable kernels includes normal kernels and is considered in some detail in Sec. 17.3.

within the image domain of K_0. With these two conditions we know, then, that $\phi = \lambda_0 K_0 \phi$ implies

$$
\begin{aligned}
H\phi &= \lambda_0 H K_0 \phi \\
&= (\lambda_0 H K_0)^* \phi \\
&= \bar{\lambda}_0 K_0^* H \phi
\end{aligned}
$$

and hence the cf's ψ of K_0^* may be represented in terms of the cf's ϕ of K_0 by $\psi = H\phi$ with

$$
(\phi, \psi) \equiv (\phi, H\phi) \neq 0 \tag{7}
$$

As a consequence of this, the $\phi_{\nu 1}$ and $\psi_{\nu m_\nu}$ of our earlier notation are such that

$$
\delta_{m_\nu, 1} = (\phi_{\nu 1}, \psi_{\nu m_\nu}) \neq 0 \tag{8}
$$

from which it follows that m_ν can only be unity. An appropriate alternative formulation of the result under consideration thus has been shown to be: A pole λ_0 of the resolvent of a given $\mathfrak{L}^2$ kernel is *simple* iff that portion K_0 of the original kernel which is responsible for the appearance of λ_0 as a characteristic value is essentially *fully symmetrizable*.†

The form of the representation (2) has one other appealing feature. It shows rather vividly that, in general, there need be no correlation whatsoever between the rank of the degenerate kernel K_0 and the number p of linearly independent characteristic functions associated with the characteristic value λ_0 (that is, the rank of the cv λ_0). For instance, when $p = 1$, the kernel K_0 as given by (2) can be rewritten as

$$
K_0(x, y) = \sum_{\mu=1}^{m} \frac{\phi_\mu(x) \overline{\psi}_\mu(y)}{\lambda_0}
$$

This is clearly a degenerate kernel of rank m and has the Fredholm determinant $D_0(\lambda) = (1 - \lambda/\lambda_0)^m$, as may be easily verified. On the other hand, by virtue of the *biorthogonality* of the ϕ_μ and ψ_ν as given by (4), K_0 has only a single characteristic function (proportional to ϕ_m) corresponding to $\lambda = \lambda_0$.

---

† Lalesco (1912, pp. 58–59) provided still another form of this result when he proved that "a necessary and sufficient condition for a pole to be simple is that to each characteristic function ϕ of the equation corresponding to this pole, there exists a solution ψ of the adjoint equation with the property that $(\phi, \psi) \neq 0$." In view of (7) and (8) above, Lalesco's formulation is seen to be an easy consequence of our analysis.

PROBLEMS

1. Give an example of a kernel $K(x,y)$ which has characteristic functions *not* representable as $\varphi(x) = r_n(x,y_0)$ in the notation of Sec. 6.1. [Northover (1970) has shown that every cf φ associated with a given cv can, however, be expressed in the form

$$\varphi(x) = \sum_{\nu=1}^{n} \int_a^b \alpha_\nu(y) r_\nu(x,y) \, dy \Bigg]$$

2. In the event that λ is a characteristic value, the Fredholm alternative asserts that solutions to the inhomogeneous equation

$$\text{(a)} \quad \varphi = f + \lambda K \varphi$$

exist iff $(f, \Omega) = 0$ for all solutions Ω of the homogeneous adjoint equation

$$\text{(b)} \quad \Omega = \bar{\lambda} K^* \Omega$$

Show that an alternate verification of this fact can be based upon the decomposition (6.3-6). *Hint*: First verify that (a) is equivalent to

$$\varphi = f + \lambda r f + \lambda K_0 \varphi$$

and then use the orthogonality of K_0 and r in arriving at the desired conclusion.

3. Consider the two kernels

$$K_1(x,y) = 1 + 3xy$$
$$K_2(x,y) = 1 + y + 3xy$$

$$-1 \le x, y \le 1$$

a. Compare the resolvents of these two kernels and verify that both K_1 and K_2 have the same Fredholm determinant.
b. Show that K_2 has only one linearly independent characteristic function associated with the characteristic value $\frac{1}{2}$ [Smithies (1962, p. 77)].
c. Find generalized characteristic functions ϕ_μ, ψ_μ of K_2, K_2^* that satisfy (6.4-4) and hence determine the representation (6.4-2) for K_2.
d. Show that K_2 is a *symmetrizable* kernel.

4. From the definitions of Sec. 6.4 for $N = \sum_{\nu=1}^{p} m_\nu$, the order of the cv λ_0 as a zero of $\tilde{D}(\lambda)$, and $n = \max [m_\nu]$, the index of λ_0, establish the inequalities

$$\frac{N}{n} \le p \le N - n + 1$$

Satisfy yourself then that $N = n$ iff $p = 1$ and $N = n + 1$ implies $p = 2$.

5. a. Use the Neumann series (3.1-4) to verify that R_{K_0} as given by (6.4-5) is the resolvent corresponding to the kernel K_0 of (6.4-2).

5.   **b.** Alternatively, show that K_0 and R_{K_0} satisfy the Fredholm identities
(3.2-5).

   c. Show that

$$\mathrm{Tr}\left(R_{K_0}\right) = -\frac{\displaystyle\sum_{\nu=1}^{p} m_\nu}{(\lambda - \lambda_0)}$$

   as is to be expected from Sec. 3.5.

6.   [Radon (1919).] Using the decomposition (6.1-1) in the expression (6.3-3),
verify that the following relations are satisfied by the $r_\nu(x,y)$, $\nu = 1, 2, \ldots, n$:

   (a) $r_\nu = -r_1 r_\nu$ $1 \le \nu \le n$

   (b) $0 = r_n r_\nu$ $2 \le \nu \le n$

   (c) $r_{\nu+1} r_\mu = r_\nu r_{\mu+1}$ $1 \le \mu, \nu \le n-1$

Deduce from (a) and (c) that $r_1 = -r_1{}^2$ and

$$r_k = -(-r_2)^{k-1} \qquad 2 \le k \le n$$

Combining this last expression with (b), we see that r_2 is what may be termed
a *nilpotent* kernel of index n. [Pogorzelski (1966, chap. 7, sec. 4) derives
relations similar to the above.]

chapter seven

Hermitian Kernels and Their Characteristic Values

7.1 THE NATURE OF HERMITIAN KERNELS

In this and the next several chapters we will be concerned principally with that subclass of $\mathfrak{L}^2$ kernels for which

$$K^*(x,y) = K(x,y)$$

i.e., *Hermitian*, or in the real case *symmetric*, kernels. As we shall see, such kernels possess a number of important properties, which collectively permit a thorough and complete analysis of linear integral equations of the second kind with these Hermitian functions as kernels.[†] Before proceeding to those considerations, however, it is worth noting that E. Schmidt (1907) has shown that linear integral equations with arbitrary real kernels can be transformed into equivalent integral equations with symmetric kernels. Generalizing to the complex case, if

$$\phi = f + \lambda K \phi$$

[†] The related theory applicable to self-adjoint differential equations is amply discussed in Ince (1956), Coddington and Levinson (1955), and Courant and Hilbert (1953).

is the integral equation of interest, we let

$$\phi = \psi - \bar{\lambda}\ K^*\psi$$

It follows then that

$$\psi = f + L\psi$$

where $L = \lambda K + \bar{\lambda}K^* - \lambda\bar{\lambda}KK^*$ is clearly Hermitian.

Although this transformation has advantages for situations in which the eigenparameter λ is given a priori, it does have the obvious drawback that λ appears in a rather awkward fashion in the new kernel L, thus rendering the technique practically useless for characteristic-value calculation purposes. Fortunately, integral equations with kernels already Hermitian occur of themselves rather frequently in applied problems, and the specialized theory of their solution is therefore of distinct practical importance.

In the next section we shall show that every nonnull Hermitian $\mathfrak{L}^2$ kernel has at least one characteristic value. For the moment, then, as we take a look at some of the beneficial properties of such Hermitian kernels, we merely assume this to be true.

Property 1   The characteristic values of Hermitian kernels are real.

Let $\phi = \lambda K\phi$ with $\|\phi\| \neq 0$. Then

$$\lambda(K\phi, \phi) = (\lambda K\phi, \phi) = (\phi, \phi)$$

so that $\lambda(K\phi, \phi)$ is positive. On the other hand, since K is Hermitian,

$$(K\phi, \phi) = (\phi, K^*\phi) = (\phi, K\phi) = \overline{(K\phi, \phi)}$$

which, with the above, implies that λ must be real.

Property 2   The characteristic values of a Hermitian kernel $K(x, y)$ and its *transpose* $K(y, x)$ [or conjugate $\overline{K(x, y)}$] are identical, while the characteristic functions may be chosen as complex conjugates.

Property 3   The characteristic functions of real symmetric kernels may be selected to be real.

This property does not remain valid in the general case (see Prob. 1).

Property 4   Characteristic functions of Hermitian kernels belonging
  to distinct characteristic values are orthogonal to one another.

This may be viewed as a special case of the more general result that if ϕ
is a characteristic function (cf) of a general $\mathfrak{L}^2$ kernel K associated with
the characteristic value λ and ψ is a cf of K^* associated with $\overline{\mu}$ where
$\mu \neq \lambda$, then ϕ and ψ are orthogonal. The proof for this follows from the
identity

$$(\phi, \psi) = (\lambda K\phi, \psi) = \lambda(K\phi, \psi) = \lambda(\phi, K^*\psi)$$
$$= \frac{\lambda}{\mu}(\phi, \overline{\mu}K^*\psi) = \frac{\lambda}{\mu}(\phi, \psi)$$

Property 5   The characteristic values of Hermitian kernels form a
  nonvoid finite or enumerable sequence (λ_ν) with no finite limit
  point. The λ_ν may be ordered in the natural way:

$$0 < |\lambda_1| \leq |\lambda_2| \leq \cdots \leq |\lambda_\nu| \leq \cdots$$

The associated characteristic functions may be chosen to be
orthonormal.

The first part of Property 5 reflects just the specialization of general
Fredholm-Carleman theory to the case of Hermitian kernels. Property 4
ensures the orthogonality of characteristic functions associated with distinct characteristic values. The finite number of such functions belonging
to the same cv can be orthonormalized by the Gram-Schmidt process
[Smithies (1962, pp. 61–62), Riesz and Sz.-Nagy (1955, pp. 67–68), and
Courant and Hilbert (1953, pp. 50–51)]. When dealing with Hermitian
kernels we shall generally assume that the cv's have been ordered in the
usual natural way and that the associated cf's have been orthonormalized.

Property 6   If $\phi_1(x)$, $\phi_2(x)$, . . . , $\phi_N(x)$ are distinct orthonormalized
  characteristic functions of the Hermitian kernel $K(x, y)$ associated,
  respectively, with the characteristic values $\lambda_1, \lambda_2, \ldots, \lambda_N$ (not necessarily all different), then the inequality

$$\sum_{\nu=1}^{N} \frac{|\phi_\nu(x)|^2}{\lambda_\nu^2} \leq \int_a^b |K(x, y)|^2 \, dy \equiv \|K\|_y^2 \tag{1}$$

is valid.

This relation (1) is just the well-known *Bessel's inequality* expressed in a form appropriate for the kernel K considered as a function of its *second* variable only (see Sec. 13.1). The standard proof follows from inspection of the one-dimensional norm

$$\left\| K(x,y) - \sum_{\nu=1}^{N} \alpha_\nu(x)\,\overline{\phi}_\nu(y) \right\|^2 \tag{2}$$

with x viewed as a parameter. Expansion and completion of the square shows that (2) has its (nonnegative) minimum value when (and only when) $\alpha_\nu(x) = \phi_\nu(x)/\lambda_\nu (1 \le \nu \le N)$, from which case the inequality (1) ensues.

If we integrate (1) with respect to x, we obtain

$$\sum_{\nu=1}^{N} \frac{1}{\lambda_\nu^{2}} \le \int_a^b \int_a^b |K(x,y)|^2 \, dx \, dy = \|K\|^2 \tag{3}$$

This relation (3) is valid independent of the number of characteristic values included in the summation. In particular, then, it provides an alternate proof of the first part of Property 5, as well as giving rise to estimates, say, of the number N of linearly independent characteristic functions corresponding to the same cv λ, viz.,

$$N \le \lambda^2 \|K\|^2 \tag{4}$$

It should be noted that when every cv is included in the summation (3), each repeated according to its rank, the inequality becomes an equality. In this case, we have from (4.3-3)

$$\sum_{\nu=1}^{\infty} \frac{1}{\lambda_\nu^{2}} = k_2 = \|K\|^2 \tag{5}$$

Analogous to this, for every regular value of K we also have from (4.4-5)

$$\sum_{\nu=1}^{\infty} \left| \frac{1}{\lambda_\nu - \lambda} \right|^2 = \sum_{\nu=1}^{\infty} \frac{1}{(\lambda_\nu - \lambda)(\lambda_\nu - \overline{\lambda})}$$

$$= \mathrm{Tr}\,[R_K(\lambda)R_K(\overline{\lambda})\,] = \mathrm{Tr}\,(R_K R_K^*) = \|R_K\|^2 \tag{6}$$

Definition [Smithies (1962, p. 113).] An orthonormal system of characteristic functions $\{\phi_\nu\}$, with associated characteristic values $\{\lambda_\nu\}$ ordered naturally as in Property 5 and such that *every* $\mathfrak{L}^2$ characteristic function ϕ of K is a finite linear combination of the functions ϕ_ν, is designated a *full characteristic system* of K.

It should be borne in mind that the set of functions $\{\phi_\nu\}$ need not be *complete*. In other words, there may well be nontrivial $\mathcal{L}^2$ functions orthogonal to all the ϕ_ν, as is obviously the case for the kernel

$$K(x,y) = \sin x \sin y \qquad 0 \le x,y \le \pi$$

for example. To handle these situations we introduce the concept of infinite cv's and say that $\lambda = \infty$ is a "characteristic value" if there exists nonnull $\phi(x)$ such that

$$\int_a^b K(x,y)\,\phi(y)\,dy = 0$$

A kernel is then called *closed* if $\lambda = \infty$ is not a "cv." Clearly, then, any degenerate kernel is not closed. If we add to the set $\{\phi_\nu\}$, associated with the finite characteristic values, all the linearly independent ϕ_μ corresponding to the "cv" $\lambda = \infty$, however, the resulting augmented set is generally complete. (We shall have more to say about these matters in a subsequent chapter.)

Property 7   All of the higher-order iterates of Hermitian kernels are themselves Hermitian. For real values of λ, the resolvent of a Hermitian kernel is Hermitian.

The composite of two Hermitian kernels, on the other hand, is *not*, in general, Hermitian.

7.2   ELEMENTARY PROOFS OF THE EXISTENCE OF CHARACTERISTIC VALUES

Once the background of previous chapters has been provided, a most elementary proof of the existence of characteristic values of Hermitian kernels follows from an examination of the traces of their second iterates. To be specific, we know that for Hermitian $\mathcal{L}^2$ kernels

$$\sum_{\nu=1}^{\infty} \left(\frac{1}{\lambda_\nu}\right)^2 = k_2 = \|K\|^2$$

[cf. (7.1-5)]. In view of this relation, therefore, all nonnull Hermitian kernels possess characteristic values. Conversely, if Hermitian K has no finite characteristic values, then it vanishes almost everywhere in its fundamental domain of definition.

A function-theoretic proof of similar simplicity was originally suggested by Kneser in a 1906 paper. His approach is based upon two simple trace inequalities that are generally valid for Hermitian kernels K, namely,

$$\text{Tr } (K^{2n}) \equiv k_{2n} \geq 0 \tag{1}$$

with equality iff $\|K\| = 0$ and

$$k_{2n}^2 \leq k_{2n-2m} k_{2n+2m} \tag{2}$$

for all positive integer $m < n$. For a moment, therefore, let us consider these two relations.

Equation (1) is a consequence of the fact that for Hermitian kernels

$$k_{2n} = \|K^n\|^2 \tag{3}$$

When the norm of K vanishes, then so also must k_{2n} since $\|K^n\| \leq \|K\|^n$ [cf. (3.5-1)]. On the other hand, if $\|K\| \neq 0$, then $\|K^n\|$ and hence k_{2n} are both nonvanishing also (see Prob. 2 at the end of this chapter). The inequality (2) follows from relation (3), the nature of the trace operator, and equation (3.5-1), viz.,

$$k_{2n}^2 \equiv [\text{Tr } (K^{2n})]^2 = [\text{Tr } (K^{n-m} K^{n+m})]^2$$
$$\leq \|K^{n-m}\|^2 \cdot \|K^{n+m}\|^2 = k_{2n-2m} k_{2n+2m}$$

Returning now to Kneser's argument, we recall that any characteristic values of a given $\mathfrak{L}^2$ kernel must appear as zeros of the modified Fredholm determinant $\tilde{D}(\lambda)$, which, from Sec. 4.1, has the representation

$$\tilde{D}(\lambda) = \exp \left(- \sum_{n=2}^{\infty} \frac{\lambda^n}{n} k_n \right)$$

at least for small λ. For a nonnull Hermitian kernel without any characteristic values, therefore, the power series in the exponential of the above expression would have to be convergent for all finite λ. In particular, then, for such a kernel the subseries

$$\sum_{n=1}^{\infty} \frac{\lambda^{2n}}{2n} k_{2n} \tag{4}$$

would have to be absolutely convergent for all λ, too.

On the other hand, for positive λ equation (4) is a series of positive terms, by (1), with a ratio of two successive terms given by

$$r_n = \lambda^2 \left(\frac{n}{n+1} \right) \frac{k_{2n+2}}{k_{2n}}$$

Moreover, in view of the inequality (2) and its implication when $m = 1$ that

$$\frac{k_{2n+2}}{k_{2n}} \geq \frac{k_{2n}}{k_{2n-2}} \geq \cdots \geq \frac{k_4}{k_2} \tag{5}$$

this ratio r_n must approach a limit r as $n \to \infty$, which is greater than or equal to $\lambda^2 k_4 / k_2$. By the ratio test for the convergence of series of positive terms, therefore, (4) is certainly divergent whenever $|\lambda| > \sqrt{k_2/k_4}$, a contradiction. We are perforce led again to the conclusion, then, that Hermitian nonnull kernels have characteristic values, and in this case our argument shows that at least one of the cv's must have a modulus less than or equal to $\sqrt{(k_2/k_4)}$.

7.3 ESTIMATION OF CHARACTERISTIC VALUES USING TRACE RELATIONS

Not every existence proof is constructive, and the two we have just given, although perhaps admirable in their simplicity, do not immediately suggest computational schemes for approximation of the characteristic values associated with Hermitian kernels. Such will not be the case, by and large, with the other existence proofs that we will discuss in some detail subsequently. On the other hand, there are a couple of straightforward estimation procedures using trace relations that are worthy of mention at this time before we move on to more difficult material.

We recall from (4.3-3) that

$$\mathrm{Tr}\,(K^n) \equiv k_n = \sum_{\nu=1}^{\infty} \left(\frac{1}{\lambda_\nu} \right)^n \tag{1}$$

with $n \geq 2$ is valid for arbitrary $\mathfrak{L}^2$ kernels K. Moreover, this relationship holds even when $n = 1$, although not generally, at least for wide classes of kernels of interest (e.g., continuous definite kernels, composite kernels, etc.). We assume that this last situation prevails, then, and consider how the set of equations (1) could be effectively used to estimate the unknown characteristic values of a given kernel in terms of the traces of the various iterates of that kernel.

One possible computational algorithm is the following:

$$\frac{1}{\lambda_1^{(1)}} = k_1$$

$$\frac{1}{\lambda_1^{(2)}} + \frac{1}{\lambda_2^{(2)}} = k_1 \qquad \left(\frac{1}{\lambda_1^{(2)}}\right)^2 + \left(\frac{1}{\lambda_2^{(2)}}\right)^2 = k_2$$

$$\sum_{\nu=1}^{3} \frac{1}{\lambda_\nu^{(3)}} = k_1 \qquad \sum_{\nu=1}^{3} \left(\frac{1}{\lambda_\nu^{(3)}}\right)^2 = k_2 \qquad \sum_{\nu=1}^{3} \left(\frac{1}{\lambda_\nu^{(3)}}\right)^3 = k_3 \qquad \text{etc.}$$

Due to the interrelation between these power sums and the coefficients of the polynomials of which they are the roots [c.f. Turnbull (1952)], this procedure is equivalent to using the truncations

$$D^{[n]}(\lambda) = \sum_{\nu=0}^{n} d_\nu \lambda^\nu$$

of the Fredholm determinant to give approximations to the characteristic values $\lambda_1, \lambda_2, \ldots, \lambda_n$.

When applied to the kernel

$$K(x,y) = \min(x,y) \qquad 0 \leq x,y \leq 1$$

for example, the above algorithm leads to

$$D^{[1]}(\lambda) = 1 - \frac{\lambda}{2}$$

$$D^{[2]}(\lambda) = 1 - \frac{\lambda}{2} + \frac{\lambda^2}{24}$$

and hence

$$\lambda_1^{(1)} = 2$$
$$\lambda_1^{(2)} = 6 - 2\sqrt{3} \doteq 2.536$$
$$\lambda_2^{(2)} = 6 + 2\sqrt{3} \doteq 9.464$$

(We use the symbol $\doteq$ to designate approximate numerical equality.) These are to be compared with the actual values of λ_ν, namely, $\lambda_1 = \pi^2/4 \doteq 2.467$ and $\lambda_2 = 9\pi^2/4 \doteq 22.207$ (see Prob. 4 at the end of this chapter).

Another simple computational scheme is given by the relations

$$\frac{1}{\lambda_1^{(1)}} = k_1$$

$$\left(\frac{1}{\lambda_1^{(2)}}\right)^2 = k_2 \qquad \frac{1}{\lambda_1^{(2)}} + \frac{1}{\lambda_2^{(2)}} = k_1$$

$$\sum_{\nu=1}^{1} \left(\frac{1}{\lambda_\nu^{(3)}}\right)^3 = k_3 \qquad \sum_{\nu=1}^{2} \left(\frac{1}{\lambda_\nu^{(3)}}\right)^2 = k_2 \qquad \sum_{\nu=1}^{3} \left(\frac{1}{\lambda_\nu^{(3)}}\right) = k_1 \qquad \text{etc.}$$

Clearly, this procedure is based upon the assumption of distinct characteristic values. If this condition is not satisfied, the method needs to be modified in the obvious manner.

When applied to the min (x, y) kernel, this alternate algorithm yields

$$\lambda_1^{(1)} = 2$$
$$\lambda_1^{(2)} \doteq \sqrt{6} \doteq 2.449$$
$$\lambda_2^{(2)} = 6 + 2\sqrt{6} \doteq 10.899$$

(see Prob. 4).

Use of the above methods is naturally not restricted to Hermitian kernels. In the general case, however, the procedure may be complicated by the complex nature of the characteristic values.

7.4   SCHMIDT'S CHARACTERISTIC-VALUE EXISTENCE PROOF

So far in this chapter we have considered a number of properties that are fundamental to the theory of Hermitian kernels and sampled several existence proofs and estimation procedures for characteristic values of such kernels. For the most part this latter material has been dependent on various trace relations and the interplay between higher-order traces and sums of reciprocal characteristic values. A classical analysis of E. Schmidt (1907), having some of these same features, will now be discussed. This approach of Schmidt not only provides another proof of the existence of characteristic values of Hermitian kernels, but it also details a constructive procedure whereby approximations of arbitrary accuracy may be obtained for at least the smallest cv and an associated cf.

As a prelude to our more general discussion and in the same vein as considerations dating back originally to Bernoulli, let us first investigate the degenerate kernel

$$K(x, y) = \sum_{\nu=1}^{N} \frac{\phi_\nu(x)\,\overline{\phi}_\nu(y)}{\lambda_\nu}$$

Here we assume that the ϕ_ν are orthonormalized $\mathfrak{L}^2$ functions and that the λ_ν are arbitrary *real* numbers satisfying

$$0 < |\lambda_1| = |\lambda_2| = \cdots = |\lambda_{N_1}| < |\lambda_{N_1+1}| \leq \cdots \leq |\lambda_N|$$

In other words, the first N_1 of the λ's have a common magnitude that is strictly less than that of any of the remaining λ's.

By construction we have

$$\phi_\nu(x) = \lambda_\nu \int_a^b K(x,y)\phi_\nu(y)\,dy \qquad \nu = 1, 2, \ldots, N$$

and $\qquad K^n(x, y) = \displaystyle\sum_{\nu=1}^{N} \frac{\phi_\nu(x)\overline{\phi}_\nu(y)}{(\lambda_\nu)^n}$

It follows, then, that

$$\lambda_1{}^{2n} K^{2n}(x, y) = \sum_{\nu=1}^{N_1} \phi_\nu(x)\,\overline{\phi}_\nu(y) + \sum_{\nu=N_1+1}^{N} \left(\frac{\lambda_1}{\lambda_\nu}\right)^{2n} \phi_\nu(x)\,\overline{\phi}_\nu(y)$$

and hence $\qquad \displaystyle\lim_{n\to\infty} \lambda_1^{2n} K^{2n}(x,y) = \sum_{\nu=1}^{N_1} \phi_\nu(x)\,\overline{\phi}_\nu(y)$ (1)

The right-hand side of this relation is a Hermitian function, which for fixed y is a linear combination of characteristic functions of K associated with the least (in magnitude) cv; by itself it is also a cf of K^2 corresponding to the cv $\lambda_1{}^2$. Applying the trace operator to (1), we obtain

$$\lim_{n\to\infty} \lambda_1{}^{2n} k_{2n} = N_1$$ (2)

Moreover, if we define c_n by

$$c_n \equiv \frac{k_{2n}}{k_{2n+2}}$$ (3)

this last result implies that

$$\lim_{n\to\infty} c_n = \lambda_1{}^2$$ (4)

Schmidt's approach to the analysis of a general Hermitian nonnull $\mathcal{L}^2$ kernel K was to investigate, in the limit of large n, the same functions as are present in (1) to (4). In particular, since

$$\infty > c_1 \geq c_2 \geq \cdots \geq c_n > 0$$

by (3), (7.2-1), and (7.2-5), we know that $\lim c_n$ exists as a nonnegative number in the general case and we can call it $\lambda_1{}^2$. Thence,

$$\frac{k_{2n}}{k_{2n+2}} \geq \lambda_1{}^2$$

for arbitrary integer n, from which it follows that

$$\infty > \lambda_1{}^2 k_2 \geq \lambda_1{}^4 k_4 \geq \cdots \geq \lambda_1{}^{2n} k_{2n} \geq 0$$

Thus,
$$\lim_{n \to \infty} \lambda_1{}^{2n} k_{2n} = N_1 \tag{5}$$

also exists. In fact $N_1 \geq 1$.†

Having established the identity of λ_1 as the result of a limiting process, we want now to show that either $+\lambda_1$ or $-\lambda_1$ (or both) is a characteristic value of the given kernel K and there is associated with it a constructible characteristic function. It will turn out later, as expected, that N_1 is the number of linearly independent cf's belonging to this (these) cv. As might be surmised from (1), we consider

$$f_n(x, y) \equiv \lambda_1{}^{2n} K^{2n}(x, y)$$

It follows then that

$$f_{n+1} = \lambda_1{}^2 f_n K^2 = \lambda_1{}^2 K f_n K = \lambda_1{}^2 K^2 f_n \tag{6}$$

---

†The argument here goes something like the following: The inequality

$$k_{2n+2m} \leq \|K^{2n}\| \cdot \|K^{2m}\| \leq \|K^n\|^2 \cdot \|K^m\|^2 = k_{2n} k_{2m}$$

and the monotonicity of the c_n imply that

$$(c_m)^n k_{2n} \geq c_m c_{m+1} \cdots c_{m+n-1} k_{2n} \geq 1$$

for every positive integer m. Hence, $\lambda_1{}^{2n} k_{2n} \geq 1$ also, from which the desired result follows.

and hence

$$|f_{n+1}(x,y) - f_{m+1}(x,y)|^2 \le \lambda_1{}^4 \|f_n - f_m\|^2 \int_a^b |K(x,z)|^2 \, dz \int_a^b |K(\tau,y)|^2 \, d\tau$$

(7)

Taking the norm, we have

$$\|f_n - f_m\|^2 = \int_a^b \int_a^b |\lambda_1{}^{2n} K^{2n}(x,y) - \lambda_1{}^{2m} K^{2m}(x,y)|^2 \, dx \, dy$$

$$= \lambda_1{}^{4n} k_{4n} - 2\lambda_1{}^{2n+2m} k_{2n+2m} + \lambda_1{}^{4m} k_{4m}$$

which tends to zero by (5) as $m,n \to \infty$. In view of the completeness of $\mathfrak{L}^2$, therefore, there exists a square integrable and Hermitian limit function $f(x,y)$ toward which the Hermitian functions $f_n(x,y)$ converge in the mean. Moreover, using (6), it results that

$$f = \lambda_1{}^2 f K^2 = \lambda_1{}^2 K f K = \lambda_1{}^2 K^2 f$$

(8)

The convergence of f_n to f is actually stronger than implied above. To be specific, (7) shows that the f_n form a Cauchy sequence in the *relative sense*.† In fact,

$$\left| f_{n+1}(x,y) - \lambda_1{}^2 \int_a^b \int_a^b K(x,z) f(z,\tau) K(\tau,y) \, dz \, d\tau \right|^2$$

$$= \lambda_1{}^4 \left| \int_a^b \int_a^b K(x,z) [f_n(z,\tau) - f(z,\tau)] K(\tau,y) \, dz \, d\tau \right|^2$$

$$\le \lambda_1{}^4 \|f_n - f\|^2 \int_a^b |K(x,z)|^2 \, dz \int_a^b |K(\tau,y)|^2 \, d\tau \qquad (9)$$

so that the series f_n is relatively uniformly convergent to $\lambda_1{}^2 K f K$. If we let $g = \lambda_1{}^2 K f K$, it is easily demonstrated that

$$g = \lambda_1{}^2 g K^2 = \lambda_1{}^2 K g K = \lambda_1{}^2 K^2 g$$

Hence by designating $f \equiv g$ if need be, we have the result that f_n is relatively uniformly convergent to f rather than merely convergent in norm.

---

†See footnote on page 26.

It should be noted that

$$\|f_n\|^2 = \lambda_1{}^{4n}\|K^{2n}\|^2 = \lambda_1{}^{4n} k_{4n}$$

so that the norms of the f_n are monotonically decreasing yet are bounded from below by $\sqrt{N_1} \geq 1$. The Hermitian limit function f is thus non-null. There exist values of y, therefore, for which $f(x,y)$ is a nonnull $\mathfrak{L}^2$ function of the single variable x. For such y, $f(x,y)$ is, by (8), a characteristic function of the kernel K^2 corresponding to the characteristic value $\lambda_1{}^2$. It is a trivial matter then to determine a nonnull cf of K itself associated with either $+\lambda_1$ or $-\lambda_1$ (see Chap. 3, Prob. 13).

7.5   THE SCHMIDT ESTIMATES FOR THE LEAST CHARACTERISTIC VALUE

In the preceding section we have shown that, given a nonnull Hermitian $\mathfrak{L}^2$ kernel K, the limit of the trace ratio k_{2n}/k_{2n+2} leads to the square of a characteristic value $\pm \lambda_1$ of K, which has associated with it a characteristic function related to

$$f(x,y) = \lim_{n \to \infty} \lambda_1{}^{2n} K^{2n}(x,y)$$

Inherent in the analysis presented, obviously, is a constructive technique for obtaining approximations to λ_1 and f. Before investigating in more detail at least the characteristic-value estimation procedure, however, there remain a few loose ends to be tidied up.

First it should be noted that $\pm \lambda_1$, as expected, is the *least* (in magnitude) characteristic value of K.   For assume there exists $\phi = \lambda K \phi$ with $|\lambda| < |\lambda_1|$. Then

$$\phi = \lambda^2 K^2 \phi = \lambda^{2n} K^{2n} \phi = \left(\frac{\lambda}{\lambda_1}\right)^{2n} f_n \phi$$

and
$$\|\phi\| \leq \left(\frac{\lambda}{\lambda_1}\right)^{2n} \|f_n\| \cdot \|\phi\|$$

As $n \to \infty$, since $\|\phi\|$ and $\|f_n\|$ are bounded and $(\lambda/\lambda_1)^{2n} \to 0$, it must actually be true that $\|\phi\| = 0$, and hence λ is not a bona fide cv.

It should also be observed that if

$$\phi = \lambda K \phi$$

with $\lambda = \pm \lambda_1$, then $\phi = f_n \phi$ for all n. This implies in the limit that

$$\phi = f\phi \tag{1}$$

We are now in a position to give an alternate representation for $f(x,y)$, much in keeping with (7.4-1). This will also serve to clarify the role of N_1 appearing in (7.4-5) in the general case. Consider all the characteristic functions $\phi_\nu(x)$ of $K^2(x,y)$ associated with the cv λ_1^2 (up to normalization, there exist only a finite number N of these functions). Assume that they are orthonormalized. Then

$$f(x,y) = \sum_{\nu=1}^{N} \alpha_\nu(y)\phi_\nu(x)$$

with

$$\alpha_\nu(y) = \int_a^b f(x,y)\,\overline{\phi}_\nu(x)\,dx$$

On the other hand, since f is Hermitian

$$\alpha_\nu(y) = \int_a^b \overline{f}(y,x)\,\overline{\phi}_\nu(x)\,dx$$

$$= \sum_{\mu=1}^{N} \overline{\phi}_\mu(y) \int_a^b \overline{\alpha}_\mu(x)\,\overline{\phi}_\nu(x)\,dx$$

$$\equiv \sum_{\mu=1}^{N} \alpha_{\nu\mu}\,\overline{\phi}_\mu(y)$$

so that

$$f(x,y) = \sum_{\mu,\nu=1}^{N} \alpha_{\nu\mu}\phi_\nu(x)\,\overline{\phi}_\mu(y)$$

It follows from (1), however, that $\phi_\nu = f\phi_\nu$. This implies that $\alpha_{\nu\mu} = \delta_{\nu\mu}$ and hence

$$f(x,y) = \sum_{\nu=1}^{N} \phi_\nu(x)\overline{\phi}_\nu(y) \tag{2}$$

The relationship (2) concisely expresses f in terms of the characteristic functions ϕ_ν of K^2 associated with the cv λ_1^2. We know, however, that if we consider all of the (orthonormalized) cf's of K itself corresponding to either $+\lambda_1$ or $-\lambda_1$, these must form a *basis* for the ϕ_ν (see Prob. 5). Thus, the above expression (2) is equally valid if the ϕ_ν are actually the orthonormalized cf's of K associated with both $\pm\lambda_1$. In view of this conclusion we have, therefore, that

$$\mathrm{Tr}\,(f) = N$$

where N is either the number of linearly independent cf's of K^2 associated with λ_1^2 or the number of such cf's of K itself belonging to $\pm\lambda_1$.

The number N so obtained is in reality equal to the N_1 given by the limit in (7.4-5). To see this we recall that (7.4-5) can be equivalently rewritten as

$$\lim_{n \to \infty} \text{Tr} \ (f_n) = N_1$$

Using relation (7.4-8) and the inequality (7.4-9), moreover, we find that

$$|\text{Tr} \ (f_n) - \text{Tr} \ (f)|$$

$$\leq \lambda_1^2 \|f_{n-1} - f\| \int_a^b \left[\int_a^b |K(x,z)|^2 \, dz \int_a^b |K(\tau,x)|^2 \, d\tau \right]^{\frac{1}{2}} dx$$

$$\leq \lambda_1^2 \|f_{n-1} - f\| \cdot \| K \|^2$$

Thus, $\text{Tr} \ (f_n)$ tends to $\text{Tr} \ (f)$ as $n \to \infty$, and hence $N = N_1$. We conclude that N_1 is also equal to the number of linearly independent cf's discussed above.

The Schmidt estimates for the least characteristic value can now be stated without any ambiguity. In view of the monotonicity of the trace ratios we have

$$\infty > \frac{k_2}{k_4} \geq \frac{k_4}{k_6} \geq \frac{k_6}{k_8} \geq \cdots \geq \lambda_1^2 \geq \cdots \geq \sqrt[3]{\frac{N}{k_6}} \geq \sqrt[2]{\frac{N}{k_4}} \geq \frac{N}{k_2} > 0 \qquad (3)$$

As the order of the traces goes up, these estimates provide increasingly improved approximations to the cv λ_1^2. For example, with the kernel

$$K(x,y) = \begin{cases} x(1-y) & x \leq y \\ y(1-x) & y \leq x \end{cases}$$

on $0 \leq x, y \leq 1$, for which $\lambda_1^2 = \pi^4 \doteq 97.41$, we obtain

$$k_2 = \frac{1}{90} \qquad k_4 = \frac{1}{9,450} \qquad \text{and} \qquad k_6 = \frac{691}{1,365(467,775)}$$

and hence $105 \geq 97.8 \geq \lambda_1^2 \geq 97.40 \geq 97.2 \geq 90$

PROBLEMS

1. **a.** Establish Properties 2, 3, and 7 of Sec. 7.1.
    b. Show that Property 3 is not true for general Hermitian kernels by analyz-

ing the integral equation

$$\phi(x) = \lambda \int_{-1}^{1} i(x - y)\phi(y)\, dy \qquad -1 \leq x, y \leq 1$$

c. Give an example of two Hermitian kernels the composite of which is non-Hermitian.

2. Use Schwarz's inequality to verify that for Hermitian kernels

$$|(K^n \phi, \psi)|^2 \leq (K^n \phi, K^n \phi)\|\psi\|^2$$
$$\leq (K^{2n}\phi, \phi)\|\psi\|^2$$
$$\leq \|K^{2n}\| \cdot \|\phi\|^2 \cdot \|\psi\|^2$$

From this result and the fact that $\|K^{n+m}\| \leq \|K^n\| \cdot \|K^m\|$, construct a recursive proof by contradiction, which demonstrates that the iterates of nonnull Hermitian kernels are themselves nonnull [Smithies (1962, p. 107)].

3. Let ϕ, ψ be arbitrary nontrivial complex-valued $\mathfrak{L}^2$ functions.
  a. Show that $(L\phi, \psi) = 0$ for all such ϕ, ψ implies $\|L\| = 0$.
  b. Use the result of part a to help demonstrate that $(K\phi, \phi)$ is real for all such ϕ iff K is Hermitian.

4. Consider the symmetric kernel

$$K(x, y) = \min(x, y) \qquad 0 \leq x, y \leq 1$$

  a. Show that $K(x, y)$ is such that $(K\phi, \phi) \geq 0$ for all $\mathfrak{L}^2$ functions ϕ.
  b. Determine a full characteristic system for K.
  c. Use the traces k_1, k_2, and k_3 to obtain estimates for the lower-order characteristic values by means of the algorithms of Sec. 7.3.

5. Let $K(x, y)$ be a nonnull Hermitian $\mathfrak{L}^2$ kernel. Verify that the totality of linearly independent characteristic functions of K associated with either $+\lambda_1$ or $-\lambda_1$ forms a basis for the characteristic functions of $K^2(x, y)$ corresponding to λ_1^2.

6. Further investigate the quality of the Schmidt estimates (7.5-3) by verifying and extending the results obtained in Sec. 7.5 with the kernel

$$K(x, y) = \begin{cases} x(1 - y) & x \leq y \\ y(1 - x) & y \leq x \end{cases}$$

on $0 \leq x, y \leq 1$. To simplify the procedure, use the facts that

$$K(x, y) = \frac{2}{\pi^2} \sum_{\nu=1}^{\infty} \frac{\sin \nu\pi x \, \sin \nu\pi y}{\nu^2}$$

and
$$\sum_{\nu=1}^{\infty} \frac{1}{\nu^{2m}} = (2\pi)^{2m} \frac{B_m}{2(2m)!}$$

where B_m is the mth Bernoulli number.

chapter eight

Other Existence Proofs for the Characteristic Values of Hermitian Kernels

8.1 INTRODUCTORY REMARKS

The approach of Schmidt discussed in the last chapter and the existence proofs we shall consider shortly generally lead to constructive procedures for the *least* (in magnitude) characteristic value and an associated characteristic function of the Hermitian kernel K under investigation. In view of the results of Sec. 6.2, however, this is not actually a restrictive situation. Assuming that we have found the least cv λ_1 and a corresponding normalized cf $\phi_1(x)$, we can then decompose K into

$$K(x,y) = \frac{\phi_1(x)\,\overline{\phi}_1(y)}{\lambda_1} + K_1(x,y)$$

and since the two kernels on the right-hand side of this relation are orthogonal, we can concentrate afresh on Hermitian K_1, extracting its least cv and associated cf. This procedure continues, at each stage giving rise to a dissection of the form

$$K(x,y) = \sum_{\nu=1}^{N} \frac{\phi_\nu(x)\,\overline{\phi}_\nu(y)}{\lambda_\nu} + K_N(x,y)$$

until all of the cv's of K have been brought forth. If this occurs for some finite value of N, then K is a degenerate kernel of rank N, of course, and K_N must perforce be a null kernel.

8.2 CHARACTERISTIC VALUES AND FUNCTIONS BY ITERATION

In 1922 Kellogg proposed use of a straightforward iterative technique based upon successive substitutions, which not only proved the existence but also led to a determination of characteristic values and characteristic functions for Hermitian kernels [see also Riesz and Sz.-Nagy (1955, pp. 240–241) and Wavre (1943 and 1944)]. Beginning with an arbitrary $\mathfrak{L}^2$ function $\psi_0(x)$ of unit norm and our kernel K, we construct

$$\psi_1(x) = \mu_1 \int_a^b K(x,y)\psi_0(y) \, dy \qquad (1)$$

where positive μ_1 is selected so that ψ_1 is of unit norm also. Continuing in like manner, then, we form $\psi_2, \psi_3, \ldots$, and, in general,

$$\psi_n(x) = \mu_n \int_a^b K(x,y)\psi_{n-1}(y) \, dy \qquad (2)$$

with $\|\psi_n\| = 1$. It turns out, as we shall see, that if ψ_0 is such that

$$\|K\psi_0\| \neq 0 \qquad (3)$$

that is, if ψ_0 is not orthogonal to the kernel K, then there exists a characteristic value μ and a corresponding normalized characteristic function ψ toward which the μ_n and ψ_n, respectively, converge.

We first note from (2) that

$$\frac{1}{\mu_n} = (K\psi_{n-1}, \psi_n) \le \|K\|$$

by Schwarz's inequality. In this same vein,

$$\frac{1}{\mu_n} = (K\psi_{n-1}, \psi_n) = (\psi_{n-1}, K\psi_n)$$

$$= \frac{1}{\mu_{n+1}} (\psi_{n-1}, \psi_{n+1}) \le \frac{1}{\mu_{n+1}}$$

The μ_n thus form a monotonically decreasing sequence bounded from below by $1/\|K\|$, and hence they converge to a positive limit μ.

For the ψ_n, making use of (2) and the normalization property, we obtain

$$\|\psi_{2n} - \psi_{2m}\|^2 = 2 - (\psi_{2m}, \psi_{2n}) - (\psi_{2n}, \psi_{2m})$$

$$= 2[1 - (\mu_1 \ldots \mu_{2m})(\mu_1 \ldots \mu_{2n})(K^{m+n}\psi_0, K^{m+n}\psi_0)]$$

$$= 2\left[1 - \frac{\mu_{m+n+1} \cdots \mu_{2m}}{\mu_{2n+1} \cdots \mu_{m+n}}\right]$$

$$\to 0 \qquad \text{as } m, n \to \infty$$

(Here we have assumed, without loss of generality, that $m \geq n$.) This argument shows that the ψ_{2n} form a Cauchy sequence in the complete space $\mathcal{L}^2$ and hence they converge, in the mean, to an $\mathcal{L}^2$ limit function ψ of unit norm.† Since in the special case $m = n + 1$ we have from the above result

$$\|\psi_{2n} - \mu_{2n+2}\,\mu_{2n+1}\,K^2\psi_{2n}\| = \|\psi_{2n} - \psi_{2n+2}\| \to 0$$

as $n \to \infty$, it follows that

$$\psi = \mu^2 K^2 \psi$$

In other words ψ, which exists as the limit of the ψ_{2n}, is a characteristic function of K^2 associated with the characteristic value μ^2. It is a simple matter then to determine characteristic functions of K itself corresponding to either $+\mu$ or $-\mu$ (or both). (See Chap. 3, Prob. 13.)

It usually occurs that the limit μ is precisely equal in magnitude to the *least* cv λ_1 of K. In this case, if we introduce the Schmidt notation of the last chapter, we have

$$\psi_{2n} = \mu_1 \ldots \mu_{2n} K^{2n} \psi_0$$

$$= \left(\frac{\mu_1}{\lambda_1}\right) \ldots \left(\frac{\mu_{2n}}{\lambda_1}\right) \lambda_1^{2n} K^{2n} \psi_0$$

$$\equiv \gamma_n f_n \psi_0$$

and hence
$$\gamma_n^2 = \frac{1}{(f_n \psi_0, f_n \psi_0)} = \frac{1}{(f_{2n}\psi_0, \psi_0)}$$

---

† Kellogg in his original paper (1922) used the Arzela-Ascoli theorem in order to derive this convergence (uniform) for continuous kernels. Riesz and Sz.- Nagy (1955, p. 241) employ an argument based upon the *completely continuous* nature of the kernels under consideration. We shall have more to say about this approach in subsequent sections.

In the limit $n \to \infty$, this gives rise to

$$\psi = \gamma f \psi_0 \qquad \text{where} \qquad \gamma^2 = \frac{1}{(f\psi_0, \psi_0)} \tag{4}$$

If $(f\psi_0, \psi_0) \neq 0$, then $\mu^2 = \lambda_1^2$ since $(\gamma_n/\gamma_{n-1}) = \mu_{2n}\mu_{2n-1}/\lambda_1^2$ and

$$\psi(x) = \gamma \sum_{\nu=1}^{N} \phi_\nu(x) \int_a^b \overline{\phi_\nu}(y) \, \psi_0(y) \, dy \tag{5}$$

which follows from the known decomposition (7.5-2) of f. Equation (5) clearly shows ψ to be a linear combination of either the cf's of K^2 associated with the cv λ_1^2 or the cf's of K itself corresponding to $\pm \lambda_1$ (depending upon the point of view adopted).

If $(f\psi_0, \psi_0) = 0$ in (4), ψ_0 unfortunately belongs to the class of functions orthogonal to all the cf's of K associated with $\pm \lambda_1$, and thus $\mu^2 = \lambda_\nu^2$ for some $\nu > 1$.[†] To avoid this case, if desirable to do so, $\psi_0(x)$ can be chosen proportional to $K(x, y_0)$ for some appropriate y_0 since then

$$(f\psi_0, \psi_0) \propto \sum_{\nu=1}^{N} \frac{|\phi_\nu(y_0)|^2}{\lambda_1^2}$$

and we know this cannot vanish for all $a \leq y_0 \leq b$. For such ψ_0 (and y_0), moreover, it follows trivially that the necessary condition (3) is automatically satisfied.

The iterative procedure discussed above has considerable popularity in engineering circles. It has also been applied profitably to certain *noncompletely continuous* symmetric transformations by Wavre (1943 and 1944) and to *symmetrizable* transformations by Zaanen (1943). Moreover, Fox and Li (1961 and 1963) and Li (1963) have used this successive substitution method with good success in their numerical investigations of a class of integral equations having so-called "complex-symmetric" kernels. Generalizations of Kellogg's iteration and similar procedures to the general non-Hermitian case, as well as proofs of the convergence of the methods, appear in Wielandt (1944) and Marek (1962).

[†] It should be noted that in practical numerical applications this situation by and large does not result. Even if ψ_0 is initially, in essence, of this class, computational errors and round-off, which build up as the iteration procedure continues, inevitably contribute sufficiently to the limit function f to ensure $(f\psi_0, \psi_0) \neq 0$.

8.3   AN EXISTENCE PROOF BASED UPON AN EXTREMAL PROPERTY

One of the more powerful techniques for the estimation of characteristic values and functions of Hermitian kernels is a procedure first used by Lord Rayleigh and later extended by Ritz, among others. Underlying this formal technique, which we shall describe in some detail in a subsequent chapter, are certain theoretical results that represent, in some sense, the extension to function spaces of the notion of matrix diagonalization. We want to discuss these results here, however, not only for the sake of completeness but also because they permit another independent proof of the existence of cv's for Hermitian kernels. Our approach will share some common ground with a 1910 paper of Riesz [see also Riesz and Sz.- Nagy (1955, pp. 231–232) and Mikhlin (1964, pp. 105ff)].†

Recall from the theory of quadratic forms that the diagonalization of

$$Q[x] = \sum_{\mu=1}^{n} \sum_{\nu=1}^{n} a_{\mu\nu} x_{\nu} \bar{x}_{\mu} \qquad a_{\mu\nu} = \bar{a}_{\nu\mu}$$

leads to

$$Q[x] = \sum_{\nu=1}^{n} \gamma_{\nu} |y_{\nu}|^2$$

where the γ_{ν} are the real *eigenvalues* of the Hermitian matrix $(a_{\mu\nu})$. In addition, the algebraically largest among these γ_{ν}, say, γ_1, satisfies the following extremal property:

$$\max \frac{Q[x]}{\|x\|^2} = \gamma_1 \tag{1}$$

We want to carry over these same ideas into function space, and thus we define the *quadratic integral form*

$$Q[\phi] \equiv \int_a^b \int_a^b K(x,y)\phi(y)\bar{\phi}(x) \, dy \, dx = (K\phi, \phi) \tag{2}$$

As we anticipate, this functional (which is real valued for Hermitian K) has a number of useful properties analogous to those well known in ma-

---

†The interested reader should also consult the dissertation of Enskog (1917) where the solution of nonhomogeneous equations with Hermitian kernels is accomplished by a *diagonalization* procedure. When appropriately modified for the characteristic-value problem, Enskog's approach is essentially equivalent to that proposed by Rayleigh and Ritz. Hecke (1922) gives a brief account of Enskog's method; Courant and Hilbert (1953, p. 156) have an even briefer description.

trix theory, and we begin our discussion by considering a few such results. For simplicity at this stage in the analysis, we assume that the Hermitian $\mathcal{L}^2$ kernel K is such that $Q[\phi] > 0$ for some ϕ.

Definition   A kernel for which $Q[\phi]$ is always positive (nonnegative, nonpositive, or negative) is called *positive* (*nonnegative, nonpositive, or negative*) *definite*.

If K is positive (nonnegative) definite, it follows then from (2) that the kernel $-K$ is negative (nonpositive) definite and conversely. The second iterate K^2, for instance, is a good example of a nonnegative definite kernel.

Property 1 (K Schwarz Inequality)   For a nonnegative (nonpositive) definite kernel

$$|(K\phi,\psi)|^2 \leq (K\phi,\phi)(K\psi,\psi) \tag{3}$$

The proof of this property proceeds in the same manner as for the general Schwarz inequality: For general complex α

$$0 \leq (K(\phi + \alpha\psi), \phi + \alpha\psi)$$

$$= (K\phi,\phi) + \alpha(K\psi,\phi) + \overline{\alpha}(K\phi,\psi) + |\alpha|^2(K\psi,\psi)$$

$$= (K\phi,\phi) + 2|\alpha| \operatorname{Re}[e^{-i\arg\alpha}(K\phi,\psi)] + |\alpha|^2(K\psi,\psi)$$

Hence $\{\operatorname{Re}[e^{-i\arg\alpha}(K\phi,\psi)]\}^2 \leq (K\phi,\phi)(K\psi,\psi)$

which, owing to the arbitrariness of α, implies the desired result (3).

Property 2   For a general Hermitian kernel, there exist *positive* constants M such that

$$Q[\phi] \leq M\|\phi\|^2$$

for all $\mathcal{L}^2$ functions ϕ.

This property is a trivial consequence of the relations

$$Q[\phi] = (K\phi,\phi) \leq \|K\| \cdot \|\phi\|^2 = \sqrt{k_2}\,\|\phi\|^2$$

If K is nonnegative definite, the least of such M, which must itself be positive, is commonly termed the *norm of the transformation* (or operator) generated by the kernel K.†

THEOREM (PREPARATION)

Let M be the least of the upper bounds of Property 2. Then *if* there exists a nontrivial $\mathfrak{L}^2$ function ϕ such that $Q[\phi] = M\|\phi\|^2$, that ϕ must be a characteristic function of the kernel K corresponding to the characteristic value $1/M$. Moreover $1/M$ is then the least positive cv of K.

Proof   (a) Consider the Hermitian kernel $M - K$, which, in view of the definition of the positive constant M, must be nonnegative definite. By the K Schwarz inequality (3) we then have

$$|(M\phi - K\phi, \psi)|^2 \leq (M\phi - K\phi, \phi)(M\psi - K\psi, \psi)$$

for arbitrary ϕ, ψ. If, however, ϕ is such that $Q[\phi] = M\|\phi\|^2$, then $(M\phi - K\phi, \phi) = 0$ and thus

$$(M\phi - K\phi, \psi) = 0 \qquad \text{for all } \psi$$

In particular, then, the choice of $\psi = M\phi - K\phi$ implies

$$\|M\phi - K\phi\|^2 = 0 \qquad \text{or} \qquad \phi = \frac{1}{M}K\phi$$

(b) Assume there exist ψ and λ such that $\psi = \lambda K\psi$ with $\lambda > 0$ and $\|\psi\| = 1$. Then

$$\lambda M \geq \lambda Q[\psi] = \lambda(K\psi, \psi) = (\psi, \psi) = 1$$

and hence $\lambda \geq \dfrac{1}{M}$ Q.E.D.

The crucial assumption in the above preparation theorem, of course, is the existence of the $\mathfrak{L}^2$ function ϕ. As the principal result of this section will show, however, this existence need not be assumed, being a consequence (albeit nonelementary) of the nature of the linear operator induced by the kernel K. For the general case, the proof will hinge on

---

† If $Q[\phi]$ can take on both signs, then the norm of the transformation is given by the least upper bound of $|Q[\phi]|/\|\phi\|^2$. We shall not have occasion to use this terminology in the sequel.

the fact that the relation

$$K\phi \equiv \int_a^b K(x,y)\,\phi(y)\,dy$$

with K in $\mathfrak{L}^2$ defines a *completely continuous* (cc) operator, i.e., an operator that transforms every infinite bounded set into a (conditionally sequentially) *compact* set (see Appendix B). The verification of this fact [details of which may be found in Riesz and Sz.-Nagy (1955, pp. 178ff) or in Mikhlin (1964, pp. 107ff), for example] involves showing that linear transformations which can be approximated in norm with arbitrary accuracy by cc linear transformations are themselves cc. Then since all linear transformations of finite rank in $\mathfrak{L}^2$ are cc, so also must be the above operator.†

We turn now to the principal result.

THEOREM (MAIN)

Let M be the least of the upper bounds of Property 2. Then *there exists* a nontrivial $\mathfrak{L}^2$ function ϕ satisfying $Q[\phi] = M\|\phi\|^2$; moreover, ϕ is such that

$$\phi = \frac{1}{M}\,K\phi$$

Proof   (*a*) In view of the nature of the positive constant M, there exists a sequence of $\mathfrak{L}^2$ functions $\{\phi_\nu\}$ such that

$$\|\phi_\nu\| = 1 \qquad \text{and} \qquad M_\nu \equiv Q[\phi_\nu] \to M \qquad \text{as } \nu \to \infty$$

Applying the K Schwarz inequality (3) with the nonnegative definite kernel $M - K$ and $\phi = \phi_\nu$, $\psi = (M - K)\phi_\nu$, we obtain

$$\|(M - K)\phi_\nu\|^2 \leq (M\phi_\nu - K\phi_\nu, \phi_\nu)((M - K)^2\phi_\nu, (M - K)\phi_\nu)$$

$$\leq \text{const}\,(M - M_\nu)$$

so that as $\nu \to \infty$

$$\phi_\nu - \frac{1}{M}\,K\phi_\nu \to 0 \qquad\qquad (4)$$

---

†Recall the important fact that the class of degenerate kernels is dense in $\mathfrak{L}^2$; see Sec. 3.3. Also see Secs. 10.2 and 13.3 where special forms of degenerate kernel approximants are discussed.

Now the linear operator generated by the $\mathfrak{L}^2$ kernel K is completely continuous so that since the $\{\phi_\nu\}$ are bounded, the sequence of $\mathfrak{L}^2$ functions given by $\{\psi_\nu\} = \{K\phi_\nu\}$ contains a mean-convergent subsequence $\{\psi_{\nu\mu}\}$. As a consequence of (4), however, this implies that the sequence $\{\phi_{\nu\mu}\}$ is itself convergent in the mean to an $\mathfrak{L}^2$ limit function, say, ϕ. It follows, then, that

$$\|\phi\| = 1 \qquad Q[\phi] = M \qquad \text{and} \qquad \phi = \frac{1}{M} K\phi$$

(b) The above argument completes the verification of the theorem in the general case. It is worth noting, on the other hand, that if the kernel K has additional structure, the proof can be simplified somewhat. For instance, if K is *continuous* in the fundamental square $a \leq x, y \leq b$, the sequence of functions $\{\psi_\nu\}$ turns out to be both equibounded and equiuniformly continuous (see Prob. 7). The familiar theorem of Arzela and Ascoli then ensures the existence of a *uniformly* convergent subsequence $\{\psi_{\nu\mu}\}$ with *continuous* limit function ψ. Moreover, ψ is nontrivial since

$$\|\psi\| = \lim \|K\phi_{\nu\mu}\| \geq \lim Q[\phi_{\nu\mu}] = M > 0$$

The proof is concluded by noting from (4) that

$$\psi_{\nu\mu} - \frac{1}{M} K\psi_{\nu\mu} \to 0$$

also, and hence the continuous limit function ψ satisfies†

$$\psi = \frac{1}{M} K\psi$$

8.4 EXTREMAL CHARACTERIZATION OF LARGER CHARACTERISTIC VALUES

At the beginning of this chapter we suggested how the larger cv's of a Hermitian kernel K could be individually analyzed at the appropriate

---

†Even in the non-Hermitian case, the cf's of continuous kernels can always be chosen to be continuous.

stage by means of the decomposition

$$K(x,y) = \sum_{\nu=1}^{n} \frac{\phi_\nu(x)\overline{\phi}_\nu(y)}{\lambda_\nu} + K_n(x,y)$$

Let us assume, therefore, that we have determined the first n *positive* cv's $0 < \lambda_1 \leq \lambda_2 \leq \cdots \leq \lambda_n$ and their corresponding orthonormalized cf's $\phi_1, \phi_2, \ldots, \phi_n$ and have made the above dissection of K. Considering now the subsidiary Hermitian kernel K_n, either $Q_n[\phi] \equiv (K_n\phi, \phi) > 0$ for some ϕ or $Q_n[\phi] \leq 0$ for all ϕ in $\mathfrak{L}^2$. In the former case, if we set

$$\frac{1}{\lambda_{n+1}} = \sup \frac{Q_n[\phi]}{\|\phi\|^2}$$

the theorems of the preceding section ensure that λ_{n+1} is the least positive cv of K_n and has a corresponding cf ϕ_{n+1} (which may be normalized). Moreover, from previous work it then follows that ϕ_{n+1} is also a cf of the original kernel K belonging to the cv λ_{n+1}, and $(\phi_{n+1}, \phi_\nu) = 0$ for $\nu = 1, 2, \ldots, n$. In other words, λ_{n+1} has a natural place as the $(n + 1)$st positive cv of the original kernel.

By introducing the class of functions $c_n(\phi_1, \phi_2, \ldots, \phi_n)$ where

$$\phi \in c_n \quad \text{implies} \quad \begin{cases} (\phi, \phi_\nu) = 0 & (\nu = 1, 2, \ldots, n) \\ \\ \|\phi\| = 1 \end{cases}$$

we can give additional structure to the above remarks. For then $Q_n[\phi] = Q[\phi]$ for ϕ in c_n, and

$$\sup_{\phi \in c_n} Q[\phi] = \sup_{\phi \in c_n} Q_n[\phi] \leq \sup_{\phi \in \mathfrak{L}^2} \frac{Q_n[\phi]}{\|\phi\|^2} = \frac{1}{\lambda_{n+1}}$$

But since ϕ_{n+1} is in c_n and $Q[\phi_{n+1}] = \|\phi_{n+1}\|^2/\lambda_{n+1} = 1/\lambda_{n+1}$, it follows that

$$\frac{1}{\lambda_{n+1}} = \sup_{\phi \in c_n} Q[\phi] \tag{1}$$

This important relation (1), which characterizes the higher-order characteristic values (and characteristic functions) as solutions of extremal problems, represents the natural extension of the matrix diagonalization discussed earlier. As a point of view, moreover, it suggests the utility of various direct methods of the variational calculus for determination of cv's and cf's. On the other hand, however, for certain purposes the relation (1) suffers from its recursive nature, requiring at each stage, as it does, knowledge of all of the lower-order characteristic functions. We shall see anon, though, how this "restriction" can be easily obviated.

8.5 THE EXTENSION TO NEGATIVE CHARACTERISTIC VALUES

In each of the two preceding sections we made the assumption, where needed, that for the kernel under consideration, $Q[\phi]$ (or $Q_n[\phi]$) was positive for some ϕ. It is evident, however, that for certain kernels (e.g., nonnegative definite kernels — see Prob. 8) this really is no assumption at all. Moreover, such an auxiliary condition could, in general, be avoided, if desired, in the following manner: Define $Q^2[\phi]$ by

$$Q^2[\phi] \equiv (K^2\phi, \phi) = \|K\phi\|^2 \tag{1}$$

Since K^2 is always nonnegative definite, the foregoing theory readily applies and in particular we find nontrivial ψ, which satisfy

$$Q^2[\psi] = \frac{1}{\lambda^2} \|\psi\|^2$$

and are thus cf's of K^2 associated with the cv λ^2. In the standard manner (see Chap. 3, Prob. 13), these ψ lead to cf's Ψ of K itself associated with the cv $\pm \lambda$. It follows, then, that

$$\frac{1}{|\lambda|} = \frac{|Q[\Psi]|}{\|\Psi\|^2} = \sup \frac{|Q[\phi]|}{\|\phi\|^2}$$

with the only difference from our previous characterization being the presence of the absolute value signs.

As our earlier analysis tends to suggest, however, there is no need to go to these absolute value signs. If at any stage, including the initial, $Q[\phi] \leq 0$ for all functions of the class in question, then attention should be shifted to the kernel $-K$ and its associated functional $-Q[\phi]$. Arguments completely analogous to those presented above then lead to the *negative* characteristic values and associated characteristic functions. For a general Hermitian kernel, therefore, we may extract step by step all the positive cv's and their corresponding cf's starting with the least; in principle, this having been done, we may in like manner extract all the negative cv's and their associated cf's.†

The question naturally arises as to whether we overlook any cv's and cf's in this extraction process. Assuming such to be the case, we let λ_0, ϕ_0 be among the "disregarded" elements. For a degenerate kernel K, it immediately follows that

$$\int_a^b K(x,y)\phi_0(y)\,dy = 0 \tag{2}$$

and hence $\lambda_0 = \infty$. In the general case, we are led to

$$\|K\phi_0\|^2 = Q^2[\phi_0] \le \frac{1}{\lambda^2}\|\phi_0\|^2$$

for a countable infinity of values $\lambda^2 \to \infty$. If ϕ_0 is truly nontrivial, then (2) must also be satisfied here, too. In our extraction process, therefore, we overlook only those cf's associated with the "characteristic value" $\lambda_0 = \infty$.

Partially imbedded in the above remarks is the following.

Auxiliary Result   [Smithies (1962, p. 118).] For a Hermitian $\mathfrak{L}^2$ kernel K, an $\mathfrak{L}^2$ function ϕ satisfies

$$\int_a^b K(x,y)\phi(y)\,dy = 0$$

iff ϕ is orthogonal to all of the characteristic functions of K associated with finite characteristic values.

The "only if" part of the proof follows from the observation that

$$K\phi = 0 \qquad \text{implies} \qquad 0 = (K\phi, \phi_\nu) = (\phi, K\phi_\nu) = \frac{(\phi, \phi_\nu)}{\lambda_\nu}$$

PROBLEMS

1. We know that nonnegative (nonpositive) definite $\mathfrak{L}^2$ kernels K are perforce Hermitian (see Chap. 7, Prob. 3). Show that this result is no longer valid gen-

---

†The particular order of extraction, of course, actually is up to the individual investigator. By alternating between considerations of K and $-K$, any possible order may be effected. Bear in mind, however, that positive (negative) definite kernels must perforce have only positive (negative) characteristic values.

erally if we only know that $(K\phi, \phi)$ is real for *real-valued* ϕ. (*Hint:* Look at rigid body rotations.)

2. Olagunju and West (1964) have shown that a necessary and sufficient condition for all the characteristic values of a given $\mathfrak{L}^2$ kernel to be real (nonnegative) is that

$$\mathrm{Tr}\,(K^n p(K)\overline{p}(K)) \geq 0$$

for all complex polynomials $p(K)$, where n is some even (odd) positive integer. Establish the necessity of this condition.

3. **a.** Determine a full characteristic system for the symmetric kernel

$$K(x,y) = \begin{cases} x(1-y) & x \leq y \\ y(1-x) & y \leq x \end{cases} \quad 0 \leq x,y \leq 1$$

b. Show that this kernel is nonnegative definite in the space of complex-valued $\mathfrak{L}^2$ functions.

4. For the kernel of Prob. 3:
a. Use Kellogg's iterative procedure, starting from the initial function $\psi_0(x) = 1$, in order to obtain estimates for the least characteristic value.
b. What second-degree polynomial would have been a better choice for $\psi_0(x)$? Why?

5. Use the Schwarz inequality, the K Schwarz inequality, and an inductive argument to verify that

$$(K\phi, \phi)^n \leq (\phi, \phi)^{n-1}(K^n \phi, \phi) \qquad n \geq 1$$

for nonnegative definite kernels K. [See Beesack (1969b) for the extension of this result to symmetric kernels which satisfy $K(x,y) \geq 0$, $a \leq x,y \leq b$.]

6. [Heinz (1951); see also Rellich (1953a).] Let a nonnegative definite kernel K and a Hermitian kernel L be such that

$$(L^2\phi, \phi) \leq (K^2\phi, \phi)$$

for all complex-valued ϕ.
a. Prove that the cv's of both the difference kernel $K - L$ as well as the sum kernel $K + L$ are positive and thus

$$|(L\phi, \phi)| \leq (K\phi, \phi)$$

b. Does there exist a related result which is valid if K is only known to be Hermitian?

7. Consider the sequence of functions given by

$$\psi_\nu(x) = \int_a^b K(x,y)\phi_\nu(y)\,dy \qquad \nu = 1, 2, 3, \ldots$$

where the kernel K is *continuous* for $a \leq x,y \leq b$ and $\|\phi_\nu\| = 1$. Show that the $\{\psi_\nu\}$ are equibounded and equiuniformly continuous.

8. Let $0 < \lambda_1 \le \lambda_2 \le \cdots \le \lambda_n$ and $\phi_1, \phi_2, \ldots, \phi_n$ be n positive cv's and associated orthonormalized cf's of a Hermitian kernel $K(x,y)$. If K is non-negative definite, verify that

$$K_n(x,y) \equiv K(x,y) - \sum_{\nu=1}^{n} \frac{\phi_\nu(x)\, \overline{\phi}_\nu(y)}{\lambda_\nu}$$

also shares this property.

chapter nine

The Rayleigh-Ritz Procedure and Related Results

9.1 ESTIMATION USING THE RAYLEIGH QUOTIENT

We have seen earlier how both the Schmidt and Kellogg procedures for establishing the existence of characteristic values and associated characteristic functions of Hermitian kernels lead naturally to estimates for those values and functions. In analogous fashion the extremal characterizations of the previous chapter give rise to valuable approximations. For instance, if for a given Hermitian kernel K we define the *Rayleigh quotient* (or *ratio*) R by

$$R[\phi] \equiv \frac{Q[\phi]}{\|\phi\|^2} = \frac{(K\phi, \phi)}{\cdot(\phi, \phi)} \tag{1}$$

we know that
$$R[\phi] \leq \frac{1}{\lambda_1}$$

where λ_1 is the least positive characteristic value of K. (We assume, as usual, that $Q[\phi]$ can take on positive values.) Hence if

$$\phi = \sum_{\nu=1}^{n} \alpha_\nu \psi_\nu$$

where $\{\psi_\nu\}$ is *any* finite orthonormal set of functions, then

$$R[\phi] = \frac{\sum_{\mu=1}^{n}\sum_{\nu=1}^{n} c_{\mu\nu}\alpha_\mu\,\overline{\alpha}_\nu}{\sum_{\nu=1}^{n}|\alpha_\nu|^2} \le \frac{1}{\lambda_1} \tag{2}$$

with

$$c_{\mu\nu} \equiv (K\psi_\mu, \psi_\nu) = \int_a^b \int_a^b K(x,y)\psi_\mu(y)\,\overline{\psi}_\nu(x)\,dy\,dx$$

The $\alpha_\nu{}^{(n)}$ that maximize the left-hand side of the inequality (2) are the components of an eigenvector of the Hermitian matrix $C \equiv (c_{\mu\nu})$ corresponding to the (algebraically) largest eigenvalue $\gamma_1{}^{(n)}$; the maximum of $R[\phi]$ is this value $\gamma_1{}^{(n)}$ itself.[†] Note especially that $\gamma_1{}^{(n)}$ provides an *upper bound* for the characteristic value λ_1; increasing the set $\{\psi_\nu\}$ to include ψ_{n+1}, moreover, leads to an improved upper-bound estimate of λ_1 in that

$$\gamma_1{}^{(n)} \le \gamma_1{}^{(n+1)} \le \frac{1}{\lambda_1}$$

We shall see subsequently (Sec. 10.3) how the "distance" between $\phi_1{}^{(n)} = \sum_{\nu=1}^{n} \alpha_\nu{}^{(n)}\psi_\nu$ and an appropriate characteristic function of K corresponding to λ_1 can be measured in terms of the difference $1/\lambda_1 - \gamma_1{}^{(n)}$. At this stage, however, it suffices to observe that from our previous analysis, if $\gamma_1{}^{(n)} \to 1/\lambda_1$ as $n \to \infty$ [which must occur if the infinite set of $\{\psi_\nu\}$ is complete (see Chap. 13)], we do know that $\phi_1{}^{(n)}$ converges in the mean to such a characteristic function.

A version of this *direct method*[‡] for the calculation of cv's and cf's was first used in various problems in acoustics by Lord Rayleigh (1870

[†] For modest n, the determination of $\gamma_1{}^{(n)}$ and the $\alpha_\nu{}^{(n)}$ is a simple algebraic exercise. Using Lagrange's method, for instance, we may solve

$$0 = \frac{\partial}{\partial\alpha_\mu}\left[(R[\phi] - \gamma)\sum_{\nu=1}^{n}|\alpha_\nu|^2\right] = \sum_{\nu=1}^{n}(c_{\mu\nu} - \gamma\delta_{\mu\nu})\overline{\alpha}_\nu, \qquad \mu = 1, 2, \ldots, n$$

with $\sum_{\nu=1}^{n}|\alpha_\nu|^2 = 1$ (see the degenerate kernel analysis of Chap. 2). Sophisticated techniques, of course, are needed when n is large.

[‡] We call a method *direct* if it reduces the task of solving the given integral equation or equivalent variational problem to that of solving a *finite* approximating system of algebraic equations.

and 1896); an improved formulation of the original technique subsequently appeared in 1899. Nine years later the Swiss physicist Ritz,
working independently, proposed essentially the same method. The successful applications reported there (1908) and elsewhere (1909) did
much to popularize the procedure which today bears the names of these
two applied scientists [see Courant (1943) and Gould (1966, pp. 79–80)
for additional historical background).]†

It is worthwhile noting at this point that often various properties of
the characteristic functions of a given integral equation are known a
priori, e.g., they have a certain number of zeros in the domain of definition, they are even or odd, they vanish at the end points of the interval,
etc. A better estimate at any given stage of the Rayleigh-Ritz procedure
will usually be obtained if the orthonormalized set of functions $\{\psi_\nu\}$ also
enjoys these same properties (see Prob. 1 at the end of this chapter for
an illustrative example).

9.2 COURANT'S APPROACH TO THE ESTIMATION PROBLEM

Complementing the Rayleigh-Ritz procedure is another direct method
proposed by Courant (1923) [see also Courant and Hilbert (1953, pp.
123–125)]. It is based upon the fact that a given Hermitian $\mathfrak{L}^2$ kernel
can be approximated in norm (with arbitrary accuracy) by degenerate
kernels which we may choose in the form

$$K_n(x,y) = \sum_{\mu=1}^{n} \sum_{\nu=1}^{n} b_{\mu\nu}{}^{(n)} \psi_\nu(x) \overline{\psi}_\mu(y) \tag{1}$$

with $b_{\mu\nu}{}^{(n)} = \overline{b_{\nu\mu}{}^{(n)}}$. We assume that the set $\{\psi_\nu\}$ is orthonormalized
so that $(\psi_\mu, \psi_\nu) = \delta_{\mu\nu}$.

An arbitrary square-integrable function ϕ can be expanded in terms
of the ψ_ν as

$$\phi = \sum_{\nu=1}^{n} \alpha_\nu \psi_\nu + r_n$$

If $\alpha_\nu = (\phi, \psi_\nu)$, then $(r_n, \psi_\nu) = 0$ for all $1 \le \nu \le n$, and thus

$$\frac{Q_n[\phi]}{\|\phi\|^2} \equiv \frac{(K_n \phi, \phi)}{(\phi, \phi)} = \frac{\displaystyle\sum_{\mu=1}^{n} \sum_{\nu=1}^{n} b_{\mu\nu}{}^{(n)} \alpha_\mu \overline{\alpha}_\nu}{\displaystyle\sum_{\nu=1}^{n} |\alpha_\nu|^2 + \|r_n\|^2} \tag{2}$$

---

† In a later section of this chapter we shall discuss an extremely general characterization of
the Rayleigh-Ritz procedure.

For all kernels K_n (except those which are not strictly nonpositive definite), the maximum

$$\frac{1}{\lambda_1{}^{(n)}} \equiv \max \frac{Q_n[\phi]}{\|\phi\|^2} \tag{3}$$

must be achieved when $\|r_n\| = 0$, and thus its determination, as in the case of the Rayleigh-Ritz method, is purely an algebraic exercise.

The quantity $\lambda_1{}^{(n)}$ defined in (3) and the associated function, which gives rise to the maximum value of the quadratic form, are approximations to the least positive characteristic value λ_1 and an appropriate characteristic function ϕ_1 of the given kernel K, respectively. It follows, moreover, reasoning carefully with the Schwarz inequality, that

$$\left| \frac{1}{\lambda_1} - \frac{1}{\lambda_1{}^{(n)}} \right| \leq \|K - K_n\| \tag{4}$$

and hence the accuracy of $\lambda_1{}^{(n)}$, as expected, is directly dependent upon the quality of approximation of K by K_n.†

9.3 INDEPENDENT INVESTIGATION OF THE LARGER CHARACTERISTIC VALUES

In the latter portion of this chapter we shall treat the Courant and Rayleigh-Ritz procedures in more comprehensive fashion. As a prelude to that discussion, let us examine again the higher-order characteristic values of Hermitian kernels. From Chap. 8 we know that these cv's can be characterized in extremal fashion, namely,‡

$$\frac{1}{\lambda_{n+1}} = \max_{\phi \, \epsilon \, c_n} Q[\phi] \tag{8.4-1}$$

where the class c_n is composed of all those functions of unit norm that are orthogonal to the characteristic functions $\phi_1, \phi_2, \ldots, \phi_n$ associated with the first n characteristic values $\lambda_1, \lambda_2, \ldots, \lambda_n$. As suggested previously, however, it is not theoretically necessary to have determined the lower-order λ's and ϕ's before progressing to λ_{n+1} and ϕ_{n+1}. To see

---

† If the existence of the least positive cv λ_1 were not already known from consideration of the other proofs presented previously, the approach of this section could be used as the basis for yet another independent proof of this fact [see Courant and Hilbert (1953, p. 125) for the case of a continuous kernel].

‡ Here and henceforth we use the notation "max" in place of "sup" whenever we know that these values are actually attained.

this, it will be helpful to extend our notion of the classes c_n. Given an arbitrary set $\{f_\nu\}$, $\nu = 1, 2, \ldots, n$, we therefore define $c_n(f_n) \equiv c_n(f_1, f_2, \ldots, f_n)$ to be that set of functions such that

$$\phi \in c_n(f_n) \quad \text{implies} \quad \begin{cases} (\phi, f_\nu) = 0 & \nu = 1, 2, \ldots, n \\ \|\phi\| = 1 \end{cases}$$

Now let

$$\mu(f_1, \ldots, f_n) \equiv \max_{\phi \in c_n(f_n)} Q[\phi] \tag{1}$$

for arbitrary $\{f_\nu\}$ and consider the set $\{\phi_\nu\}$, $\nu = 1, 2, \ldots, n+1$, of orthonormalized characteristic functions associated with the first $(n+1)$ *positive* cv's.† We next form

$$\Phi(x) = \sum_{\nu=1}^{n+1} \alpha_\nu \phi_\nu(x)$$

and determine (as can always be done) the coefficients α_ν so that $\Phi \in c_n(f_n)$. The following inequalities and equalities then result:

$$\mu(f_1, \ldots, f_n) \geq Q[\Phi] = \sum_{\nu=1}^{n+1} \frac{|\alpha_\nu|^2}{\lambda_\nu}$$

$$\geq \frac{1}{\lambda_{n+1}} \sum_{\nu=1}^{n+1} |\alpha_\nu|^2 = \frac{1}{\lambda_{n+1}}$$

On the other hand, from (8.4-1) and (1),

$$\mu(\phi_1, \ldots, \phi_n) = \frac{1}{\lambda_{n+1}}$$

and thus it must be true that

$$\frac{1}{\lambda_{n+1}} = \min_{\{f_1, \ldots, f_n\}} \max_{\phi \in c_n(f_n)} Q[\phi] \tag{2}$$

This result (2) is the celebrated *Weyl-Courant minimax characterization* of positive cv's of Hermitian kernels [see Weyl (1912a) and

---

† If a Hermitian kernel does not possess a kth positive (negative) cv, we assume that $\lambda_k = \infty$ and consider the associated cf ϕ_k to be null.

Courant (1920)]. A corresponding relationship, either with the roles of min and max interchanged or with Q replaced by $-Q$, is valid for the negative characteristic values.

Although (2) is generally an inappropriate formulation for computational purposes, it does lend itself well to various theoretical investigations (some of which have a quantitative flavor). For example, let us assume that we have a sequence of (not necessarily degenerate) square-integrable Hermitian kernels K_n converging in the mean to a given Hermitian $\mathfrak{L}^2$ kernel K. Using Schwarz's inequality, we find

$$|Q[\phi] - Q_n[\phi]| = |(K\phi - K_n\phi, \phi)|$$
$$\leq \|K - K_n\|$$

or
$$Q[\phi] \leq Q_n[\phi] + \|K - K_n\|$$
$$Q_n[\phi] \leq Q[\phi] + \|K - K_n\| \tag{3}$$

for functions ϕ of unit norm. If $\phi \in c_k(\phi_k^{(n)})$ where $\phi_\nu^{(n)}$, $\nu = 1, 2, \ldots, k$, are orthonormalized cf's of K_n corresponding to the first k positive cv's, $\lambda_1^{(n)}, \ldots, \lambda_k^{(n)}$, then the first inequality of (3) leads to

$$Q[\phi] \leq \frac{1}{\lambda_{k+1}^{(n)}} + \|K - K_n\|$$

By the minimax result (2) we then have

$$\frac{1}{\lambda_{k+1}} \leq \frac{1}{\lambda_{k+1}^{(n)}} + \|K - K_n\|$$

where λ_{k+1} is the $(k+1)$st positive cv of K. In analogous fashion we obtain from the other inequality in (3)

$$\frac{1}{\lambda_{k+1}^{(n)}} \leq \frac{1}{\lambda_{k+1}} + \|K - K_n\|$$

and hence†
$$\left| \frac{1}{\lambda_{k+1}} - \frac{1}{\lambda_{k+1}^{(n)}} \right| \leq \|K - K_n\| \tag{4}$$

The inequality expressed in this relation, which generalizes the result (9.2-4) given earlier for the least cv in the case of degenerate approximating kernels, is equally valid for the negative cv's as well as the

---

† See Chap. 15, Prob. 6, for a substantial refinement of this inequality.

positive. (Here we have temporarily abandoned our usual ordering in terms of modulus.) It measures the "distance" between two characteristic values in terms of the "proximity" of the two kernels. Moreover, if $\|K - K_n\| \to 0$ as assumed, then (4) clearly implies the important and fundamental result that

$$\lim_{n \to \infty} \frac{1}{\lambda_k{}^{(n)}} = \frac{1}{\lambda_k} \qquad uniformly \text{ in } k \qquad (5)$$

Although no relation analogous to (4) is uniformly valid in the non-Hermitian situation, this last result does carry over, in essence, to the general case [see Dunford and Schwartz (1963, sec. 11.9), Yamamoto (1968), or Gohberg and Krein (1969, sec. 1.5), for example].

9.4 A GENERALIZED RAYLEIGH-RITZ PROCEDURE

The Weyl-Courant minimax characterization (9.3-2) unfortunately is not very practical for explicit determination of even upper or lower bounds for the larger characteristic values of Hermitian kernels. For instance, already in the case of the second cv, (9.3-2) becomes

$$\frac{1}{\lambda_2} = \min_{f_1} \ \max_{\phi \, \epsilon \, c_1 \, (f_i)} \ Q\,[\phi]$$

Since we cannot explore, in practice, all of the space $\mathfrak{L}^2$ of interest, we usually content ourselves with some more realistic proper subspace $S \subset \mathfrak{L}^2$. But then

$$\min_{f_1 \, \epsilon \, S} \geq \min_{f_1}$$

whereas

$$\max_{\phi \, \epsilon \, c_1 (f_1) \cap S} \leq \max_{\phi \, \epsilon \, c_1 (f_1)}$$

and we are faced with the delicate task of determining the cumulative effect of these compensating changes.

There is, however, a characterization complementary to that of Weyl and Courant, which obviates these difficulties. In order to derive this result, let $\{f_\nu\}$, $\nu = 1, 2, \ldots, n$ be n *arbitrary* orthonormal functions spanning a subspace $S^{(n)}$ of the space of interest and consider

$$\min_{\phi \, \epsilon \, S^{(n)}} R\,[\phi] \equiv \min_{\phi \, \epsilon \, S^{(n)}} \frac{Q\,[\phi]}{\|\phi\|^2}$$

Now if ϕ_ν, $\nu = 1, 2, \ldots, n$ are orthonormalized cf's associated with the first n *positive* cv's λ_ν, $\nu = 1, 2, \ldots, n$, of a given Hermitian kernel K, the following relationships are then valid:

$$\min_{\phi \in S^{(n)}} R[\phi] \leq \min_{\phi \in S_{(n)} \cap c_{n-1}(\Phi_{n-1})} Q[\phi]$$

$$\leq \max_{\phi \in S_{(n)} \cap c_{n-1}(\Phi_{n-1})} Q[\phi]$$

$$\leq \max_{\phi \in c_{n-1}(\Phi_{n-1})} Q[\phi]$$

$$= \frac{1}{\lambda_n}$$

On the other hand, if $f_\nu = \phi_\nu$, $\nu = 1, 2, \ldots, n$, then

$$\min_{\phi \in S^{(n)}} R[\phi] = \frac{1}{\lambda_n}$$

It follows, therefore, that among all possible choices of the f_ν,

$$\frac{1}{\lambda_n} = \max_{\{f_1, f_2, \ldots, f_n\}} \min_{\phi \in S^{(n)}} R[\phi] \tag{1}$$

Here, then, is an alternate characterization of the nth positive characteristic value of a Hermitian kernel. It is usually credited to Poincaré, who touched on related results for differential operators in an 1890 paper [see also Pólya (1954), Pólya and Schiffer (1953–1954, pp. 276–279), Hersch (1961), Weinberger (1962, pp. 74–76), and Stenger (1966)].[†] Like the Weyl-Courant result (9.3-2), this max-min formulation is independent of any explicit reference to preceding cv's or cf's. In addition, similar to (9.3-2), it has its analogue for negative characteristic values, too. One new feature, however, is now present. If we have *two closed linear spaces S and S' of interest* with $S' \subset S$ and we consider the nature of the linear transformation associated with a given kernel

---

[†] Fischer (1905) stated the result (1) in precisely this way for the case of real quadratic forms in finite-dimensional spaces. He also deduced what is equivalent to a finite-dimensional version of the Weyl-Courant characterization (9.3-2). Stenger shows how both the Poincaré and Weyl-Courant characterizations can be viewed as separate special cases of an even more general variational principle.

when restricted to each of these domains, then clearly for the positive cv's,

$$\frac{1}{\lambda_n}\bigg|_{S'} \le \frac{1}{\lambda_n}\bigg|_{S} \qquad (2)$$

since the maximum in (1) is taken over a larger collection of elements in the case of S than with S'. (Here it is assumed, of course, that $S^{(n)} \subset S' \subset S$.) This suggests that the new Poincaré formulation will allow more practical manipulation, and such is indeed the case, as we shall see shortly.

The inequality (2) shows that if we *increase* the space under consideration, we *decrease* the positive characteristic values (*increase* the negative cv's). Conversely, if the action of the linear transformation generated by a given kernel is restricted to a *smaller* domain, the positive cv's *increase*. To denote this dependency upon the space in question, we superscript the λ_n, for example, $\lambda_n^{(0)}$. Having done this, we may write a more general version of (1):

$$\frac{1}{\lambda_k'} = \max_{\{f_1, \ldots, f_k\} \in S'} \min_{\phi \in S(k)} R[\phi] \qquad (1')$$

Here, now, the λ_k' refer to the first k positive characteristic values of the restriction (projection) of the kernel K to (onto) the space of interest S'.

With the max-min characterization $(1')$ in mind, let us see what happens when we choose $S^{(n)}$, the space spanned by n arbitrary (but given) orthonormal functions, to be the *practical space under consideration*. In this case we have for each $k \le n$

$$\frac{1}{\lambda_k^{(n)}} = \max_{\{f_1, \ldots, f_k\} \in S^{(n)}} \min_{\phi \in S(k)} R[\phi] \qquad (3)$$

Moreover, in view of the monotonicity property (2), if $S^{(n)}$ is enlarged to $S^{(n+1)}$, $S^{(n+2)}$, etc., by the inclusion of additional orthonormal functions, (3) gives rise to

$$\frac{1}{\lambda_k^{(n)}} \le \frac{1}{\lambda_k^{(n+1)}} \le \cdots \le \frac{1}{\lambda_k} \qquad (4)$$

the last inequality in this chain being a consequence of the fact that all the $S^{(\nu)}$ are proper subspaces of $\mathfrak{L}^2$.

When $k = 1$, the relationship (3) reduces to

$$\frac{1}{\lambda_1^{(n)}} = \max_{f_1 \in S(n)} \min_{\phi \in S(1)} R[\phi]$$

$$= \max_{\phi \in S(n)} R[\phi]$$

This is precisely what in Sec. 9.1 we termed the *Rayleigh-Ritz approximant to the least positive cv of the given kernel K.* Moreover, we saw there that $1/\lambda_1^{(n)}$ was given by the algebraically largest *eigenvalue* of the $n \times n$ Hermitian matrix

$$C \equiv (c_{\mu\nu}) = ((K\psi_\mu, \psi_\nu))$$

where the ψ_ν are the orthonormal functions spanning $S^{(n)}$. Pursuing this analogy further, if we order the eigenvalues $\gamma_\nu^{(n)}$ of C so that

$$\gamma_1^{(n)} \geq \gamma_2^{(n)} \geq \cdots \geq \gamma_n^{(n)}$$

then it follows for general k from (3) that

$$\frac{1}{\lambda_k^{(n)}} = \max_{\nu_1, \ldots, \nu_k} \min_{\alpha_{\nu_1}, \ldots, \alpha_{\nu_k}} \frac{\displaystyle\sum_{i=1}^{k}\sum_{j=1}^{k} c_{\mu_i\nu_j} \alpha_{\mu_i} \overline{\alpha_{\nu_j}}}{\displaystyle\sum_{j=1}^{k} \|\alpha_{\nu_j}\|^2}$$

$$= \max_{\nu_1, \ldots, \nu_k} \min_{\beta_{\nu_1}, \ldots, \beta_{\nu_k}} \frac{\displaystyle\sum_{j=1}^{k} \gamma_{\nu_j}^{(n)} |\beta_{\nu_j}|^2}{\displaystyle\sum_{j=1}^{k} |\beta_{\nu_j}|^2}$$

$$= \max_{\nu_1, \ldots, \nu_k} \min_{1 \leq j \leq k} \gamma_{\nu_j}^{(n)}$$

$$= \gamma_k^{(n)}$$

We may call these eigenvalues of C, therefore, the *Rayleigh-Ritz approximants to the higher-order characteristic values of K.*† Note,

---

† Aronszajn (1948) has proposed a further generalization of the Rayleigh-Ritz procedure. It will be discussed in Chap. 11 how his more comprehensive approach appears as a natural counterpart to A. Weinstein's classical method of approximation.

particularly, that (except for some of the higher-order $\gamma_k^{(n)}$, which may turn out to be negative) *each* of the eigenvalues of C gives rise to an *upper bound* for a corresponding cv of K.

9.5 SOME USEFUL INEQUALITIES

To bring this chapter to a close, let us sample a series of general results which, by and large, can be easily deduced from the Weyl-Courant and/or Poincaré characterizations. For the most part, these date back to Schwarz (1885), Weyl (1912a, 1912b, and 1915) and Courant (1920), although we do include the more recent contributions of Aronszajn (1948) and Stenger (1967). The kernels throughout are square integrable and Hermitian, and we assume, as usual, that their cv's are ordered in the natural way, namely, according to increasing magnitude. The subclasses of positive and negative cv's are presumed to be similarly ordered.

1. Let $K = K' + K''$ and denote the cv's associated with K, K', and K'' by λ_ν, λ'_ν, and λ''_ν, respectively. Then from (9.3-2)

$$\frac{1}{\lambda_{\nu+\mu-1}} \leq \frac{1}{\lambda'_\nu} + \frac{1}{\lambda''_\mu}$$

for the positive cv's and

$$\frac{1}{\lambda_{\nu+\mu-1}} \geq \frac{1}{\lambda'_\nu} + \frac{1}{\lambda''_\mu}$$

for the negative cv's. Reasoning carefully, it follows from these two relations that

$$\left| \frac{1}{\lambda_{\nu+\mu-1}} \right| \leq \left| \frac{1}{\lambda'_\nu} \right| + \left| \frac{1}{\lambda''_\mu} \right| \tag{1}$$

regardless of the sign of the individual cv's. For $\mu = \nu = 1$, this result becomes

$$\left| \frac{1}{\lambda_1} \right| \leq \left| \frac{1}{\lambda'_1} \right| + \left| \frac{1}{\lambda''_1} \right|$$

2. If K'' above is of finite rank N and has p positive cv's and $N - p$ negative cv's, then

$$\frac{1}{\lambda_{\nu+p}} \leq \frac{1}{\lambda'_\nu} \qquad \nu = 1, 2, \ldots$$

for the positive cv's and

$$\frac{1}{\lambda_{\nu+N-p}} \geq \frac{1}{\lambda'_\nu} \qquad \nu = 1, 2, \ldots$$

for the negative cv's.

3. If K'' above is *nonnegative* definite, then $Q[\phi] \geq Q'[\phi]$ and

$$\frac{1}{\lambda_\nu} \geq \frac{1}{\lambda'_\nu} \qquad \nu = 1, 2, \ldots \tag{2}$$

for all the cv's (both positive and negative). If K'' is *nonpositive* definite, the corresponding result is

$$\frac{1}{\lambda_\nu} \leq \frac{1}{\lambda'_\nu}$$

4. Now let two closed linear spaces S and S' be such that $S' \subset S$. Define $S'' \equiv \{f : f \in S, f \notin S'\}$ and denote the positive (or negative) cv's associated with K and its restrictions to the smaller spaces S' and S'' by λ_ν, λ'_ν, and λ''_ν, respectively. As we have already noted in (9.4-2),

$$\left| \frac{1}{\lambda'_\nu} \right| \leq \left| \frac{1}{\lambda_\nu} \right| \qquad \nu = 1, 2, \ldots \tag{3} \dagger$$

In addition, however, if S'' is n dimensional, the following inequalities are also valid:

$$\left| \frac{1}{\lambda_{\nu+n}} \right| \leq \left| \frac{1}{\lambda'_\nu} \right| \qquad \nu = 1, 2, \ldots \tag{4}$$

[Cauchy (1829) actually appears to have been the first to observe these inequalities, at least for real symmetric transformations on finite-dimensional spaces; see also Hamburger and Grimshaw (1951, pp. 73–78).]

5. Under the assumptions of 4, a simple argument (see Prob. 7) shows that the several Rayleigh ratios $R[\phi]$, $R[\phi']$, and $R[\phi'']$, where ar-

---

† This is an example of the well-known physical principle that whenever new *constraints* are imposed, *characteristic frequencies* increase. One important instance of this behavior occurs when the domain of definition of the kernel K is *decreased*, for example, $a \leq a' \leq x, y \leq b' \leq b$.

bitrary $\phi = \alpha\phi' + \beta\phi''$ with $\phi' \epsilon S'$, $\phi'' \epsilon S''$, and

$$\|\phi\| = \|\phi'\| = \|\phi''\| = 1$$

satisfy

$$R[\phi] + \frac{1}{\lambda_{\bar{1}}} \leq R[\phi'] + R[\phi''] \tag{5}$$

$$R[\phi] + \frac{1}{\lambda_{\bar{1}}^+} \geq R[\phi'] + R[\phi''] \tag{5'}$$

Here we have designated the least (in magnitude) negative (positive) cv of K in S by $\lambda_{\bar{1}}$ ($\lambda_{\bar{1}}^+$). Alternately using the two characterizations (9.3-2) and (9.4-1), we obtain for the positive cv's[†]

$$\frac{1}{\lambda_{\nu+\mu-1}} + \frac{1}{\lambda_{\bar{1}}} \leq \frac{1}{\lambda_{\nu}'} + \frac{1}{\lambda_{\mu}''} \leq \frac{1}{\lambda_{\nu+\mu}} + \frac{1}{\lambda_{\bar{1}}^+} \tag{6}$$

while for the negative cv's

$$\frac{1}{\lambda_{\nu+\mu-1}} + \frac{1}{\lambda_{\bar{1}}^+} \geq \frac{1}{\lambda_{\nu}'} + \frac{1}{\lambda_{\mu}''} \geq \frac{1}{\lambda_{\nu+\mu}} + \frac{1}{\lambda_{\bar{1}}} \tag{6'}$$

These two se s of inequalities can be written compactly as

$$\left| \frac{1}{\lambda_{\nu+\mu-1}} \right| \leq \left| \frac{1}{\lambda_{\nu}'} \right| + \left| \frac{1}{\lambda_{\mu}''} \right| + \left| \frac{1}{\lambda_{\bar{1}}^{\pm}} \right| \leq \left| \frac{1}{\lambda_{\nu+\mu}} \right| + \left| \frac{1}{\lambda_{\bar{1}}^+} \right| + \left| \frac{1}{\lambda_{\bar{1}}} \right| \tag{7}$$

where we assume that $\lambda_{\bar{1}}^+$ ($\lambda_{\bar{1}}$) appears in the inner inequality whenever we are concerned with the negative (positive) cv's. Obvious special cases occur if K is either nonnegative or nonpositive definite. There is also a further refinement of (7), which is valid in these cases whenever the linear space S is finite dimensional (see Prob. 7).

PROBLEMS

1. For the familiar kernel

$$K(x,y) = \begin{cases} x(1-y) & x \leq y \\ y(1-x) & y \leq x \end{cases}$$

[†] Using a *functional equation* satisfied by the Rayleigh quotient R, Diaz and Metcalf (1968) established these and more general inequalities (see Prob. 8).

on $0 \leq x,y \leq 1$, use the Rayleigh-Ritz procedure with simple polynomials in order to obtain estimates for the least characteristic value. Try first without using the known fact that all characteristic functions of K must vanish at $x = 0, 1$.

2. Verify the inequality (9.2-4) which suggests the accuracy that is to be obtained using Courant's approach to characteristic value estimation.

3. Satisfy yourself that for Hermitian kernels K_N of finite rank, the two criteria (9.3-2) and (9.4-1) are reciprocally related in the following sense: If the characteristic values are ordered according to *algebraic* value, the Poincaré characterization of the nth cv of K_N is equivalent to the Weyl-Courant characterization of the $(N - n + 1)$st cv of $- K_N$.

4. [Lax (1962).] Let S_n be a subspace of $\mathfrak{L}^2$ with the property that, given any other subspace $S^{(n-1)}$ of dimension $(n - 1)$, there is a nontrivial element of S_n orthogonal to $S^{(n-1)}$. Show that, in view of (9.3-2),

$$\min_{\phi \, \epsilon \, S_n} R[\phi] \leq \frac{1}{\lambda_n}$$

for the nth positive cv of any Hermitian kernel (with an analogous relation holding for the negative cv's). (In his paper Lax suggests how this result might be used as the basis for rather practical cv estimation procedures.)

5. Write the min (x,y) kernel as

$$K(x,y) \equiv \min (x,y) = K'(x,y) + xy$$

where $K'(x,y)$ is the kernel of Prob. 1. Use the inequalities of Sec. 9.5 and the results above to obtain estimates for the cv's of this kernel.

6. **a.** Let a and c be any nonnegative constants and b a complex number such that

$$f(z) = az\bar{z} + bz + \overline{bz} + c$$

is nonnegative for every complex z. Show that

$$b\bar{b} \leq ac$$

and hence $\qquad f(z) \leq (1 + z\bar{z})(a + c)$

[See Hamburger and Grimshaw (1951, p. 76).]

b. As in 4 and 5 of Sec. 9.5, let

$$\phi = \alpha \phi' + \beta \phi'' \qquad \text{with} \quad \begin{matrix} \phi' \, \epsilon \, S' \\ \phi'' \, \epsilon \, S'' \end{matrix}$$

and

$$\|\phi\| = \|\phi'\| = \|\phi''\| = 1$$

Verify by direct computation that

$$0 \leq \frac{1}{|\beta|^2} \left(R[\phi] - \frac{1}{\lambda_{\bar{1}}} \right)$$

$$= \left| \frac{\alpha}{\beta} \right|^2 \left(R[\phi'] - \frac{1}{\lambda_{\bar{1}}} \right) + \frac{\alpha}{\beta} (K\phi', \phi'')$$

$$+ \left(\overline{\frac{\alpha}{\beta}} \right) (\phi'', K\phi') + R[\phi''] - \frac{1}{\lambda_{\bar{1}}}$$

and then, employing the result of part a, deduce that

$$R[\phi] + \frac{1}{\lambda_{\bar{1}}} \leq R[\phi'] + R[\phi'']$$

c. In analogous fashion demonstrate that

$$R[\phi] + \frac{1}{\lambda_{\bar{1}}^{+}} \geq R[\phi'] + R[\phi'']$$

7. Satisfy yourself that in the case of finite dimension n, if $R[\phi]$ is always of one sign, then $\lambda_{\bar{1}}^{\pm}$ can be replaced by $\lambda_{\bar{n}}^{\mp}$ in the results of Prob. 6.

8. [Diaz and Metcalf (1968).] Let n be a positive integer, $\{\psi_\nu\}$ $(1 \leq \nu \leq n)$ any finite orthonormal set of functions, and $(\alpha_{\mu\nu})$ $(1 \leq \mu, \nu \leq n)$ an array of complex scalars. Using the definition (9.1-1) of the Rayleigh quotient, show that for nontrivial functions $\phi_\mu = \sum\limits_{\nu=1}^{n} \alpha_{\mu\nu} \psi_\nu$, the functional equation

$$\sum_{\mu=1}^{n} R[\phi_\mu] = \sum_{\nu=1}^{n} R[\psi_\nu]$$

is valid whenever

$$\sum_{\mu=1}^{n} \frac{\alpha_{\mu i} \overline{\alpha_{\mu j}}}{\sum\limits_{\nu=1}^{n} |\alpha_{\mu\nu}|^2} = \delta_{ij}$$

(This relation expresses the invariance, under unitary transformations, of the sum of the Rayleigh quotients of n arbitrary orthonormal functions.)

chapter ten

Error Bounds for the Characteristic Values and Characteristic Functions of Hermitian Kernels

10.1 THE GENERAL QUALITY OF APPROXIMATION

In Chap. 9 we established a rather comprehensive result concerning the accuracy with which the (unknown) positive and negative characteristic values of a given Hermitian $\mathfrak{L}^2$ kernel K could be approximated by the (assumed known) cv's of "nearby" Hermitian $\mathfrak{L}^2$ kernels K_n, namely,

$$\left| \frac{1}{\lambda_\nu} - \frac{1}{\lambda_\nu^{(n)}} \right| \leq \| K - K_n \| \qquad (9.3\text{-}4')$$

for all $\nu = 1, 2, \ldots$. Unfortunately, this fundamental relation (which could equally well have been proved using the inequalities of Sec. 9.5) has, in general, no characteristic function counterpart.† On the other hand, analogous relations do exist when the approximating kernels K_n

---

†With the exception of some partial results such as those of Mysovskih (1959) and Atkinson (1967), there also appears to be no direct analogue of the error estimate (9.3-4) that is generally valid in the non-Hermitian situation. See Chap. 15, Prob. 6, however, for not only an appropriate analogue, but also a significant generalization of (9.3-4) in the case of normal and hence Hermitian kernels.

are selected in certain specialized fashions, e.g., à la Courant or Rayleigh. In this chapter we shall explore the more important among these special cases, indicating also, where appropriate, how error bounds that improve upon (9.3-4') may be derived.

Before moving on to these interesting topics, however, let us note a simple result that is valid in the frequently encountered case when the approximating kernel K_n is of finite rank n. We have then from (9.3-4') that

$$\left| \frac{1}{\lambda_{n+\nu}} \right| \le \| K - K_n \| \equiv \epsilon_n \tag{1}$$

for all $\nu = 1, 2, \ldots$, from which we may conclude that the higher-order cv's of the given kernel K are all located *outside* the open interval $(-1/\epsilon_n, 1/\epsilon_n)$.

The relation (1) also may be viewed in converse fashion. For example, setting $\nu = 1$ gives

$$\| K - K_n \| \ge \left| \frac{1}{\lambda_{n+1}} \right| \tag{2}$$

a meaningful indication of precisely how well one can hope to approximate K with a degenerate kernel of rank n.†

10.2 ANOTHER LOOK AT THE COURANT PROCEDURE

If, in Courant's approach to the estimation problem (see Sec. 9.2), we take the $b_{\mu\nu}{}^{(n)}$ of the expansion

$$K_n(x,y) = \sum_{\mu=1}^{n} \sum_{\nu=1}^{n} b_{\mu\nu}{}^{(n)} \psi_\nu(x) \overline{\psi}_\mu(y) \tag{9.2-1'}$$

to be the *Fourier coefficients* of the Hermitian kernel K in terms of the orthonormalized functions $\{\psi_\nu\}$, we then obtain the *best* (in the mean-square sense) approximation to K of the form given (see Chap. 13).

---

† Actually, as will be clear from the analysis of the next section,

$$\| K - K_n \| \ge \left[\sum_{\nu=1}^{\infty} \left(\frac{1}{\lambda_{n+\nu}} \right)^2 \right]^{\frac{1}{2}}$$

Note, moreover, that since

$$b_{\mu\nu}^{(n)} = (K\psi_\mu, \psi_\nu) = c_{\mu\nu} \tag{1}$$

in this case, *the entire Courant procedure becomes completely equivalent to that of Rayleigh and Ritz.* The net result of this interplay is a set of two-sided bounds for the characteristic values λ_ν of K, namely,[†]

$$\left| \frac{1}{\lambda_\nu^{(n)}} \right| \le \left| \frac{1}{\lambda_\nu} \right| \le \left| \frac{1}{\lambda_\nu^{(n)}} \right| + \epsilon_n \qquad \nu = 1, 2, \ldots \tag{2}$$

Here, as usual, the $\lambda_\nu^{(n)}$ are the positive (or negative) cv's of K_n, which in this case are the reciprocals of the eigenvalues of the matrix $C = (c_{\mu\nu})$, algebraically ordered, and $\epsilon_n = \|K - K_n\|$. The relation (2) clearly shows that as $\epsilon_n \to 0$, the Rayleigh-Ritz approximants $\lambda_\nu^{(n)}$ approach the corresponding cv's λ_ν of K.

In view of the nature of the coefficients (1), the error term ϵ_n appearing in (2) can be put into a number of more practical forms, for example,

$$\epsilon_n^2 \equiv \|K - K_n\|^2 = \|K\|^2 - \|K_n\|^2$$

$$= \|K\|^2 - \sum_{\mu=1}^{n} \sum_{\nu=1}^{n} |c_{\mu\nu}|^2$$

$$= k_2 - \mathrm{Tr}\,(CC^*)$$

$$= \sum_{\nu=1}^{\infty} \left(\frac{1}{\lambda_\nu}\right)^2 - \sum_{\nu=1}^{n} \left(\frac{1}{\lambda_\nu^{(n)}}\right)^2 \tag{3}$$

It follows from (3), using (2), that ϵ_n^2 has the lower bound

$$\epsilon_n^2 \ge \sum_{\nu=n+1}^{\infty} \left(\frac{1}{\lambda_\nu}\right)^2 \tag{4}$$

(see the footnote on page 132). Thus if the rank n of K_n is assumed fixed but the ψ_ν are allowed to vary, the best one can do occurs when the $\psi_\nu (1 \le \nu \le n)$ span the subspace generated by the first n characteristic functions of K. In this case, $\lambda_\nu^{(n)} = \lambda_\nu$ ($1 \le \nu \le n$) and equality prevails

---

[†] From matrix considerations Wielandt (1954) has derived a weaker version of this expression.

in (4). It is interesting to note also that as a consequence of the above relations,

$$\|K_n\|^2 \leq \|K\|^2$$

with equality iff $\epsilon_n = 0$. This last situation will eventually occur as $n \to \infty$ if the ψ_ν are complete in $\mathcal{L}^2$ (see Chap. 13).†

Using ideas similar to those employed heretofore, Trefftz (1933) has shown how the upper half of the inequality (2) can be easily, yet substantially, improved. Since $(\lambda_\nu)^2 \leq (\lambda_\nu^{(n)})^2$ for $1 \leq \nu \leq n$, it follows that

$$\sum_{\mu=1}^{n} \left(\frac{1}{\lambda_\mu^{(n)}}\right)^2 - \left(\frac{1}{\lambda_\nu^{(n)}}\right)^2 \leq \sum_{\mu=1}^{n} \left(\frac{1}{\lambda_\mu}\right)^2 - \left(\frac{1}{\lambda_\nu}\right)^2$$

for each $\nu \leq n$. Thus,

$$\left(\frac{1}{\lambda_\nu}\right)^2 - \left(\frac{1}{\lambda_\nu^{(n)}}\right)^2 \leq \sum_{\mu=1}^{n} \left(\frac{1}{\lambda_\mu}\right)^2 - \sum_{\mu=1}^{n} \left(\frac{1}{\lambda_\mu^{(n)}}\right)^2$$

$$\leq \sum_{\mu=1}^{\infty} \left(\frac{1}{\lambda_\mu}\right)^2 - \sum_{\mu=1}^{n} \left(\frac{1}{\lambda_\mu^{(n)}}\right)^2$$

$$= \|K - K_n\|^2 \equiv \epsilon_n^2$$

and hence

$$\left|\frac{1}{\lambda_\nu^{(n)}}\right| \leq \left|\frac{1}{\lambda_\nu}\right| \leq \left[\left(\frac{1}{\lambda_\nu^{(n)}}\right)^2 + \epsilon_n^2\right]^{\frac{1}{2}} \tag{5}$$

This is a tighter inequality than (2) and less sensitive to increasing ν.

10.3 THE NATURE OF RAYLEIGH-RITZ APPROXIMATIONS TO CHARACTERISTIC FUNCTIONS

Let us denote the characteristic values and characteristic functions of a given Hermitian kernel K by λ_ν and ϕ_ν, respectively, and the corresponding Rayleigh-Ritz approximants (with respect to the subspace spanned by orthonormal $\psi_1, \psi_2, \ldots, \psi_n$) by $\lambda_\nu^{(n)}$ and $\phi_\nu^{(n)}$. Thus from Sec. 10.2 we have that not only

$$\phi_\nu = \lambda_\nu K \phi_\nu \qquad \|\phi_\nu\| = 1$$

---

† Completeness in the complement of the null space of K is actually sufficient, of course.

but also $$\phi_\nu^{(n)} = \lambda_\nu^{(n)} K_n \phi_\nu^{(n)} \qquad \|\phi_\nu^{(n)}\| = 1$$

where $$K_n(x,y) = \sum_{\mu=1}^{n} \sum_{\nu=1}^{n} (K\psi_\mu, \psi_\nu)\psi_\nu(x)\overline{\psi}_\mu(y)$$

Using an approach motivated by the work of Koehler (1953), we want to show now how the "distance" between $\phi_\nu^{(n)}$ and an appropriate ϕ_ν can be measured in terms of $\epsilon_n = \|K - K_n\|$. We begin by treating the case $\nu = 1$, the argument for higher-order ν being of the same general nature.

Assume that the least cv of K is positive (the analysis is easily modified if $\lambda_1 < 0$) and has *multiplicity* k. Form the function

$$\phi \equiv \phi_1^{(n)} - \sum_{\mu=1}^{k} (\phi_1^{(n)}, \phi_\mu)\phi_\mu$$

and note its orthogonality to the first k cf's of K. It follows, therefore, that

$$(K\phi, \phi) \le \frac{1}{\lambda_{k+1}} (\phi, \phi)$$

or since $(K\phi_1^{(n)}, \phi_1^{(n)}) = 1/\lambda_1^{(n)}$,†

$$\frac{1}{\lambda_1^{(n)}} - \frac{1}{\lambda_1} \sum_{\mu=1}^{k} |(\phi_1^{(n)}, \phi_\mu)|^2 \le \frac{1}{(\lambda_{k+1})} \left[1 - \sum_{\mu=1}^{k} |(\phi_1^{(n)}, \phi_\mu)|^2 \right]$$

Hence, $$\sum_{\mu=1}^{k} |(\phi_1^{(n)}, \phi_\mu)|^2 \ge 1 - e_1 \qquad (1)$$

---

† It is important to recognize that not only is $(K_n\phi_\nu^{(n)}, \phi_\nu^{(n)}) = 1/\lambda_\nu^{(n)}$ but this relation also is equally valid with K_n replaced by K. This fact, which underlies Rayleigh-Ritz analysis, can be verified in several ways. The easiest, perhaps, is to introduce the projection operator P_n, which transforms elements of $\mathcal{L}^2$ into their components in the subspace generated by $\psi_1, \psi_2, \ldots, \psi_n$ [see, e.g., Riesz and Sz.-Nagy (1955, pp. 266ff)] Since $K_n = P_n K P_n$ and P_n is symmetric, we then have

$$1 = (\phi_\nu^{(n)}, \phi_\nu^{(n)}) = \lambda_\nu^{(n)}(K_n\phi_\nu^{(n)}, \phi_\nu^{(n)})$$
$$= \lambda_\nu^{(n)}(P_n K P_n \phi_\nu^{(n)}, \phi_\nu^{(n)})$$
$$= \lambda_\nu^{(n)}(K P_n \phi_\nu^{(n)}, P_n \phi_\nu^{(n)})$$
$$= \lambda_\nu^{(n)}(K \phi_\nu^{(n)}, \phi_\nu^{(n)})$$

from which the desired result follows.

where
$$e_1 = \frac{\dfrac{1}{\lambda_1} - \dfrac{1}{\lambda_1^{(n)}}}{\dfrac{1}{\lambda_1} - \dfrac{1}{\lambda_{k+1}}} \geq 0 \tag{2}$$

Note particularly that from (10.2-2), say, for large enough n, e_1 is less than a constant (independent of n) times ϵ_n. As $\epsilon_n \to 0$, e_1 becomes less than unity and indeed tends to zero itself. In the sequel we assume $e_1 < 1$.

Owing to the multiplicity of the cv λ_1, the cf's $\phi_1, \phi_2, \ldots, \phi_k$ are only uniquely determined modulo a unitary transformation. In other words, a perfectly satisfactory set of initial cf's would be $\tilde{\phi}_\nu$, $1 \leq \nu \leq k$, where, for example,

$$\tilde{\phi}_1 = \frac{\displaystyle\sum_{\mu=1}^{k} (\phi_1^{(n)}, \phi_\mu) \phi_\mu}{\left[\displaystyle\sum_{\mu=1}^{k} |(\phi_1^{(n)}, \phi_\mu)|^2 \right]^{\frac{1}{2}}}$$

In this case, using (1), we would have

$$(\phi_1^{(n)}, \tilde{\phi}_1) \geq \sqrt{1 - e_1} \geq 1 - e_1 \geq 0$$

We assume, therefore, that the original cf's ϕ_ν, $1 \leq \nu \leq n$, have been selected so that

$$(\phi_\nu^{(n)}, \phi_\nu) \geq \sqrt{1 - e_\nu}$$

where e_ν has all of the features analogous to e_1. (An inductive argument shows this to be possible for the higher-order ν, appropriate care being taken if some of the corresponding λ_ν are negative.) It follows, then, that

$$\|\phi_\nu - \phi_\nu^{(n)}\|^2 = 2 \left[1 - (\phi_\nu^{(n)}, \phi_\nu) \right]$$
$$\leq 2e_\nu \tag{3}$$

and this constitutes the desired result.†

Once the bound (3) has been established for the mean-square error, we are in a position to investigate the absolute difference $|\phi_\nu(x) -$

---

† See Weinberger (1960) for generalizations of this result. Also see Bückner (1952, p. 80) for another type of bound, applicable when $k = 1$, $n = 1$. Lastly see Vainikko (1964 and 1965) for general error bounds of both a priori and a posteriori type.

$\phi_\nu^{(n)}(x)|$. In view of the nature of these characteristic functions, we have

$$| \phi_\nu(x) - \phi_\nu^{(n)}(x)| = |\lambda_\nu K \phi_\nu(x) - \lambda_\nu^{(n)} K_n \phi_\nu^{(n)}(x)|$$

$$\leq |\lambda_\nu - \lambda_\nu^{(n)}| \cdot |K \phi_\nu(x)|$$

$$+ |\lambda_\nu^{(n)}| \cdot |K \phi_\nu(x) - K \phi_\nu^{(n)}(x)|$$

$$+ |\lambda_\nu^{(n)}| \cdot |(K' - K_n) \phi_\nu^{(n)}(x)|$$

$$\leq |\lambda_\nu^{(n)}| \left\{ \|K\|_y \left[|\lambda_\nu| \cdot \left| \frac{1}{\lambda_\nu} - \frac{1}{\lambda_\nu^{(n)}} \right| + \|\phi_\nu - \phi_\nu^{(n)}\| \right] \right.$$

$$\left. + \|K - K_n\|_y \right\}$$

where we have used the obvious notation (see Appendix B)

$$\|A\|_y = \left[\int_a^b |A(x, y)|^2 \, dy \right]^{\frac{1}{2}}$$

If $\epsilon_n \to 0$, the above result shows the convergence of the Rayleigh-Ritz approximants to the cf's to be *relatively uniform* (see Sec. 3.1), considerably better than the expected mean square. Moreover, if K, K_n are continuous, the convergence is uniform.[†]

10.4 KELLOGG'S APPROACH ONCE MORE

In Sec. 8.2 we analyzed the method of successive substitutions as proposed by Kellogg (1922) and verified that under certain minimal restrictions it was indeed a viable technique for the determination of cv's and cf's of Hermitian kernels. We want now to derive bounds for the estimation error associated with this approach.[‡] Certain of the results appear to have been given first by Collatz (1939 and 1940) [see also Iglisch (1941) and Collatz (1960)], who adapted earlier work of Temple (1928a and b) to the integral equations setting.

[†] In his 1931 memoir N. M. Krylov presents a number of sharp error bounds for the Rayleigh-Ritz approximants to the cv's and cf's associated with certain linear ordinary differential equations with continuous coefficients. These results naturally carry over, then, to the restricted class of continuous Hermitian kernels related thereto.

[‡] It should be understood that Kellogg's straightforward technique is not the only successful iterative procedure. Wiarda (1930), Carrier (1948), Bückner (1949), and Lanczos (1950), among others, have all proposed variants that occasionally are superior to the approach of Kellogg. Any detailed analysis of these other methods, however, is beyond the scope of this book.

We begin with an arbitrary $\mathfrak{L}^2$ function ψ_0 satisfying only $\| K\psi_0 \| \neq$ 0 for the given Hermitian kernel K under consideration and form the sequence $\psi_1, \psi_2, \ldots$, using the relation

$$\psi_n = K\psi_{n-1} \qquad n = 1, 2, \ldots$$

In contradistinction to our earlier analysis, we choose *not* to normalize the ψ's at each stage of the procedure. If we define the so-called "Schwarz constants" (1885) a_n by the relation

$$a_n \equiv (\psi_{n-m}, \psi_m) = (\psi_n, \psi_0) \qquad (1)$$

it follows that $a_{2n} > 0$ for all n and

$$a_{2n}{}^2 \leq a_{2n-2} a_{2n+2} \qquad n = 1, 2, \ldots \qquad (2)$$

In a similar manner it can be verified that

$$a_{2n+1}^2 \leq a_{2n} a_{2n+2} \qquad n = 0, 1, \ldots \qquad (3)$$

Under the assumption that the a_n for odd n do not vanish, the quotient

$$r_n \equiv \frac{a_{n-1}}{a_n} \qquad (4)$$

gives rise to a well-defined ratio termed the *Schwarz quotient*. We have, then, from (3), that

$$|r_{2n+1}| \geq |r_{2n+2}|$$

while from (2) we obtain

$$r_1 r_2 \geq r_3 r_4 \geq \cdots \geq r_{2n-1} r_{2n} \geq \cdots$$

It is not necessarily true, however, that $|r_{2n}| \geq |r_{2n+1}|$ (see the example of Sec. 10.6 and Probs. 3 and 4 at the end of this chapter).

Recalling our original analysis of Sec. 8.2, we see that the product

$$r_{2n-1} r_{2n} = \frac{a_{2n-2}}{a_{2n}} = \frac{(\psi_{n-1}, \psi_{n-1})}{(K^2 \psi_{n-1}, \psi_{n-1})}$$

plays the same role here as $\mu_n{}^2$ did there. These products monotonically decrease, converging in the limit of large n to the square of a characteristic value of the given kernel K. In the sequel we shall assume that this cv is the least one, namely λ_1.

Now form

$$\tilde{\psi}_0 = \psi_0 - \sum_{\nu=1}^{k} (\psi_0, \phi_\nu) \phi_\nu$$

where $\{\phi_\nu\}$, $\nu = 1, 2, \ldots, k$, are orthonormalized cf's of the kernel K associated with the first k cv's and successively generate, in a manner duplicating the above, $\tilde{\psi}_n$, $\tilde{a}_n$, and $\tilde{r}_n$. Note particularly that

$$\tilde{\psi}_n = \psi_n - \sum_{\nu=1}^{k} \frac{(\psi_0, \phi_\nu)}{\lambda_\nu{}^n} \phi_\nu \tag{5}$$

from which it readily follows that the $\tilde{\psi}_n$ are orthogonal to the cf's $\phi_1, \phi_2, \ldots, \phi_k$. If the kernel K is *nonnegative definite*, we have

$$\frac{1}{\tilde{r}_{2n+1}} = \frac{\tilde{a}_{2n+1}}{\tilde{a}_{2n}} = \frac{(K\tilde{\psi}_n, \tilde{\psi}_n)}{(\tilde{\psi}_n, \tilde{\psi}_n)} \le \frac{1}{\lambda_{k+1}}$$

In fact, the $\tilde{r}_n$ converge monotonically from above to λ_{k+1} as $n \to \infty$ (Prob. 3).

We are now in a position to derive the first of the desired results. For the moment take K to be nonnegative definite and assume

$$0 < \lambda_1 = \lambda_2 = \cdots = \lambda_k < \lambda_{k+1}$$

As a consequence of (5) we have

$$\tilde{\psi}_n - \lambda_1 K \tilde{\psi}_n = \psi_n - \lambda_1 K \psi_n$$

and hence

$$a_n - \lambda_1 a_{n+1} = (\psi_n - \lambda_1 K \psi_n, \psi_0)$$

$$= \left(\tilde{\psi}_n - \lambda_1 K \tilde{\psi}_n, \tilde{\psi}_0 + \sum_{\nu=1}^{k} (\psi_0, \phi_\nu) \phi_\nu \right)$$

$$= (\tilde{\psi}_n - \lambda_1 K \tilde{\psi}_n, \tilde{\psi}_0)$$

$$= \tilde{a}_n - \lambda_1 \tilde{a}_{n+1}$$

Therefore

$$\frac{1}{\lambda_{k+1}} \geq \frac{1}{\tilde{r}_n} \geq \frac{1}{\tilde{r}_n} \frac{1/\lambda_1 - 1/\tilde{r}_{n+1}}{1/\lambda_1 - 1/\tilde{r}_n}$$

$$= \frac{\tilde{a}_n - \lambda_1 \tilde{a}_{n+1}}{\tilde{a}_{n-1} - \lambda_1 \tilde{a}_n} = \frac{a_n - \lambda_1 a_{n+1}}{a_{n-1} - \lambda_1 a_n}$$

$$= \frac{1}{r_n} \frac{1/\lambda_1 - 1/r_{n+1}}{1/\lambda_1 - 1/r_n}$$

When $r_n < \lambda_{k+1}$ (which it eventually must become as $n \to \infty$), this leads
to

$$\frac{1}{\lambda_1} - \frac{1}{r_{n+1}} \leq \frac{1/r_{n+1} - 1/r_n}{\lambda_{k+1}/r_n - 1} \tag{6}$$

Relation (6) gives a set of upper bounds to go along with the lower
bounds $1/r_{n+1}$. These upper bounds are expressed for the most part in
terms of quantities readily computable during the course of the successive substitutions procedure. [See Collatz (1940) for generalizations to
higher-order cv's.] For practical purposes λ_{k+1} can be replaced, if
desired, by a lower bound, as long as that lower bound remains greater
than r_n. A possible candidate is

$$\left(k_2 - \frac{k}{r_n^2}\right)^{-\frac{1}{2}}$$

since $$\frac{1}{\lambda_{k+1}} \leq \sqrt{k_2 - \frac{k}{\lambda_1^2}} \leq \sqrt{k_2 - \frac{k}{r_n^2}}$$

follows as a consequence of (7.1-5). Note, however, that in order to be
able to employ this lower bound, the relation

$$\frac{k}{r_n^2} \leq k_2 \leq \frac{k+1}{r_n^2}$$

must hold. Fortunately, these inequalities are often satisfied in practical
problems with well-separated characteristic values.

    If in the above analysis we had concerned ourselves with the ker-
nel K^2 rather than K, the end result would have been

$$\frac{1}{\lambda_1^2} - \frac{1}{\mu_{n+1}^2} \leq \frac{1/\mu_{n+1}^2 - 1/\mu_n^2}{\lambda_{k+1}^2/\mu_n^2 - 1} \tag{7}$$

Here, in keeping with our earlier observation, we have written $\mu_n{}^2$ for the product $r_{2n-1} r_{2n}$. This relation (7) is the analogue of (6) for general Hermitian kernels.

Equations (6) and (7) show quite clearly that the cv error bounds associated with the method of Kellogg are sensitive functions of the ratio $|\lambda_{k+1}/\lambda_1|$. If this ratio is large, as it often is, for example, in various problems in mechanics, the procedure can be rapidly convergent. Similar remarks apply to the mean-square error of the characteristic function estimates, as we shall see in the next section.

10.5 ERROR BOUNDS FOR ITERATIVE APPROXIMATIONS TO CHARACTERISTIC FUNCTIONS

Under the assumption that the Kellogg iterative procedure leads to the least cv λ_1 and an associated cf ψ, we have from Sec. 8.2 and (7.4-8)

$$
\begin{aligned}
\| \psi - \psi_{2n} \|^2 &= 2 - (\psi, \psi_{2n}) - (\psi_{2n}, \psi) \\
&= 2 - (\gamma f \psi_0, \gamma_n f_n \psi_0) - (\gamma_n f_n \psi_0, \gamma f \psi_0) \\
&= 2 - \gamma \gamma_n (f \psi_0, \psi_0) - \gamma_n \gamma (\psi_0, f \psi_0) \\
&= 2 \left(1 - \frac{\gamma_n}{\gamma} \right)
\end{aligned}
$$

To obtain characteristic function error bounds, then, we need only estimate the difference $1 - (\gamma_n/\gamma)$ in terms of quantities easily computed at any given stage of the process. Unfortunately, to do this necessitates two results that we have not as yet established, namely, the following:

1. Given an orthonormalized set of square-integrable functions $\{\phi_\nu\}$, the relation

$$
\sum_{\nu=1}^{\infty} |(f, \phi_\nu)|^2 \leq \|f\|^2
$$

   is valid for any function $f \in \mathfrak{L}^2$.†

2. If the ϕ_ν of 1 are the characteristic functions associated with all of the finite characteristic values l_ν of a Hermitian $\mathfrak{L}^2$ kernel K, then the square-integrable functions

$$
g_n \equiv \sum_{\nu=1}^{n} \frac{(f, \phi_\nu) \phi_\nu}{\lambda_\nu}
$$

---

† This result is commonly termed *Bessel's inequality*; one special case was given earlier as Property 6 of Sec. 7.1.

with $f \epsilon \mathfrak{L}^2$, are mean convergent to the $\mathfrak{L}^2$ function $g \equiv Kf$. Moreover, in this case,

$$\sum_{\nu=1}^{\infty} \frac{|(f,\phi_\nu)|^2}{\lambda_\nu^2} = \|g\|^2$$

For the present, we shall assume these results to be valid; further discussion and their proof are more appropriately postponed until Chap. 13.

As in the preceding section, we take the Hermitian kernel K under consideration to be one for which

$$|\lambda_1| = |\lambda_2| = \cdots = |\lambda_k| < |\lambda_{k+1}|$$

If we designate the orthonormalized cf's of K as usual by ϕ_ν, then it follows from results 1 and 2 that

$$\sum_{\nu=1}^{\infty} |(\psi_0, \phi_\nu)|^2 \leq 1 \tag{1}$$

$$\gamma_n^2 \sum_{\nu=1}^{\infty} \left(\frac{\lambda_1}{\lambda_\nu}\right)^{4n} |(\psi_0, \phi_\nu)|^2 = 1 \tag{2}$$

and

$$\gamma^2 \sum_{\nu=1}^{k} |(\psi_0, \phi_\nu)|^2 = 1 \tag{3}$$

since ψ_0, $\psi_{2n} \equiv \gamma_n f_n \psi_0$ and $\psi \equiv \gamma f \psi_0$ all have unit norm.

Coarse estimates for $1 - (\gamma_n/\gamma)$ may be easily obtained from the three relations (1), (2), and (3). For instance, (2) implies

$$\frac{1}{\gamma_n^2} = \sum_{\nu=1}^{k} |(\psi_0, \phi_\nu)|^2 + \sum_{\nu=k+1}^{\infty} \left(\frac{\lambda_1}{\lambda_\nu}\right)^{4n} |(\psi_0, \phi_\nu)|^2$$

from which we deduce

$$\frac{1}{\gamma_n^2} \leq \frac{1}{\gamma^2} + \left(\frac{\lambda_1}{\lambda_{k+1}}\right)^{4n}$$

and hence

$$\frac{1}{\gamma_n} \leq \frac{1}{\gamma} + \left(\frac{\lambda_1}{\lambda_{k+1}}\right)^{2n}$$

or

$$\left(1 - \frac{\gamma_n}{\gamma}\right) \leq \gamma_n \left(\frac{\lambda_1}{\lambda_{k+1}}\right)^{2n} = \frac{\mu_1 \mu_2 \cdots \mu_{2n}}{\lambda_{k+1}^{2n}} \tag{4}$$

Sharper error bounds, which are more in keeping with (10.4-7), can be derived by an approach similar to that used by Trefftz (1933). Since $\psi_{2n+1} \equiv \mu_{2n+1} K \psi_{2n}$, with μ_{2n+1} chosen so that ψ_{2n+1} has norm unity, it follows from (1), (2), (3), and result 2 that

$$
\left(\frac{1}{\mu_{2n+1}}\right)^2 = \gamma_n^2 \sum_{\nu=1}^{\infty} \left(\frac{\lambda_1}{\lambda_\nu}\right)^{4n} \frac{|(\psi_0, \phi_\nu)|^2}{\lambda_\nu^2}
$$

$$
= \frac{\gamma_n^2}{\lambda_1^2 \gamma^2} + \gamma_n^2 \sum_{\nu=k+1}^{\infty} \left(\frac{\lambda_1}{\lambda_\nu}\right)^{4n} \frac{|(\psi_0, \phi_\nu)|^2}{\lambda_\nu^2}
$$

$$
\leq \frac{\gamma_n^2}{\lambda_1^2 \gamma^2} + \frac{1}{\lambda_{k+1}^2} \left(1 - \frac{\gamma_n^2}{\gamma^2}\right)
$$

Algebraic manipulation then leads to

$$
1 - \frac{\gamma_n^2}{\gamma^2} \leq \lambda_{k+1}^2 \frac{1/\lambda_1^2 - 1/\mu_{2n+1}^2}{\lambda_{k+1}^2/\lambda_1^2 - 1} \tag{5}
$$

which, since $1 - (\gamma_n/\gamma) \leq 1 - (\gamma_n/\gamma)^2$, is the desired result. If we had started initially with ψ_n rather than ψ_{2n}, the corresponding result would have been

$$
\|\psi - \psi_n\|^2 \leq 2\lambda_{k+1}^2 \frac{1/\lambda_1^2 - 1/\mu_{n+1}^2}{\lambda_{k+1}^2/\lambda_1^2 - 1} \tag{6}
$$

In either case, the dependence of the error bound upon the ratio $|\lambda_{k+1}/\lambda_1|$, as suggested earlier, is clearly displayed. †

As was true for the Rayleigh-Ritz procedure, Kellogg's approach actually leads to somewhat better covergence than that implied by (6). For completeness, we briefly investigate this point under the assumption that the limit function ψ is itself a cf of K associated with the cv $\lambda_1 > 0$. (The modifications that need be made in order to handle the general case should be obvious.) We find from our earlier analysis that

$$
|\psi(x) - \psi_{n+1}(x)| = |\lambda_1 K\psi - \mu_{n+1} K\psi_n|
$$

$$
\leq (\mu_{n+1} - \lambda_1)|K\psi| + \mu_{n+1}|K(\psi - \psi_n)|
$$

$$
\leq \|K\|_\nu \{(\mu_{n+1} - \lambda_1) + \mu_{n+1}\|\psi - \psi_n\|\}
$$

The convergence of ψ_n to ψ, therefore, is relatively uniform in general and uniform whenever $\|K\|_\nu$ is bounded.

---

† In both (5) and (6), λ_1^2 can be replaced by an appropriate upper bound such as μ_{2n}^2 or μ_n^2 as long as that upper bound remains less than λ_{k+1}^2. Analogous caution must be exercised in replacing λ_{k+1}^2 by lower bounds.

10.6 A CLASSICAL EXAMPLE

In his 1940 paper Collatz discussed some of the numerical implications of the preceding iterative characteristic-value error bounds as they pertained to the kernel

$$K(x,y) = |x - y| \qquad -\tfrac{1}{2} \le x, y \le \tfrac{1}{2}$$

This example is interesting because the Hermitian kernel K has precisely one positive cv, which turns out to be the least, and a doubly infinite set of negative cv's.

If we start with $\psi_0(x) \equiv 1$, we find

$$\psi_1(x) = x^2 + \tfrac{1}{4} \qquad \text{and} \qquad \psi_2(x) = \frac{x^4}{6} + \frac{x^2}{4} + \frac{3}{32}$$

It follows that $a_0 = 1$, $a_1 = \tfrac{1}{3}$, $a_2 = \tfrac{7}{60}$, $a_3 = \tfrac{17}{420}$, and $a_4 = \tfrac{319}{22,680}$ while

$$r_1 = 3$$

$$r_2 = \tfrac{20}{7} = 2.8571. \ . \ .$$

$$r_3 = \tfrac{49}{17} = 2.8823 . \ . \ .$$

$$r_4 = \tfrac{918}{319} = 2.8777. \ . \ .$$

and

$$\mu_1 = 2.9277. \ . \ .$$

$$\mu_2 = 2.8800. \ . \ .$$

The r_n are converging (albeit *not* monotonically) to the least cv $\lambda_1 \doteq 2.8784$, as are the μ_n. Using (10.4-7) with $n = 1$ and the approximate lower bound 4.6571 for $|\lambda_2|$ gives

$$2.85 \le \lambda_1 \le 2.88$$

If the more exact value $|\lambda_2| \doteq 4.9348$ had been used in (10.4-7), we would have obtained

$$2.86 \le \lambda_1 \le 2.88$$

In this latter case (10.5-6) leads to

$$\|\psi - \psi_1\|^2 \le 0.0035$$

This example serves equally well to illustrate the numerical accuracy of the error bounds associated with Rayleigh-Ritz approximation. We relegate these matters, however, to the problems at the end of this chapter.

PROBLEMS

1. **a.** Verify (9.3-4′) using the inequalities of Sec. 9.5.
   b. Establish the bounds (10.2-2).
2. **a.** Use the Courant method to obtain estimates for the least characteristic value of the kernel

$$K(x,y) = \begin{cases} x(1-y) & x \leq y \\ y(1-x) & y \leq x \end{cases}$$

   on $0 \leq x,y \leq 1$. Let the ψ_ν be polynomials of degree $\nu - 1$.
   b. How well can K be approximated using only ψ_1 and ψ_2?
   c. Verify that the best approximation from part b is inferior to an appropriate approximation of K using the first two characteristic functions

$$\phi_1(x) = \sqrt{2} \ \sin \pi x \quad \text{and} \quad \phi_2(x) = \sqrt{2} \ \sin 2\pi x$$

3. **a.** Construct a simple example wherein successive substitution leads to Schwarz quotients r_1, r_2, and r_3 satisfying $r_1 \geq r_2$ but $r_2 < r_3$.
   b. Show that this behavior cannot occur if the given Hermitian kernel is *nonnegative definite*; i.e., using the K Schwarz inequality (8.3-3), verify that $r_{2n} \geq r_{2n+1} > 0$ whenever $\|K\psi_0\| \neq 0$, and hence

$$r_1 \geq r_2 \geq \cdots \geq r_n \geq \cdots > 0$$

4. Consider the kernel

$$K(x,y) = |x - y| \qquad -\tfrac{1}{2} \leq x,y \leq \tfrac{1}{2}$$

   a. Determine the characteristic values and associated characteristic functions of K.
   b. Apply the Kellogg procedure, starting from the initial function $\psi_0(x) = x$. (This will lead to estimates of λ_2.)
   c. How would you go about constructing an iterative procedure that generated approximations to λ_3?
5. For the kernel of Prob. 4:
   a. Use the Rayleigh-Ritz procedure with simple polynomials in order to obtain estimates for the least characteristic value.
   b. Discuss some of the implications of the bounds (10.2-2), (10.2-5), and (10.3-3) as they pertain to this kernel when the results obtained in part a are utilized.

6. Denote the cv's and cf's of a given Hermitian kernel K by λ_ν and ϕ_ν, respectively, and the corresponding Rayleigh-Ritz approximants (with respect to the subspace spanned by orthonormal $\psi_1, \psi_2, \ldots, \psi_n$) by $\lambda_\nu^{(n)}$ and $\phi_\nu^{(n)}$. Show from first principles that

$$\cdot(K\phi_\nu^{(n)}, \phi_\nu^{(n)}) = \frac{1}{\lambda_\nu^{(n)}} \qquad \nu = 1, 2, \ldots, n$$

7. **a.** Discuss the suitability of $(k_2 - k/r_n^2)^{-\frac{1}{2}}$ as a practical replacement for λ_{k+1} in (10.4-6).

    b. Can this or a similar quantity be analogously employed in (10.5-5) and (10.5-6)? Why?

chapter eleven

Intermediate Operators and the Method of Weinstein and Aronszajn

11.1 LOWER BOUNDS (ONCE OVER LIGHTLY)

In preceding sections we have already encountered a number of procedures whereby *lower* bounds for the characteristic values of a given Hermitian $\mathfrak{L}^2$ kernel K can be obtained. To the list may also be added, for example, various other simplifications and/or improvements suggested by Trefftz (1933)† and Bertram (1957 and 1959), results based on the convexity theorems of Wing (1967a) and Boland (1969a and b) or the enclosure theorem of Collatz (1941 and 1960), and bounds that follow from more sophisticated techniques such as the method of transplantation [see Pólya and Schiffer (1953–1954)].‡ Unfortunately, it is usually the case with all of these procedures that the lower bounds so obtained turn out to be relatively crude when compared with *upper* bounds derived using comparable labor.

---

† A special case was noted earlier in Sec. 10.4.

‡ Collatz (1939) has a good collection of numerical examples for a number of cv estimation procedures.

We want now to consider various ramifications of an approach first suggested by A. Weinstein (1935 and 1937), which by and large do not suffer from the deficiency noted above. The genesis for these outgrowths can be found in the work of Aronszajn (1948 and 1951), but their development must be primarily credited to Bazley (1961) and Bazley and Fox (1961, 1962a, and 1966) [see also Gould (1966), Weinberger (1959 and 1962), Weinstein (1966), and Fichera (1966)]. For convenience we concern ourselves with approximating (from below) the *positive* characteristic values of the given Hermitian kernel K since, in analogy with, say, the Rayleigh-Ritz estimation procedure, corresponding results for the negative cv's may be obtained by replacing K by $-K$ throughout.

11.2 INTERMEDIATE COMPARISON OPERATORS

Lower bounds are intrinsically of a nature different from that of upper bounds. For example, it is clear from our analysis in Chap. 9 that, say, Rayleigh-Ritz upper bounds are determined solely by solving related finite-dimensional problems. Inspection of the max-min characterization (9.4-1), however, shows that this feature cannot be shared by lower bounds in general if the underlying space has infinite dimension. Alternatively expressed, additional information concerning the given kernel is needed before valid "direct methods" for arbitrarily accurate lower bounds may be established.

This additional information takes many forms. In the iterative bounds (10.4-7), for instance, we see it as the term λ_{k+1}^2; in the Rayleigh-Ritz inequalities (10.2-5) it appears as the mean-square error $\|K - K_n\|^2$; in the approach of Weinstein it enters in through explicit knowledge of the characteristic values and characteristic functions of closely related *base* and *intermediate* problems.

Weinstein (1935 and 1937) observed that, given a characteristic-value problem, there often is a "nearby" problem, called a *base problem*, which can be easily and explicitly resolved and which has characteristic values uniformly *no greater than* the cv's of the original problem of interest.† By linking these two problems together with a series of *intermediate* problems, which can be solved in terms of the results of the base problem and which yield progressively better lower bounds for the cv's of the original problem, we obtain a viable estimation procedure.‡

---

† In his original work Weinstein obtained such solvable base problems by relaxing the boundary conditions associated with differential operators.

‡ See Weinberger (1952a) for an analysis of the best accuracy attainable at any given stage of the procedure. In this regard also see Stenger (1969).

In 1950 Aronszajn suggested that there are actually at least two distinct ways in which the operators T_0 and T of the base (and intermediate) and original problems, respectively, can be related in order that the conditions for Weinstein's procedure are satisfied: (1) T_0 is the *extension* of T to a *larger* space of functions; and (2) the difference operator $T_0 - T$ is *nonnegative definite* (see Sec. 9.5). Weinstein's original considerations were in keeping with 1. His intermediate operators were the restrictions of T_0 to the domain of T_0 less finite-dimensional components of the difference space $S_0 \ominus S$. †

Following some general comments of a fundamental nature in the next section, the remainder of this chapter will deal largely with problems of type 2 above. For definiteness we will assume that we know completely the characteristic values $\lambda_\nu^{(0)}$, characteristic functions $\phi_\nu^{(0)}$, and hence resolvent R_0 of a Hermitian $\mathfrak{L}^2$ base kernel K_0 whose cv's satisfy

$$\lambda_\nu^{(0)} \leq \lambda_\nu \tag{1}$$

where the λ_ν are the unknown cv's of the given Hermitian kernel in which we are interested. We shall show how sequences of intermediate kernels K_n, which give rise to families of intermediate comparison operators, may be practically constructed such that

1. The cv's $\lambda_\nu^{(n)}$ of each K_n can be computed in terms of the $\lambda_\nu^{(0)}$ and $\phi_\nu^{(0)}$.
2. $\lambda_\nu^{(0)} \leq \lambda_\nu^{(n)} \leq \lambda_\nu^{(n+1)} \leq \lambda_\nu$.

In view of (9.3-4), if $\| K - K_n \| \to 0$, then $\lambda_\nu^{(n)} \to \lambda_\nu$.

11.3   THEORETICAL BASIS FOR THE WEINSTEIN-ARONSZAJN METHOD

Consider for a moment that we are interested in solving the nonhomogeneous integral equation

$$\phi = f + \lambda_0 L \phi \tag{1}$$

where L is an arbitrary $\mathfrak{L}^2$ kernel. From our analysis in earlier chapters

---

† Aronszajn (1948) proposes a *generalized Rayleigh-Ritz procedure* as the complementation of Weinstein's approach. If the operator T_0 of the base problem is the restriction of T to a smaller space of functions, then a set of upper bounds can be obtained. In this case an appropriate sequence of intermediate operators (providing progressively sharper bounds) is given by the extensions of T_0 to the domain of T_0 plus finite-dimensional components of the difference space $S \ominus S_0$. The *ordinary* Rayleigh-Ritz method then corresponds to the special situation when S_0 contains only the zero function.

we can deduce the specific forms that the solution must take in the two possible cases:

Case 1 λ_0 is a regular value of L. In this case the solution of (1) is given uniquely by

$$\phi = f + \lambda_0 R_L f \qquad (2)$$

where R_L is the resolvent of L evaluated at $\lambda = \lambda_0$.

Case 2 λ_0 is a cv of L. Here it is convenient to introduce the notation of Sec. 6.4, denoting by L_0 that portion of L which is responsible for the appearance of the cv λ_0, namely,

$$L_0(x,y) = \frac{1}{\lambda_0} \sum_{\nu=1}^{p} \sum_{\mu=1}^{m_\nu} \phi_{\nu\mu}(x)\, \overline{\psi}_{\nu\mu}(y)$$

Then $\mathfrak{L}^2$ solutions of (1) exist iff $(f,\psi_{\nu m_\nu}) = 0$ for each of the p cf's of the adjoint kernel L^* corresponding to the cv $\overline{\lambda}_0$. If this condition is satisfied, the (nonunique) $\mathfrak{L}^2$ solutions have the form

$$\phi = f + \lambda_0 R_{L'} f + \sum_{\nu=1}^{p} \left[\alpha_\nu \phi_{\nu 1} - \frac{1}{\lambda_0} \sum_{\mu=1}^{m_\nu - 1} (f,\psi_{\nu\mu}) \phi_{\nu,\mu+1} \right] \qquad (3)$$

where the α_ν are arbitrary constants, the $\phi_{\nu\mu}$ are the (perhaps generalized) cf's of L associated with the cv λ_0, and $R_{L'}$ is the resolvent of the kernel $L' \equiv L - L_0$ evaluated at $\lambda = \lambda_0$. (See Prob. 1 at the end of this chapter.)†

We are now ready to state the following result, which, for the special case of Hermitian kernels, forms the foundation for the approaches of Weinstein and Aronszajn.

THEOREM 11.3

Let two (not necessarily Hermitian) $\mathfrak{L}^2$ kernels K and L differ by a degenerate kernel. Then the characteristic values of one may be

---

† Recall that $R_{L'} = R_L - R_{L_0}$, owing to the orthogonality of L_0 and L'.

computed in terms of the cv's, cf's (including generalized), and
hence resolvent of the other (and its adjoint).

Proof   This is, of course, nothing more than an alternative declaration
of the fundamental notion underlying Schmidt's method of solving
Fredholm integral equations of the second kind. We have from (1)
and (2) above, for instance, that the solution of

$$\phi = \lambda_0 K \phi \tag{4}$$

where

$$K(x, y) = L(x, y) + \sum_{\nu=1}^{n} A_\nu(x)\, \overline{B}_\nu(y)$$

may be expressed as

$$\phi = \lambda_0 \sum_{\nu=1}^{n} (A_\nu + \lambda_0 R_L A_\nu)(\phi, B_\nu)$$

whenever λ_0 *is not* a cv of L. In other words, in this case the solution of (4) is equivalent to the resolution of an equation with the
degenerate kernel

$$\sum_{\nu=1}^{n} [A_\nu(x) + \lambda_0 R_L A_\nu(x)]\, \overline{B}_\nu(y)$$

Using the notation of Chap. 3 wherein

$$A_\nu(\lambda) \equiv A_\nu + \lambda R_L A_\nu$$

$$c_{\mu\nu}(\lambda) \equiv (A_\nu(\lambda), B_\mu)$$

we see that the cv's of K that are *distinct* from the cv's of L are
given then by the zeros of the analytic function

$$\det (\delta_{\mu\nu} - \lambda c_{\mu\nu}(\lambda)) \tag{5}$$

When λ_0 is a cv of L, we effect the decomposition $L = L_0 + L'$
and determine from (3) that if

$$\sum_{j=1}^{n} (A_j, \psi_{\nu m_\nu})(\phi, B_j) = 0 \qquad \nu = 1, 2, \ldots, p \tag{6}$$

then solutions of (4) have the form

$$\phi = \lambda_0 \sum_{j=1}^{n} \left[A_j + \lambda_0 R_{L'} A_j - \frac{1}{\lambda_0} \sum_{\nu=1}^{p} \sum_{\mu=1}^{m_\nu - 1} (A_j, \psi_{\nu\mu}) \phi_{\nu, \mu+1} \right] (\phi, B_j)$$
$$+ \sum_{\nu=1}^{p} \alpha_\nu \phi_{\nu 1}$$

Taking the inner product of this relation with each of the B_i and rearranging, we obtain

$$\sum_{j=1}^{n} \left[\delta_{ij} - \lambda_0 c'_{ij}(\lambda_0) + \sum_{\nu=1}^{p} \sum_{\mu=1}^{m_\nu - 1} (A_j, \psi_{\nu\mu})(\phi_{\nu, \mu+1}, B_i) \right] (\phi, B_j)$$
$$- \sum_{\nu=1}^{p} \alpha_\nu (\phi_{\nu 1}, B_i) = 0 \qquad (7)$$

for $1 \leq i \leq n$, where

$$c'_{ij}(\lambda) \equiv (A_j + \lambda R_{L'} A_j, B_i)$$

It follows, then, from (6) and (7) that any λ, which is a cv of L, is also a cv of K if and only if the $(n + p) \times (n + p)$ determinant of the coefficients vanishes, that is,†

$$\det \begin{pmatrix} \delta_{ij} - \lambda c'_{ij} + \sum_{\nu=1}^{p} \sum_{\mu=1}^{m_\nu - 1} (A_j, \psi_{\nu\mu})(\phi_{\nu, \mu+1}, B_i) & \vdots & (\phi_{\nu 1}, B_i) \\ ------------------------------ & \vdots & --------- \\ (A_j, \psi_{\nu m_\nu}) & \vdots & 0 \end{pmatrix} = 0 \qquad (8)$$

In summary, in order to compute the cv's of K, we first find the zeros of the analytic function (5). Those among these zeros that *are not* in the spectrum of L are cv's of K. Then we solve (8) for λ. Roots of this equation that *are* cv's of L are also in the spectrum of K. In both cases the computation involves only a priori knowledge of the cv's, cf's, and resolvent of L and its adjoint, and the proof of the theorem is thus complete.

11.4 CLASSICAL CONSTRUCTION PROCEDURES

As mentioned earlier, the Weinstein-Aronszajn approach necessitates construction of intermediate kernels K_n "between" the Hermitian base kernel K_0 and the given kernel K. In addition, the K_n are to be such

---

†This general relation simplifies considerably, of course, in the special cases of interest wherein the kernels in question are Hermitian.

that their cv's $\lambda_\nu^{(n)}$ can be computed in terms of the cv's $\lambda_\nu^{(0)}$ and the cf's $\phi_\nu^{(0)}$ of the base kernel K_0. In view of the results of the preceding section, we see that this last condition is satisfied if the difference kernel $K_n - K_0$ is degenerate. For simplicity, then, we shall call a given method *a construction procedure* in *the sense of Weinstein and Aronszajn* if it leads to intermediate kernels for which

(a) $\lambda_\nu^{(0)} \leq \lambda_\nu^{(n)} \leq \lambda_\nu^{(n+1)} \leq \lambda_\nu$ for each n, ν
(b) $K_n - K_0$ is degenerate
(c) $\|K - K_n\| \to 0$ as $n \to \infty$

Condition (c) ensures that the procedure is capable of providing arbitrarily accurate lower bounds for the cv's λ_ν of the kernel K of interest.

As we have done in previous chapters, let us denote by $S^{(n)}$ the space spanned by n linearly independent functions and by P_n the projection operator onto this space. The original method of Weinstein consisted essentially of a construction that we may represent symbolically as

$$K_n \equiv K_0 - P_n K_0 - K_0 P_n + P_n K_0 P_n$$

or equivalently $K_n = K_0 - P_n K_0$

or $\qquad\qquad K_n = K_0 - K_0 P_n \qquad\qquad\qquad\qquad (1)$

(see Prob. 3).

In Aronszajn's alternative approach (see Sec. 11.2), the assumption is that

$$K = K_0 - D \qquad\qquad\qquad\qquad (2)$$

where D is *nonnegative definite*. The intermediate kernels are constructed using projections in the Hilbert space generated by the inner product $[f, g] \equiv (Df, g)$. In order to simplify the discussion we introduce a set $\{\psi_\nu\}$ of $\mathfrak{L}^2$ functions that is *complete* (see Chap. 13) in the complement of the null space of D and is orthonormalized under the new inner product; i.e.,

$$(D\psi_\nu, \psi_\mu) = \delta_{\mu\nu} \qquad\qquad\qquad\qquad (3)$$

The Aronszajn intermediate kernels then take the form

$$K_n = K_0 - D\widetilde{P}_n D \qquad \text{where } \widetilde{P}_n = \sum_{\nu=1}^{n} \psi_\nu(x)\overline{\psi}_\nu(y)$$

or $$K_n(x,y) = K_0(x,y) - \sum_{\nu=1}^{n} (D\psi_\nu)(x) \cdot \overline{(D\psi_\nu)}(y) \qquad (4)$$

It is obvious that these kernels satisfy condition b above. We leave the verifications of conditions a and c as exercises for the interested reader (see Prob. 4).

The characteristic values of the intermediate kernels K_n of (4) can be computed in terms of the cv's and cf's of the base kernel K_0 as detailed in Theorem 11.3. Those cv's of K_n that are distinct from the cv's of K_0 are given by

$$\det\left(\delta_{\mu\nu} + \lambda(D\psi_\nu + \lambda R_{K_0} D\psi_\nu, D\psi_\mu)\right) = 0 \qquad (5)$$

And those λ that are cv's of K_0 are also cv's of K_n iff

$$\det\left(\frac{\delta_{\mu\nu} + \lambda(D\psi_\nu + \lambda R_{K_0'} D\psi_\nu, D\psi_\mu)}{(D\psi_\nu, \phi_i^{(0)})} \,\middle|\, \frac{(\phi_i^{(0)}, D\psi_\mu)}{0}\right) = 0 \qquad (6)$$

where K_0' is the kernel remaining after subtracting from K_0 that portion which is responsible for the appearance of the cv λ, that is,

$$K_0'(x,y) = K_0(x,y) - \frac{1}{\lambda}\sum_{i=1}^{p} \phi_i^{(0)}(x)\,\overline{\phi}_i^{(0)}(y) \qquad (7)$$

11.5   THE METHOD OF SPECIAL CHOICES

Determination of the cv's from (11.4-5) and (11.4-6) presents some practical difficulties in general since it is necessary to solve for the zeros of various meromorphic functions. As Bazley (1961) observed [see also Bazley and Fox (1961) and Fox and Rheinboldt (1966)], however, a special choice of the ψ_ν of the previous section suffices to reduce the computational problem to that of finding the roots of polynomials.

In *the method of special choices* we construct the intermediate kernels K_n so that the degenerate kernels $(K_n - K_0)$ are orthogonal to the cf's $\phi_\mu^{(0)}$ of K_0 *for all but a finite number of* μ. In other words, we select the ψ_ν so that not only is (11.4-3) satisfied, but also

$$D\psi_\nu = \sum_{\mu=1}^{N(n)} \alpha_{\nu\mu}\phi_\mu^{(0)} \qquad \nu = 1, 2, \ldots, n \qquad (1)$$

The complex constants $\alpha_{\nu\mu}$ are assumed to be known quantities, as are the fixed integers $N(n)$. Substituting in the earlier expressions, (11.4-5) becomes

$$\det\left(\delta_{\mu\nu} + \lambda \sum_{k=1}^{N} \alpha_{\nu k}\,\overline{\alpha}_{\mu k}\,\frac{\lambda_k^{(0)}}{\lambda_k^{(0)} - \lambda}\right) = 0 \qquad (2)$$

while (11.4-6) takes on the new form

$$\det\begin{pmatrix} \delta_{\mu\nu} + \lambda \sum_{k=1}^{N}{}' \alpha_{\nu k}\,\overline{\alpha}_{\mu k}\,\dfrac{\lambda_k^{(0)}}{\lambda_k^{(0)} - \lambda} & \vdots & \overline{\alpha}_{\mu i} \\ \hline \alpha_{\nu i} & \vdots & 0 \end{pmatrix} = 0 \qquad (3)$$

In (3), Σ' signifies that those terms associated with the cv of K_0 under investigation do not appear in the summation. Elsewhere in the expression, *only* those α's that *are* associated with this cv appear. Clearly, solving (2), (3) now involves finding the zeros only of polynomials in λ, as expected.

Actually, in the present case it is a little easier to determine the cv's of the K_n in an equivalent and more direct fashion rather than as solutions of the above equations (2) and (3). We first rewrite (11.4-4), using (1), as

$$K_n(x,y) = K_0(x,y) - \sum_{i,j=1}^{N} \sum_{\mu=1}^{n} \alpha_{\mu i}\,\overline{\alpha}_{\mu j}\,\phi_i^{(0)}(x)\,\overline{\phi}_j^{(0)}(y) \qquad (4)$$

recognizing the summation therein as a degenerate kernel *of rank N* of the form

$$\sum_{i=1}^{N} \phi_i^{(0)}(x)\,\overline{\psi}_i(y) \qquad \text{with} \qquad \psi_i(y) = \sum_{j=1}^{N} \sum_{\mu=1}^{n} \alpha_{\mu j}\,\overline{\alpha}_{\mu i}\,\phi_j^{(0)}(y)$$

It follows, then, from (11.3-5) directly, that the cv's of K_n that are *distinct* from the cv's of K_0 are given by the zeros of the $N \times N$ determinant

$$\det\left(\delta_{\mu\nu} + \frac{\lambda\lambda_\nu^{(0)}}{\lambda_\nu^{(0)} - \lambda} \sum_{i=1}^{n} \alpha_{i\nu}\,\overline{\alpha}_{i\mu}\right) \qquad (5)$$

and it is obvious from this expression that there are precisely N such zeros.

The N characteristic functions corresponding to the above cv's are linear combinations of the $\phi_i^{(0)}$, $1 \le i \le N$, and hence, taken as a whole, they span the same N-dimensional space as the $\phi_i^{(0)}$. Additional characteristic values of K_n, which, on the one hand, must be *identical* with cv's of K_0, now, on the other hand, in view of this observation are constrained to have cf's associated with them that are orthogonal to all the $\phi_i^{(0)}$, $1 \leq i \leq N$. A glance at (4), then, indicates that these additional cv's are perforce restricted to the higher-order λ's of K_0, namely, $\lambda_{N+1}^{(0)}$, $\lambda_{N+2}^{(0)}$, ...

In summary, once the characteristic-value problem with the base kernel K_0 has been computationally resolved, the method of special choices necessitates the selection of the complete "orthonormal" set $\{\psi_\nu\}$ in such a way that (1) is satisfied for each ν. N of the cv's are then given in terms of the $\alpha_{\nu\mu}$ by the zeros of (5). When these are ordered together with the cv's $\lambda_{N+1}^{(0)}$, $\lambda_{N+2}^{(0)}$, . . . , carried over from K_0, we obtain a monotonically increasing sequence of values, which provides the desired improved lower bounds to the cv's of K.†

11.6 LOWER BOUNDS BY TRUNCATION OF THE BASE OPERATOR

The practical construction techniques à la Weinstein-Aronszajn currently in vogue by and large involve variants of the projection methods outlined earlier, a special choices approach, and/or some form of a truncation procedure proposed by Weinberger (1959). Stated most simply, Weinberger's technique involves replacing the operator generated by the base kernel K_0 with a truncated operator of the form

$$K_0^{(N)} \phi \equiv \sum_{\nu=1}^{N} (\phi, \phi_\nu^{(0)}) \frac{\phi_\nu^{(0)}}{\lambda_\nu^{(0)}} + \frac{1}{\lambda_{N+1}^{(0)}} \left[\phi - \sum_{\nu=1}^{N} (\phi, \phi_\nu^{(0)}) \phi_\nu^{(0)} \right] \qquad (1)$$

An appropriate sequence of *intermediate operators* is then given, using (11.4-4), say, by

$$K_n^{(N)} \phi \equiv K_0^{(N)} \phi - D\tilde{P}_n D\phi$$

$$= K_0^{(N)} \phi - \sum_{\nu=1}^{n} (\phi, D\psi_\nu) D\psi_\nu \qquad (2)$$

[Here it is assumed that the $\lambda_\nu^{(0)}$ appearing in (1) are all positive.] Strictly speaking, the *truncated base operator* and hence the intermedi-

---

† Ordering of the cv's is an important step in the Bazley procedure; otherwise the *persistence* of certain cv's may obviate condition a of Sec. 11.4 for various n,ν [see Weinstein (1935, 1937, 1966) and Fox and Rheinboldt (1966)].

ate operators are not completely continuous; thus they have no antecedent kernels, i.e., kernels by which they are generated. On the other
hand, the $K_n^{(N)}$ do lead to characteristic values $\lambda_\nu^{(n,N)}$ satisfying the two
sets of inequalities

$$\lambda_\nu^{(n,N)} \leq \lambda_\nu^{(n,N+1)} \leq \lambda_\nu^{(n)} \leq \lambda_\nu$$

and

$$\lambda_\nu^{(n,N)} \leq \lambda_\nu^{(n+1,N)} \leq \lambda_\nu$$

(see Prob. 6). In analogy with the method of special choices, moreover,
the resolution of these arbitrarily accurate lower bounds is merely an algebraic exercise.

To verify this last observation we first note that if ϕ is in the
complement of the finite-dimensional space $S_{n,N}$ spanned by the
$\phi_\nu^{(0)}$ $(1 \leq \nu \leq N)$ and the $D\psi_\nu$ $(1 \leq \nu \leq n)$ combined, then

$$K_n^{(N)} \phi = \frac{1}{\lambda_{N+1}^{(0)}} \phi$$

Thus $\lambda_{N+1}^{(0)}$ appears as a cv of generally infinite multiplicity, as is to be expected from a classical result of Weyl (1909) [see also Riesz and Sz.-
Nagy (1955, p. 367)]. In order to determine additional cv's of the intermediate operators, therefore, we need only consider $\phi \in S_{n,N}$. Moreover, without loss of generality we can assume that $\lambda_N^{(0)}$ is strictly less
than $\lambda_{N+1}^{(0)}$.

It follows, then, from the form of (2) that (11.4-5) and (11.4-6) will
suffice to determine the desired remaining cv's. We need only replace
$R_{K_0} D\psi_\nu$ by the expression†

$$\sum_{k=1}^{N} (D\psi_\nu, \phi_k) \frac{(\lambda_{N+1} - \lambda_k)}{(\lambda_k - \lambda)(\lambda_{N+1} - \lambda)} \phi_k + \frac{D\psi_\nu}{\lambda_{N+1} - \lambda}$$

and $R_{K_0'} D\psi_\nu$ by the same expression with the terms corresponding to the
cv under investigation omitted. Alternatively, we can substitute the expansion

$$\phi = \sum_{\nu=1}^{N} \alpha_\nu \phi_\nu + \sum_{k=1}^{n} \beta_k D\psi_k \tag{3}$$

---

† In this expression and elsewhere in the remainder of this section, the (0) superscript is
implied where appropriate but suppressed.

with undetermined coefficients, into (2), thence obtaining

$$K_n^{(N)}\phi = \sum_{\nu=1}^{N}\left[\frac{\alpha_\nu}{\lambda_{N+1}} + (\phi, \phi_\nu)\left(\frac{1}{\lambda_\nu} - \frac{1}{\lambda_{N+1}}\right)\right]\phi_\nu$$

$$+ \sum_{k=1}^{n}\left[\frac{\beta_k}{\lambda_{N+1}} - (\phi, D\psi_k)\right]D\psi_k$$

The characteristic-value problem for the intermediate operator on $S_{n,N}$ thus has the form

$$\sum_{\nu=1}^{N}(\phi,\phi_\nu)\left(\frac{1}{\lambda_\nu} - \frac{1}{\lambda_{N+1}}\right)\phi_\nu - \sum_{k=1}^{n}(\phi, D\psi_k)D\psi_k + \phi\left(\frac{1}{\lambda_{N+1}} - \frac{1}{\lambda}\right) = 0 \qquad (4)$$

Taking the inner product of this equation with the spanning functions $\phi_\nu (1 \le \nu \le N)$ and $D\psi_k (1 \le k \le n)$, we are led to the conclusion that the cv's which we seek are also given by the zeros of the $(N + n) \times (N + n)$ determinant

$$\det\left(\begin{array}{c|c}\delta_{\mu\nu}(1/\lambda_\nu - 1/\lambda) & -(D\psi_k, \phi_\mu) \\ \hline (\phi_\nu, D\psi_i)(1/\lambda_\nu - 1/\lambda_{N+1}) & \delta_{ik}(1/\lambda_{N+1} - 1/\lambda) - (D\psi_k, D\psi_i)\end{array}\right) \quad (5)$$

In addition to clearly displaying the simplified form that the characteristic-value calculations assume in Weinberger's truncation method, this last result (5) has many of the same advantages, from a computational point of view, as did the alternative expression (11.5-5), suggested in connection with the method of special choices. In employing (5) to determine the cv's associated with the intermediate problems, we obviate using several different formulas as is the case with (11.4-6) or (11.5-3). We also avoid whatever numerical instabilities may be attendant with (11.4-5) and (11.5-2) in the neighborhood of the base cv's. To be sure, however, we may have in (5) a larger matrix than need be; i.e., a number of the zeros of the determinant may turn out to be precisely equal to λ_{N+1}. On the other hand, since this unnecessary feature can only occur if the dimension of $S_{n,N}$ is strictly less than $(n + N)$, it can be sidestepped by recognizing any linear dependence of the ϕ_ν and the $D\psi_k$ while forming (5) from (4).

11.7 GENERAL REMARKS

As is to be expected, the methods outlined above can be generalized and extended in a number of ways and the results used in various fashions. For instance, Bazley (1961) observed that the technique of special choice can also be employed to obtain *upper* and hence two-sided error bounds for the cv's of Hermitian operators [see also Weinberger (1952b), Weinstein (1953), and Fox and Rheinboldt (1966)]. Similarly, Weinberger (1959 and 1962) discusses how more general truncation procedures can be developed which, when interplayed with Rayleigh-Ritz results, give sharp error bounds not only for the approximate cv's but also for related cf's [compare Weinberger (1960)]. Bazley and Fox (1962b) likewise consider the characteristic-function-approximation question.

Fox and Rheinboldt in their 1966 paper (which contains numerous references to the recent literature†) describe a generalized truncation procedure that is often applicable in practical situations. Meanwhile Fichera (1966) and Stakgold (1969) have also developed useful extensions of the classical construction techniques. Finally Bazley and Fox (1966) have demonstrated the intimate relationship between various of the seemingly different outgrowths of the basic Weinstein-Aronszajn method. For numerical results the above-mentioned paper of Fox and Rheinboldt (and other articles mentioned therein), the book of Gould (1966), or the review of Weinstein (1966) should be consulted. ‡

PROBLEMS

1. Let λ_0 be a characteristic value of the $\mathfrak{L}^2$ kernel L and assume that the portion of L which gives rise to the cv λ_0 may be expressed as

$$L_0(x,y) = \frac{1}{\lambda_0} \sum_{\nu=1}^{p} \sum_{\mu=1}^{m_\nu} \phi_{\nu\mu}(x)\, \overline{\psi}_{\nu\mu}(y)$$

in the notation of Sec. 6.4. Show that solutions of

$$\phi = f + \lambda_0 L\phi \tag{1}$$

exist iff $(f, \psi_{\nu m_\nu}) = 0$ for each of the p cf's of L^* associated with the cv $\overline{\lambda}_0$. If this condition is satisfied, deduce that the *general* solution of (1) is the sum of a linear combination of the $\phi_{\nu 1}$ and the *particular* solution

$$\phi = f + \lambda_0 R_{L'} f - \frac{1}{\lambda_0} \sum_{\nu=1}^{p} \sum_{\mu=1}^{m_\nu - 1} (f, \psi_{\nu\mu})\phi_{\nu,\mu+1}$$

Here $R_{L'}$ is the resolvent of the kernel $L' \equiv L - L_0$ evaluated at $\lambda = \lambda_0$.

---

† An earlier review paper of Diaz (1958) also presents a representative bibliography.
‡ Also see Wilson (1965) and the several papers following this one in the same journal.

2. If $\{A_j\}$ $(j = 1, 2, \ldots, n)$ is a set of linearly independent $\mathfrak{L}^2$ functions, verify that the set $\{A'_j\}$ where

$$A'_j = A_j + \lambda_0 R_{L'} A_j - \frac{1}{\lambda_0} \sum_{\nu=1}^{p} \sum_{\mu=1}^{m_\nu-1} (A_j, \psi_{\nu\mu}) \phi_{\nu,\mu+1}$$

is also linearly independent. Then show that $(A'_j, \psi_{\nu 1}) = 0$ for each $1 \leq j \leq n$, $1 \leq \nu \leq p$. (The notation is that of Prob. 1.)

3. Let K be the linear transformation generated by an arbitrary Hermitian $\mathfrak{L}^2$ kernel and let P be the projection operator onto a given subspace of $\mathfrak{L}^2$. Show that the three transformations

$$\begin{array}{ll} \text{(a)} & K - PK - KP + PKP \\ \text{(b)} & K - PK \\ \text{(c)} & K - KP \end{array}$$

all have the same cv's in position and multiplicity.

4. Assume that the square-integrable Hermitian kernels K, K_0, and D are such that $K = K_0 - D$ with D nonnegative definite. Let the set $\{\psi_\nu\}$ of $\mathfrak{L}^2$ functions be complete in the complement of the null space of D and satisfy $(D\psi_\nu, \psi_\mu) = \delta_{\nu\mu}$. Finally, define the intermediate kernels K_n $(n = 1, 2, \ldots)$ by

$$K_n(x,y) = K_0(x,y) - \sum_{\nu=1}^{n} (D\psi_\nu)(x) \cdot \overline{(D\psi_\nu)}(y)$$

a. Show that

$$(K_0\phi, \phi) \geq (K_n\phi, \phi) \geq (K_{n+1}\phi, \phi) \geq (K\phi, \phi)$$

and hence that the positive cv's of the various kernels satisfy

$$\lambda_\nu^{(0)} \leq \lambda_\nu^{(n)} \leq \lambda_\nu^{(n+1)} \leq \lambda_\nu \qquad \text{for each } n, \nu$$

b. Assuming that

$$\|D\|^2 = \sum_{\nu=1}^{\infty} \|D\psi_\nu\|^2 = \sum_{\mu,\nu=1}^{\infty} |(D\psi_\nu, D\psi_\mu)|^2$$

[which follows since $\|D\|$ is also the *Hilbert-Schmidt* norm of the operator generated by the kernel D in the Hilbert space formed with the inner product $[f,g] \equiv (Df,g)$; see Sec. 14.1; see also Dunford and Schwartz (1963, Sec. 11.6) and Schatten (1960, Chap. 2)], verify that

$$\|K - K_n\| \to 0 \quad \text{as } n \to \infty$$

5. Establish (11.4-5), (11.4-6), (11.5-2), and (11.5-3).

6. Let $K_0^{(N)}\phi$ and $K_n^{(N)}\phi$ be defined as in Sec. 11.6.
   a. Show that the cv's and associated cf's of the truncated base operator may be given by

$$\begin{cases} \lambda_\nu^{(0)}, \phi_\nu^{(0)} & 1 \leq \nu \leq N \\ \lambda_{N+1}^{(0)}, \phi_\nu^{(0)} & N+1 \leq \nu \end{cases}$$

   b. Using *Bessel's inequality* as given in Sec. 10.5, demonstrate that

$$\left(K_n^{(N+1)}\phi - K_n^{(N)}\phi, \phi \right) \leq 0$$

   and hence deduce that the positive characteristic values of these intermediate operators satisfy

$$\lambda_\nu^{(n, N)} \leq \lambda_\nu^{(n, N+1)} \leq \lambda_\nu^{(n)} \leq \lambda_\nu$$

   (The analogous monotonicity in n, for fixed N, has essentially already been verified in Prob. 4 above.)
7. a. Verify that (11.6-4) follows from the assumed representation (11.6-3).
   b. Show that λ_{N+1} occurs as a p-fold zero of (11.6-5) iff the dimension of $S_{n,N}$ equals $n + N - p$.

chapter twelve

Parametric Dependence of Characteristic Values and Characteristic Functions

12.1 ON THE NATURE OF REGULAR PERTURBATIONS

As a natural adjunct to the approximation-theoretic material of earlier chapters we want now to discuss a number of results from *linear operator perturbation theory* that also are relevant to the practical side of cv and cf estimation. Our concern will be with the (approximate) solution of Fredholm integral equations whose kernels differ only slightly from simpler kernels for which the cv problem can presumably be completely solved. To be specific, we shall generally assume, for simplicity, that we are dealing with a kernel $K(x,y;\varepsilon)$ which depends on a complex "smallness" or perturbation parameter ε in such a way that not only is

    1. $K(x,y;\varepsilon) \in \mathcal{L}^2$ as a function of x, y for $|\varepsilon| < \varepsilon_0$,

but, in fact,

    2. $K(x,y;\varepsilon) = K_0(x,y) + \varepsilon K_1(x,y) + \varepsilon^2 K_2(x,y) + \cdots$ for $|\varepsilon| < \varepsilon_0$ (1)†

---

† Here explicitly the meaning of the convergent infinite series is that

$$\lim_{n \to \infty} \left\| K - \sum_{\nu=0}^{n} \varepsilon^\nu K_\nu \right\| = 0 \qquad \text{for } |\varepsilon| < \varepsilon_0$$

As is the custom elsewhere, we shall often refer, then, to $K(x,y;\varepsilon)$ as depending *analytically* on the parameter ε. This notion, of course, can be generalized to domains of the complex ε plane other than the circular disks about the origin with which we will be concerned. Similar remarks also apply to series of functions of one independent variable and to ordinary infinite series, as well as to asymptotic expansions of the form (1).

where K_0, K_1, K_2, ..., are themselves square integrable in the fundamental domain.

The formal analysis of problems such as the above dates back at least to Rayleigh (1894) with more recent contributions, of a pioneering nature, having been made by Schrödinger (1926) in quantum-mechanical applications. It remained, however, for Rellich (1937–1942) to settle rigorously the various mathematical questions underlying the Rayleigh-Schrödinger approach. The work of Rellich (see also 1950 and 1953b) with various refinements and/or extensions suggested principally by Sz.-Nagy (1946–1947 and 1951) and Kato (1949b, 1950, 1951, and 1966) forms the basis for the treatment of perturbation theory presented herein.†

One anticipates that, under conditions 1 and 2, the cv's and associated cf's of K themselves have power series that are convergent for small enough ε. Such turns out to be essentially the case, as we shall see, whenever K is Hermitian for real ε, a result that generalizes, in part, to bounded operators on arbitrary separable Hilbert spaces. Although slightly less is often true if the Hermitian character is lacking, the perturbation problem we are concerned with here still remains, even then, of a nonsingular or regular nature; a small change in the perturbation parameter ε generally results in correspondingly small changes in the cv's of K (and often in the cf's as well). The precise nature of these resulting changes is focal to the ensuing discussion.‡

12.2 THE ANALYTICITY OF THE RESOLVENT

As we have often suggested (albeit sometimes implicitly), one way to obtain information regarding the characteristic values of a given kernel is to study the related resolvent kernel of which the cv's appear as pole singularities. The following theorem is, then, a natural first step in the application of such an approach to the investigation of the cv's of $\mathfrak{L}^2$ kernels satisfying (12.1-1) (compare Theorem 5.4).

THEOREM 12.2

Let the arbitrary $\mathfrak{L}^2$ kernel $K(x, y; \varepsilon)$ depend analytically on ε, say, for $|\varepsilon| < \varepsilon_0$. The associated resolvent kernel $R(\lambda, \varepsilon) \equiv$

---

† Also see Meyer zur Capellen (1933), whose short note preceded Rellich's series of fundamental papers; Hölder (1937); and Bückner (1952, pp. 84ff). The broad applicability of a related perturbation technique to Fredholm integral equations of the first kind is surveyed in Kanwal (1970).

‡ For the Hermitian case we already know from our earlier work [see, for example, (9.3-4)] that small changes (in the sense of the norm of the kernel) are going to give rise to small changes in the reciprocals of the cv's.

$R_K(x, y; \lambda, \varepsilon)$ is, then, for small enough ε, an analytic function of ε in each closed bounded region Δ of the complex λ plane that does *not* contain any of the characteristic values of the *unperturbed* kernel $K(x, y; 0)$.

Proof   We begin by making the decomposition

$$K(x, y; \varepsilon) = K(x, y; 0) + \varepsilon L(x, y; \varepsilon)$$
$$\equiv K_0 + \varepsilon L$$

and associating with each of the kernels K and K_0 its corresponding resolvent kernel, namely, $R(\lambda, \varepsilon)$ and $R(\lambda, 0)$, respectively. Substitution of the above dissection into the first of the Fredholm identities (3.2-5) leads to

$$R(\lambda, \varepsilon) = K + \lambda(K_0 + \varepsilon L)R(\lambda, \varepsilon)$$

from which it follows, operating (symbolically) on both sides with $I + \lambda R(\lambda, 0)$ and rearranging, that

$$R(\lambda, \varepsilon) = \tilde{K} + \lambda \varepsilon \tilde{L} R(\lambda, \varepsilon) \tag{1}$$

where
$$\tilde{K} = K + \lambda R(\lambda, 0) K \tag{2}$$
$$\tilde{L} = L + \lambda R(\lambda, 0) L$$

This expression (1), which has the form of an integral equation for $R(\lambda, \varepsilon)$ with kernel $\varepsilon \tilde{L}$, is meaningful (as far as λ is concerned) for all λ distinct from the cv's of both K_0 and K.

    It is clear from (2) that both $\tilde{K}$ and $\tilde{L}$ are analytic functions of ε for $|\varepsilon| < \varepsilon_0$. Moreover, for λ in a given closed bounded subset Δ of the resolvent set of K_0, $R(\lambda, 0)$ and hence $\tilde{L}$ must have finite upper bounds (in norm). As a consequence, ε can be chosen small enough so that the Neumann series *for the resolvent* $\rho(\lambda, \varepsilon)$ of $\varepsilon \tilde{L}$ converges (relatively) uniformly. We conclude, therefore, that under these conditions ρ depends analytically on ε since each iterate of $\varepsilon \tilde{L}$ obviously has this property.

    For small enough ε and $\lambda \in \Delta$, then, the solution of (1), viewed as an equation for $R(\lambda, \varepsilon)$, can be expressed as

$$R(\lambda, \varepsilon) = \tilde{K} + \lambda \rho(\lambda, \varepsilon) \tilde{K}$$

From this we immediately deduce the desired behavior for $R(\lambda, \varepsilon)$.

The theorem we have just proved establishes the fact that the analytic character of a given $\mathfrak{L}^2$ kernel K commonly carries over to its resolvent. From our knowledge of the effect of analytic perturbations even in spaces of finite dimension, however, we do not expect such to be the case in general as far as the characteristic values (and characteristic functions) of K are concerned.† On the other hand, as the results of the next section show, the prevalent situation relative to cv's is really not so bad at all and, in fact, essentially parallels the behavior encountered in the finite-dimensional case.

12.3   THE ALGEBRAIC DEPENDENCE OF THE CHARACTERISTIC VALUES

We know from Theorem 12.2 that the characteristic values of $\mathfrak{L}^2$ kernels satisfying (12.1-1) are, for small enough ε, restricted to neighborhoods of the cv's of the unperturbed kernel K_0. To make this notion more precise, consider an arbitrary (but fixed) cv λ_0 of $K_0 \equiv K(x,y;0)$ and surround λ_0 (in the λ plane) with a *simple closed contour* C, which contains *no other* cv of K_0. In keeping with Secs. 6.3 and 6.4 we split off from K_0 that portion of the kernel which is responsible for the appearance of the cv λ_0 and similarly decompose $R(\lambda,0)$. In view of the meromorphic character of $R(\lambda,0)$, it follows, then—by direct calculation from (6.4-2) to (6.4-4), if you wish—that

$$-\frac{1}{2\pi i}\oint_C r(\mu,0)\,d\mu \equiv -\frac{1}{2\pi i}\oint_C \mathrm{Tr}\,[R(\mu,0)-K_0]\,d\mu$$

$$=\sum_{\nu=1}^{p} m_\nu \tag{1}$$

This contour integration thus gives rise to the multiplicity‡ N of the cv λ_0 of K_0.

Mindful of Theorem 12.2, we can in like fashion define

$$r(\lambda,\varepsilon) \equiv \mathrm{Tr}\,[R(\lambda,\varepsilon)-K]$$

---

† For example, the 2×2 matrix

$$\begin{pmatrix} 2 & 1-\varepsilon \\ -1 & 0 \end{pmatrix}$$

has the cv's $\lambda = 1/(1\pm\sqrt{\varepsilon})$.

‡ *Multiplicity* here refers, of course, to the *algebraic* multiplicity of λ_0.

and calculate, along the identical contour C,†

$$N(\varepsilon) \equiv -\frac{1}{2\pi i} \oint_C r(\mu, \varepsilon) \, d\mu \tag{2}$$

On the one hand, we know that the value of $N(\varepsilon)$ should be precisely equal to the number of cv's of K, counting multiplicity, within C (compare remarks of Sec. 3.5). On the other hand, for small enough ε, $N(\varepsilon)$ must depend analytically on ε since this property, shared by both $R(\lambda, \varepsilon)$ and K, is invariant under the trace operation as well as the contour integration. The end result is that, for small ε, $N(\varepsilon)$ must equal N and we have thus proved the following:

THEOREM 12.3-1

Let the $\mathfrak{L}^2$ kernel $K(x, y; \varepsilon)$ depend analytically on ε as in (12.1-1), and consider a simple closed contour C which encloses a single arbitrary cv of $K(x, y; 0)$, of multiplicity N, and no others. Counting multiplicity, $K(x, y; \varepsilon)$ itself has precisely N cv's within C for all small enough ε.

Considerably more information can be obtained from this approach by recognizing that the integral relation (2) is actually only one of a series of such expressions. For example, recall from (3.5-7) that the modified Fredholm determinant $\tilde{D}(\lambda)$ of a given $\mathfrak{L}^2$ kernel K can be uniquely expressed as

$$\tilde{D}(\lambda) \equiv \exp\left[-\int_0^\lambda \mathrm{Tr}\,(R_K - K)\,d\mu\right]$$

It follows, then, introducing the dependence upon ε and assuming that ε is small enough to ensure the validity of Theorem 12.3-1, that

$$\begin{aligned}
N^k(\varepsilon) &\equiv -\frac{1}{2\pi i} \oint_C \mu^k r(\mu, \varepsilon)\, d\mu \\
&= \frac{1}{2\pi i} \oint_C \mu^k \frac{\partial \tilde{D}(\mu, \varepsilon)/\partial \mu}{\tilde{D}(\mu, \varepsilon)}\, d\mu \qquad k = 0, 1, 2, \ldots \\
&= \sum_{\nu=1}^N \lambda_\nu^{\,k}(\varepsilon)
\end{aligned} \tag{3}$$

---

† The contour integral employed in (1) and (2) had its origin, as far as perturbation theory is concerned, in the work of Sz.-Nagy (1946–1947) and Kato (1949b); see also Kato (1966, p. 67).

This last relation is, of course, a direct consequence of the residue calculus and the fact that the cv's of a given kernel generally appear, with appropriate multiplicity, as the zeros of the modified Fredholm determinant.

By the same argument as used previously, we thus know from (3) that the power sums $N^k(\varepsilon)$ of the cv's of K depend analytically on ε for small enough values of this perturbation parameter. Hence, if we form the polynomial $p_N(\lambda,\varepsilon)$ which has the $\lambda_\nu(\varepsilon)$ ($\nu = 1, 2, \ldots, N$) as roots, viz.,

$$\prod_{\nu=1}^{N} [\lambda - \lambda_\nu(\varepsilon)] \equiv p_N(\lambda,\varepsilon) \equiv \sum_{\nu=0}^{N} c_\nu(\varepsilon)\lambda^\nu$$

the coefficients $c_\nu(\varepsilon)$ of the alternative representation must enjoy this same behavior since, by Newton's formulas [see Turnbull (1952)], they are just sum-and-product combinations of the $\Sigma \lambda_\nu^k$. This last conclusion permits the immediate deduction that the $\lambda_\nu(\varepsilon)$ themselves are each *algebraic* functions of ε. Stated more formally, we have the following theorem.

THEOREM 12.3-2

Under the assumptions of Theorem 12.3-1, for small enough ε the characteristic values of $K(x,y;\varepsilon)$ within C consist of branches of one or several analytic functions which have at most algebraic singularities of order $1 \leq p - 1 \leq N - 1$ at $\varepsilon = 0$.†

Before we move on to consideration of the important special case of Hermitian K, let us review precisely what has been established in the analysis presented so far. We have seen firstly that analytic dependence of a given $\mathfrak{L}^2$ kernel K upon the perturbation parameter ε gives rise to the same sort of behavior in the resolvent R_K, wherever R_K has meaning. Secondly, each characteristic value λ_0 of the *unperturbed* kernel K_0 is continuable into one or several cv's $\lambda_\nu(\varepsilon)$ of the *perturbed* kernel K. These $\lambda_\nu(\varepsilon)$, moreover, are analytic functions in the neighborhood of $\varepsilon = 0$ and have the representation, typically,

$$\lambda_\nu(\varepsilon) = \lambda_0 + \varepsilon^{1/p}\lambda_{\nu1} + \varepsilon^{2/p}\lambda_{\nu2} + \cdots \tag{4}$$

with integer $p \geq 1$. The expression (4) explicitly exhibits the nature of the branch singularity at the origin $\varepsilon = 0$.

---

† Note that if $N = 1$, we already have the result that $\lambda(\varepsilon)$ is analytic in ε.

It should be noted, however, that the $\lambda_\nu(\varepsilon)$ discussed above, even when taken in aggregate for all possible choices of λ_0, need not exhaust the cv's of K, independent of how small a value the perturbation parameter assumes. For example (see Prob. 1), the kernel

$$K(x,y;\varepsilon) = \min(x,y) - (1+\varepsilon)xy \qquad 0 \le x,y \le 1$$

has, for $\varepsilon > 0$, the negative cv $\lambda(\varepsilon) = -\mu^2$ where μ is the unique positive root of the equation $\tanh \mu = \varepsilon\mu/(1+\varepsilon)$. This $\lambda(\varepsilon)$, which tends to $-\infty$ as $\varepsilon \to 0$, does not appear as the direct continuation of any of the cv's $\lambda_n = n^2 \pi^2$ associated with the unperturbed kernel $K(x,y;0)$. In an extreme situation, K_0 could be a kernel having no cv's at all, while K possesses, for $\varepsilon \ne 0$, a countably infinite spectrum.† In such cases, of course, all of the cv's of K must increase without bound as $\varepsilon \to 0$.

12.4   THE SPECIAL CASE OF HERMITIAN K

When the square-integrable kernel $K(x,y;\varepsilon)$, depending analytically on the smallness parameter ε, has the further property that it is *Hermitian for real ε*, then the remarkable result follows that the *perturbed characteristic values and characteristic functions of K are themselves analytic functions of ε for real values of this parameter*. Actually, these fundamental conclusions hold true even for the considerably larger class of *semisimple or essentially fully symmetrizable* kernels (which includes normal and hence Hermitian kernels as a subclass). We choose to carry out our discussion in this more general framework (see Sec. 6.4),‡ the various proofs turning out to be only moderately more complicated.

In analogy with the analysis of the previous section, we shall make substantial use of an appropriate contour integral of the resolvent kernel of K. To that end we again consider an arbitrary (but fixed) cv λ_0 of $K(x,y;0)$ and enclose it with a simple closed contour C, within which is contained no other cv of K_0. We then form the function§

$$\Pi(x,y;\varepsilon) \equiv -\frac{1}{2\pi i} \oint_C R(x,y;\mu,\varepsilon) \, d\mu$$

where R, of course, is the resolvent of the given $\mathcal{L}^2$ kernel K, and recog-

---

† One way to construct such K is by means of orthogonal kernels (see Prob. 2).

‡ Although it is obvious that Hermitian $K(x,y;\varepsilon)$ implies that $K_0, K_1, K_2, \ldots$, of the series (12.1-1) are also Hermitian and conversely, more general conditions on K are not so easily expressed in terms of the kernels K_ν.

§ Contour integrals such as these play an important role in various function-theoretic investigations in general operator theory [see, e.g., Dunford and Schwartz (1958, chap. 7)].

nize (from Theorem 12.2) that Π depends analytically on ε for small enough values of this perturbation parameter. If $K(x,y;\varepsilon)$ is a semisimple kernel for real ε, which we shall assume to be the case for the remainder of this section, then R has only simple poles in the λ plane. Our earlier considerations (Sec. 6.4), therefore, lead to the direct determination of Π as

$$\Pi(x,y;\varepsilon) = \sum_{\nu=1}^{N} \phi_\nu(x,\varepsilon)\overline{\psi}_\nu(y,\varepsilon) \qquad \varepsilon \text{ real} \tag{1}$$

In this expression ϕ_ν and ψ_ν are linearly independent square-integrable cf's of the kernels K and K^* associated with the cv's $\lambda_\nu(\varepsilon)$ and $\overline{\lambda}_\nu(\varepsilon)$ (arising from λ_0 and $\overline{\lambda}_0$), respectively. Moreover,

$$(\phi_\nu(\varepsilon),\psi_\mu(\varepsilon)) = \delta_{\nu\mu} \tag{2}$$

We now form the composite kernel ΠK given by

$$\Pi K(x,y;\varepsilon) \equiv \int_a^b \Pi(x,z;\varepsilon)K(z,y;\varepsilon)\,dz$$

$$= \sum_{\nu=1}^{N} \frac{\phi_\nu(x,\varepsilon)\overline{\psi}_\nu(y,\varepsilon)}{\lambda_\nu(\varepsilon)} \qquad \varepsilon \text{ real} \tag{3}$$

Similar to Π itself, this also is a degenerate kernel, for real ε, depending analytically on the perturbation parameter. Note, moreover, that

$$\Pi K(x,y;0) = \frac{1}{\lambda_0}\Pi(x,y;0)$$

It follows, then, that the relation

$$H(x,y;\varepsilon) \equiv \frac{1}{\varepsilon}\left[\Pi K(x,y;\varepsilon) - \frac{1}{\lambda_0}\Pi(x,y;\varepsilon)\right]$$

$$= \sum_{\nu=1}^{N} \frac{\phi_\nu(x,\varepsilon)\overline{\psi}_\nu(y,\varepsilon)}{h_\nu(\varepsilon)} \qquad \varepsilon \text{ real} \tag{4}$$

where $\qquad \dfrac{1}{h_\nu(\varepsilon)} \equiv \dfrac{1}{\varepsilon}\left[\dfrac{1}{\lambda_\nu(\varepsilon)} - \dfrac{1}{\lambda_0}\right]$

$$\tag{5}$$

although we may not at first expect it, actually defines a kernel $H(x,y;\varepsilon)$ *analytic in ε* for ε small enough. In addition, and this provides the rationale for the entire approach, H is an appropriate kernel to consider in precisely the same manner as we have done with K since, for real ε, it is semisimple also.

In this construction (which Kato terms *the reduction process*), then, what we have accomplished is to shift from consideration of the kernel K and its cv's $\lambda_\nu(\varepsilon)$ to consideration of the kernel H and its cv's $h_\nu(\varepsilon)$. However, since by (5), $1/h_\nu(\varepsilon)$ is the *differential quotient* of $1/\lambda_\nu(\varepsilon)$ relative to the origin $\varepsilon = 0$, information concerning the behavior of the $h_\nu(\varepsilon)$ for small ε provides valuable information regarding the *derivative* of $\lambda_\nu(\varepsilon)$ in this same region.

In carrying out the reduction process to this point we have made two implicit assumptions, which deserve mention. Firstly, we have assumed that none of the original cv's $\lambda_\nu(\varepsilon)$ are identically equal to λ_0. If there are some with this property, then they, of course, are a priori analytic in ε and are removed from further consideration by the construction (5). We have also assumed that some *splitting* of the cv's $\lambda_\nu(\varepsilon)$ actually occurs as ε moves away from the origin; in other words, we suppose that not all of the $\lambda_\nu(\varepsilon)$ $\nu = 1, 2, \ldots, N$, are equal. If the converse is, in fact, the case, then the analyticity of the $\lambda_\nu(\varepsilon)$ already follows directly from the analyticity of Π and ΠK as given by (1) and (3).

The reduction process is a step-by-step procedure. Having constructed $H(x,y;\varepsilon)$, it now has to be analyzed as $K(x,y;\varepsilon)$ was above. An apparent complication can arise, however, if some of the $h_\nu(\varepsilon)$ are such that

$$\lim_{\varepsilon \to 0} h_\nu(\varepsilon) = \infty$$

This inconvenience, which is caused by missing terms in the expansion of the related $\lambda_\nu(\varepsilon)$ about the origin, can be avoided by redefining H with the alternative expression

$$H \equiv \frac{1}{\varepsilon}\left[\Pi K - \left(\frac{1}{\lambda_0} + \alpha\varepsilon\right)\Pi \right]$$

where the constant α is appropriately chosen. This change has no effect on our line of reasoning, and we shall assume that it is made whenever needed.

It should be clear, then, that the procedure embodied in the steps described above can be repeated for as long as necessary, at each stage giving rise to information regarding the derivative of the cv's associated with the kernel of the stage immediately preceding. By tracing back the process to the original cv's $\lambda_\nu(\varepsilon)$, we are able to detail the precise behavior of the higher derivatives of these functions. It results, naturally, that all of these derivatives are continuous at the origin $\varepsilon = 0$, and a formal power series in ε emerges for each of the λ_ν. We note, however, that although the process is continuable ad infinitum, we need not carry it to this extreme. The splitting of the cv's, which is what necessitates the various constructions in the first place, must end after a finite number (less than N) of steps. At that stage we will have not only the desired analyticity of the λ_ν but their full expansion as well. This concludes the proof of the following theorem.

THEOREM 12.4

Let the $\mathfrak{L}^2$ kernel $K(x,y;\varepsilon)$ depend analytically on ε as in (12.1-1). If K is semisimple for real ε, then those of its cv's that arise from perturbation of the cv's of the undisturbed kernel $K(x,y;0)$ are themselves analytic in ε for small enough real values of this parameter.

It is worth observing that if we only concern ourselves with *Hermitian* K, the proof of this particular theorem can be reduced to just a few lines. This is a consequence of the *reality* of the cv's of K and the fact that a function with a true algebraic singularity at the origin $\varepsilon = 0$ cannot be real for both positive and negative real values of ε [see Rellich (1953b)].

12.5 THE ANALYTICITY OF CHARACTERISTIC FUNCTIONS OF SEMISIMPLE KERNELS

In the preceding section we proved the analytic dependence of the perturbed cv's of semisimple kernels that are analytic functions of a smallness parameter ε. In the sequel we want to establish the corresponding result for the associated cf's, namely, the following theorem.

THEOREM 12.5

Under the assumptions of Theorem 12.4, if λ_0 is a cv of $K(x,y;0)$ of multiplicity N, then there exist N linearly independent cf's $\phi_\nu(x,\varepsilon)$ of $K(x,y;\varepsilon)$ and N linearly independent cf's $\psi_\nu(x,\varepsilon)$ of $K^*(x,y;\varepsilon)$, corresponding to the cv's $\lambda_\nu(\varepsilon)$ and $\overline{\lambda}_\nu(\varepsilon)$ (arising from λ_0 and $\overline{\lambda}_0$), respectively. These cf's can be chosen so that, for small enough real ε,† not only do they satisfy the general biorthonormality rela-

---

†Throughout this section, for simplicity of expression we suppress the phrase "for small enough real ε" (or its equivalents) wherever possible.

tions (12.4-2) but they also are analytic functions of the perturbation parameter.†

Proof   The argument proceeds by induction on the multiplicity N. For the case $N = 1$ we have from (12.4-1) that

$$\Pi(x,y;\varepsilon) = \phi_1(x,\varepsilon)\overline{\psi}_1(y,\varepsilon) \qquad \text{with} \qquad (\phi_1(\varepsilon), \psi_1(\varepsilon)) = 1$$

(The x dependence of the various factors within this and succeeding inner products should be clear.) Composing this expression with $\phi_{10}(y) \equiv \phi_1(y,0)$, we obtain

$$\Phi(x,\varepsilon) \equiv \int_a^b \Pi(x,y;\varepsilon)\phi_{10}(y) \, dy$$

$$= \phi_1(x,\varepsilon)(\phi_{10}, \psi_1(\varepsilon)) \tag{1}$$

In analogous fashion, using $\psi_{10}(x) \equiv \psi_1(x,0)$, we form

$$\Psi(y,\varepsilon) \equiv \int_a^b \psi_{10}(x)\overline{\Pi}(x,y;\varepsilon) \, dx$$

$$= \psi_1(y,\varepsilon)(\psi_{10}, \phi_1(\varepsilon)) \tag{2}$$

We now note that Φ and Ψ are cf's of K and K^* corresponding to λ_1 and $\overline{\lambda}_1$, respectively, since ϕ_1, ψ_1 have these properties. Moreover, they are both analytic in ε owing to the analyticity of Π. It follows, then, that the inner product

$$(\Phi(\varepsilon), \Psi(\varepsilon)) = (\phi_{10}, \psi_1(\varepsilon))(\phi_1(\varepsilon), \psi_{10})$$

is also an analytic function of ε. The leading term in its power series is unity, however, and thus, for small enough ε, its negative square root $Q(\varepsilon)$ is well defined and analytic in ε. As a consequence, the functions

$$\tilde{\Phi}(x,\varepsilon) = \Phi(x,\varepsilon)Q(\varepsilon) \qquad \text{and} \qquad \tilde{\Psi}(x,\varepsilon) = \Psi(x,\varepsilon)\overline{Q}(\varepsilon)$$

have all the behavior promised in the statement of the theorem.
     Let us now assume the theorem to be true for all multiplicities less than or equal to $N - 1$ and investigate its validity for the next

---

† As the classical example discussed in Prob. 3 shows, analyticity of the kernel cannot be replaced by infinite differentiability without affecting the validity of this result.

higher multiplicity N. At the first stage of the reduction process, the kernels H and H^* as determined from (12.4-4) have new cv's h_ν and $\bar{h}_\nu$, given by (12.4-5), but the *same* cf's ϕ_ν and ψ_ν as K and K^*, respectively. If splitting has occurred and not all of the derivatives $\lambda_\nu'(0)$ are equal, i.e., not all the $h_\nu(0)$ are the same, then the multiplicity of any given cv h_ν of H is at most $N - 1$. In this case, since the theorem is assumed true, its application to the kernel H would therefore yield the desired results. We need only consider, then, the alternate case in which no splitting occurs (to first order) in the premier reduction. In fact, repetitions of this same argument applied at subsequent stages in the reduction process show that actually we only need consider the case when *no splitting occurs at all*, i.e., when the $\lambda_\nu(\varepsilon)$, $\nu = 1, 2, \ldots, N$, are all identical.

We consider the case, therefore, wherein $\lambda_\nu(\varepsilon) = \lambda(\varepsilon)$, $\nu = 1, 2, \ldots, N$. Using the expression (12.4-1) for Π, we form, in analogy with (1) and (2), the N new functions

$$\Phi_\nu(x,\varepsilon) \equiv \int_a^b \Pi(x,y;\varepsilon)\phi_{\nu 0}(y) \, dy$$

$$= \sum_{\mu=1}^N \phi_\mu(x,\varepsilon)(\phi_{\nu 0}, \psi_\mu(\varepsilon)) \tag{3}$$

$$\Psi_\nu(y,\varepsilon) \equiv \int_a^b \psi_{\nu 0}(x) \, \overline{\Pi}(x,y;\varepsilon) \, dx$$

$$= \sum_{\mu=1}^N \psi_\mu(y,\varepsilon)(\psi_{\nu 0}, \phi_\mu(\varepsilon)) \tag{4}$$

Here we have used the shorthand notation $\phi_{\nu 0}$ and $\psi_{\nu 0}$ to designate the values of the biorthogonal functions ϕ_ν and ψ_ν, respectively, at the origin $\varepsilon = 0$. We observe that the Φ_ν and Ψ_ν, $\nu = 1, 2, \ldots, N$, have a number of the desired properties: (1) they are analytic in ε since Π enjoys this behavior; (2) they are linear combinations of cf's of K and K^* belonging to the *same* cv's $\lambda(\varepsilon)$ and $\bar{\lambda}(\varepsilon)$, respectively, and hence this characteristic feature of the ϕ_ν and ψ_ν carries over to them also; (3) taken together, they form two sets of nontrivial square-integrable functions that are linearly independent for small enough real ε.† The only important property the Φ_ν, Ψ_ν generally lack is the biorthonormality of (12.4-2), and this situation we proceed to rectify.

---

† Note that $\Phi_\nu(x,0) = \phi_{\nu 0}$, $\Psi_\nu(y,0) = \psi_{\nu 0}$.

Let us look at the $N \times N$ matrix $P(\varepsilon)$ whose entries are given by

$$P_{\mu\nu}(\varepsilon) \equiv (\Phi_\mu, \Psi_\nu)$$

$$= \sum_{k=1}^{N} (\phi_{\mu 0}, \psi_k(\varepsilon))(\phi_k(\varepsilon), \psi_{\nu 0})$$

Since the Φ_μ and Ψ_ν are analytic in ε, so also are the elements $P_{\mu\nu}$. We note, moreover, that for $\varepsilon = 0$ we have $P_{\mu\nu} = \delta_{\mu\nu}$. It follows, then, that the decomposition

$$P(\varepsilon) = I + \varepsilon \, \Delta(\varepsilon) \tag{5}$$

is valid, where I is the identity matrix in N space and $\Delta(\varepsilon)$ is bounded analytic. As a consequence of (5), for small enough ε the matrix $Q(\varepsilon) \equiv P^{-\frac{1}{2}}(\varepsilon)$ is well defined and analytic in ε (see Prob. 4). If we now construct

$$\widetilde{\Phi}_\nu(x, \varepsilon) \equiv \sum_{k=1}^{N} Q_{\nu k}(\varepsilon) \Phi_k(x, \varepsilon)$$

$$\widetilde{\Psi}_\nu(x, \varepsilon) \equiv \sum_{k=1}^{N} \overline{Q}_{k\nu}(\varepsilon) \Psi_k(x, \varepsilon)$$

where the $Q_{\mu\nu}(\varepsilon)$ are the elements of $Q(\varepsilon)$, then these new functions have *all* the behavior desired for the validity of the theorem and this concludes our proof.

12.6   TWO-PARAMETER PERTURBATIONS

The analysis of perturbation problems naturally becomes more complicated when the kernels in question depend upon more than one independent perturbation parameter.† For example [see Rellich (1950, p. 611)], the 2×2 matrix

$$A = \begin{pmatrix} 1 + 2\varepsilon_1 & \varepsilon_1 + \varepsilon_2 \\ \varepsilon_1 + \varepsilon_2 & 1 + 2\varepsilon_2 \end{pmatrix}$$

has the cv's $\lambda = 1/[1 + \varepsilon_1 + \varepsilon_2 \pm \sqrt{2}\,(\varepsilon_1{}^2 + \varepsilon_2{}^2)^{\frac{1}{2}}]$. Even though A is

---

† For an interesting study of two-parameter problems as they occur in *singular* perturbation theory see O'Malley (1965 and 1968).

Hermitian for real ε_1 and ε_2 and analytic jointly in the two variables in the neighborhood of the origin $\varepsilon_1 = \varepsilon_2 = 0$, the cv's obviously cannot be expanded in a regular two-variable power series.

There is a special case of two-parameter perturbations that is worth mentioning, however, and this occurs when the second parameter is actually not independent of the first but rather is related to it by an equation such as

$$\varepsilon_2 = \varepsilon_1{}^\alpha \quad \text{with} \quad \alpha > 0 \tag{1}$$

To be more specific, we consider an $\mathscr{L}^2$ kernel $K(x,y;\varepsilon)$, which has the decomposition

$$K(x,y;\varepsilon) \equiv L(x,y;\varepsilon) + \varepsilon^\alpha M(x,y;\varepsilon) \tag{2}$$

where $\alpha > 0$, M is continuous in ε in the neighborhood of the origin, and L depends analytically upon ε as in (12.1-1). By a procedure analogous to that employed in the proof of Theorem 12.2, we can show that

$$R_K = R_L + \varepsilon^\alpha (\tilde{M} + \lambda \rho \tilde{K}) \tag{3}$$

In this expression R_K, R_L, and $\varepsilon^\alpha \rho$ are the resolvents of K, L, and $\varepsilon^\alpha \tilde{M}$, respectively, where $\tilde{M} = M + \lambda R_L M$ and $\tilde{K} = K + \lambda R_L K$. Note particularly that $\tilde{M}$ and hence ρ (for small enough ε) are bounded in any given closed bounded subset of the resolvent set of L. The following theorem may then be easily established.

THEOREM 12.6-1

Under the assumptions associated with the decomposition (2), for small enough ε the resolvent kernel R_K differs from an analytic function of ε at most by a term of order $O(\varepsilon^\alpha)$† in each closed bounded region Δ of the complex λ plane that does *not* contain any of the cv's of the unperturbed kernel $K(x,y;0)$.

The analysis of Sec. 12.4 can also be extended to yield meaningful results in the present situation, as, for example, in the following theorem.

---

† The symbols O and o indicate, as usual, familiar order relations. If $g(\varepsilon)$ is a nonvanishing real or complex-valued function of a variable ε that ranges over a set R of real or complex space, then $f(\varepsilon) = O(g(\varepsilon))$ as $\varepsilon \to 0$ is defined to mean that $f(\varepsilon)/g(\varepsilon)$ is bounded as $\varepsilon \to 0$ in R. Similarly, $f(\varepsilon) = o(g(\varepsilon))$ if $f(\varepsilon)/g(\varepsilon) \to 0$ as $\varepsilon \to 0$ in R.

THEOREM 12.6-2

If in addition to the assumptions of the previous theorem we also assume that $L(x,y;\varepsilon)$ is semisimple for real ε, then those of the cv's of $K(x,y;\varepsilon)$ that arise from perturbation of the cv's of the undisturbed kernel $K(x,y;0)$ differ from analytic functions (for small enough real ε) at most by terms of the order $O(\varepsilon^\alpha)$. In particular, these cv's of K are continuously differentiable at least $[\alpha]$ times.†

Proof   In view of the results of the previous theorem, choosing an arbitrary (but fixed) cv λ_0 of $K(x,y;0)$, the reduction process detailed in Sec. 12.4 can be applied to the kernel K at least $[\alpha]$ successive times. At the $[\alpha]$ th stage we have a kernel H given for ε real by

$$H(x,y;\varepsilon) = \sum_{\nu=1}^{N} \frac{\phi_\nu(x,\varepsilon)\bar{\psi}_\nu(y,\varepsilon)}{\tilde{h}_\nu(\varepsilon)} + \varepsilon^{\alpha-[\alpha]} O(1) \qquad (4)$$

In this expression, ϕ_ν and ψ_ν are N analytic cf's of L and L^*, respectively, and $\varepsilon^{[\alpha]}$ times the $(1/\tilde{h}_\nu)$ are the remainders, beginning with the $[\alpha]$ th terms, of the power-series developments for the reciprocals of the related analytic cv's l_ν. Furthermore, the function of x,y and ε symbolized by the large $O(1)$ i continuous in ε.

Except for a similar factor $\varepsilon^{[\alpha]}$, the reciprocals $(1/h_\nu)$ of the cv's of H itself are the *tails* of the expansions for the reciprocals of the N cv's‡ λ_ν of the original kernel K, which arise from perturbation of λ_0 (see Prob. 5). Moreover, for small enough real ε and integer k, we have from (4), in analogy with (12.3-3), that

$$\sum_{\nu=1}^{N} h_\nu{}^k(\varepsilon) = \sum_{\nu=1}^{N} \tilde{h}_\nu{}^k(\varepsilon) + \varepsilon^{\alpha-[\alpha]} O(1) \qquad (5)$$

In other words, the power sums of essentially the *tails* differ from analytic functions by remainder terms of the form given.

From relation (5) we can draw the desired conclusion that (with reordering if necessary) the h_ν satisfy

$$h_\nu(\varepsilon) = \tilde{h}_\nu(0) + \varepsilon^{\alpha-[\alpha]} O(1) \qquad (6)$$

$\nu = 1, 2, \ldots, N$. For integer α, the result is immediate. In the more general case, we know that (6) is valid unless two or more of

---

† $[\alpha]$ = greatest integer less than or equal to α.

‡ Theorem 12.3-1 regarding the invariance of algebraic multiplicity is equally valid under the weaker assumption of continuity in ε.

the $\tilde{h}_\nu(0)$ are identical (see Prob. 6). If the latter situation prevails, we can consider those special $h_\nu(\varepsilon)$ apart from the rest, in which case we only need analyze an expression such as

$$\sum_{\nu=1}^{M} h_\nu{}^k(\varepsilon) = M + \varepsilon^{\alpha-[\alpha]} O(1) \qquad (5')$$

with integer $M \leq N$. Here without loss of generality, we have assumed $\tilde{h}_\nu(0) = 1$ for $1 \leq \nu \leq M$. Referring back to (4) and setting $\varepsilon = 0$, we are thus concerned with the perturbation of a degenerate kernel of rank M, all of whose cv's are unity. From a matrix point of view, this is completely equivalent to analyzing the effect of perturbation of the identity operator in M space, and for this transformation it is well known that the perturbation carries over uniformly to the characteristic values (see Prob. 7). It follows, then, that (6) must actually be valid for all $1 \leq \nu \leq N$, which conclusion, when traced back through the reduction process to the cv's $\lambda_\nu(\varepsilon)$ of the original kernel K, yields the desired results.   Q.E.D.

An important special case of the above theorem is the following [compare Kato (1966, p. 121)]:

COROLLARY

If the $\mathfrak{L}^2$ kernel $K(x,y;\varepsilon)$ depends analytically upon the perturbation parameter ε as in (12.1-1) and $K_0 \equiv K(x,y;0)$ is semisimple (normal, Hermitian) for real ε, then those of the cv's of K that arise from perturbation of the cv's of K_0 are continuously differentiable functions for small enough real ε.

Before leaving this section, it is worth mentioning that the "regular" behavior exhibited by the cv's of kernels having the form (2) with L semisimple unfortunately does not carry over to the cf's in general. For example, the simple 2×2 matrix $\begin{pmatrix} 1 & \varepsilon \\ 0 & 1 \end{pmatrix}$, with cv's $\lambda_{1,2} = 1$, is such that either it or its adjoint has at least one (generalized) cf with a singularity, as $\varepsilon \to 0$, of at least the order $O(\varepsilon^{-\frac{1}{2}})$.

12.7 APPLICATIONS

As an appropriate conclusion to our discussion of analytic perturbation theory, we want to consider in somewhat more detail the situation, which actually occurs quite often in practice, wherein the square-integrable kernel $K(x,y;\varepsilon)$ satisfying (12.1-1) is Hermitian for real ε. To be specific, we assume K is given by

$$K(x,y;\varepsilon) = K_0(x,y) + \varepsilon K_1(x,y) + \varepsilon^2 K_2(x,y) + \cdots$$

for $|\varepsilon| < \varepsilon_0$ where the K_ν are $\mathfrak{L}^2$ and Hermitian in a common fundamental domain. We begin by singling out a cv λ_0 of multiplicity N belonging to the unperturbed kernel K_0. This cv gives rise to corresponding cv's $\lambda_\nu(\varepsilon)$ and associated cf's $\phi_\nu(x,\varepsilon)$ of K, which satisfy†

$$\lambda_\nu(\varepsilon) = \lambda_0 + \sum_{\mu=1}^{\infty} \lambda_{\nu\mu}\varepsilon^\mu$$

$$\phi_\nu(x,\varepsilon) = \sum_{\mu=0}^{\infty} \phi_{\nu\mu}(x)\varepsilon^\mu \tag{1}$$

Substitution of the series (1) and the expansion for K into the equation $\phi_\nu = \lambda_\nu K \phi_\nu$ should allow the recursive calculation of the various coefficients.

Performing the operation suggested above and equating like powers of ε on both sides of (1), we find

$$\phi_{\nu 0} = \lambda_0 K_0 \phi_{\nu 0} \tag{2}$$

$$\phi_{\nu 1} = \lambda_{\nu 1} K_0 \phi_{\nu 0} + \lambda_0 K_1 \phi_{\nu 0} + \lambda_0 K_0 \phi_{\nu 1} \tag{3}$$

$$\phi_{\nu 2} = \lambda_{\nu 2} K_0 \phi_{\nu 0} + \lambda_{\nu 1} K_1 \phi_{\nu 0} + \lambda_{\nu 1} K_0 \phi_{\nu 1}$$
$$+ \lambda_0 K_2 \phi_{\nu 0} + \lambda_0 K_1 \phi_{\nu 1} + \lambda_0 K_0 \phi_{\nu 2} \tag{4}$$

. . . .

Each of these expressions, of course, actually denotes N separate equations corresponding to the N possible values of ν. Moreover, supplementing (2) to (4) are the additional relations obtained from the assumed

---

† Here and throughout this chapter we have considered various perturbation expansions without giving explicit conditions for their convergence. Where applicable, simple estimates of convergence radii follow from the existence of nonnegative constants a and c such that

$$\|K_\nu\| \le ac^{\nu-1} \qquad \nu = 1, 2, \ldots$$

which result itself is a consequence of the assumed convergence of (12.1-1). Since these and related error estimates are amply discussed in various of the references cited, we leave any further consideration of these matters to the interested reader.

orthonormality of the ϕ_ν, namely,

$$(\phi_{\nu0}, \phi_{\mu0}) = \delta_{\nu\mu}$$

$$(\phi_{\nu1}, \phi_{\mu0}) + (\phi_{\nu0}, \phi_{\mu1}) = 0 \qquad 1 \le \mu, \nu \le N \qquad (5)$$

$$(\phi_{\nu2}, \phi_{\mu0}) + (\phi_{\nu1}, \phi_{\mu1}) + (\phi_{\nu0}, \phi_{\mu2}) = 0$$

$$\cdots\cdots\cdots\cdots\cdots\cdots\cdots\cdots\cdots$$

In the solution of a typical problem, the first step is to determine N orthonormal cf's of the equation $\phi = \lambda_0 K_0 \phi$. If we call these functions ψ_i, the $\phi_{\nu0}$ satisfying (2) and (5) must be linear combinations, say, of the form

$$\phi_{\nu0} = \sum_i \alpha_{\nu i} \psi_i \qquad \text{with} \qquad \sum_i |\alpha_{\nu i}|^2 = 1$$

Equation (3) then becomes

$$\phi_{\nu1} = \frac{\lambda_{\nu1}}{\lambda_0} \sum_{i=1}^{N} \alpha_{\nu i} \psi_i + \lambda_0 \sum_{i=1}^{N} \alpha_{\nu i} K_1 \psi_i + \lambda_0 K_0 \phi_{\nu1} \qquad (6)$$

which, taking the inner product with ψ_j on both sides and rearranging, yields

$$\sum_{i=1}^{N} \alpha_{\nu i} \left[(K_1 \psi_i, \psi_j) + \frac{\lambda_{\nu1}}{\lambda_0^2} \delta_{ij} \right] = 0 \qquad j = 1, 2, \ldots, N \qquad (7)$$

The vanishing of the determinant of the bracketed coefficient matrix in (7) determines the N possible values of $\lambda_{\nu1}$. These in turn lead to the $\alpha_{\nu i}$ and hence to the $\phi_{\nu0}$. Equation (3) [or (6)] is then an integral equation with known inhomogeneity from which the first-order perturbations $\phi_{\nu1}$ can be computed, and so forth. Higher-order terms follow in analogous fashion.†

As an example, let us consider the nonhomogeneous vibrating string of unit length mentioned earlier in Sec. 1.1. If the linear mass density has the form $1 + \varepsilon\rho(x)$, then the characteristic-value problem is

---

† Rellich (1940, pp. 358–359) suggests using the fact that the $\phi_{\nu\mu}$ can in general be chosen such that $(\phi_{\nu0}, \phi_{\nu\mu})$ is *real* for all μ. This leads to some simplification in the resulting expressions for $\lambda_{\nu\mu}$, etc. Also see Bückner (1952, pp. 87ff).

$\phi = \lambda K \phi$ with

$$K(x,y;\varepsilon) = [1 + \varepsilon\rho(y)] \begin{cases} x(1-y) & 0 \le x \le y \le 1 \\ y(1-x) & 0 \le y \le x \le 1 \end{cases}$$

The zeroth-order solutions are $\sqrt{2}\sin n\pi x$ corresponding to $\lambda_n = n^2\pi^2$. It easily follows, then, that

$$\frac{1}{\lambda_n(\varepsilon)} = \frac{1}{n^2\pi^2}\left[1 + 2\varepsilon\int_0^1 \rho(y)\sin^2 n\pi y\, dy + O(\varepsilon^2)\right] \tag{8}$$

and we observe the well-known phenomenon that an *increase* in density leads to a *decrease* in the eigenfrequencies. If ρ is constant all along the string, the $1/\lambda_n(\varepsilon)$ turn out to be linear in ε since the $O(\varepsilon^2)$ remainder term vanishes identically for this case (see Prob. 8).

As in the above example, the general analysis is simplified considerably whenever the multiplicity N of the cv under investigation is unity. In particular, we have from (7) that

$$\lambda_1 = -\lambda_0^2(K_1\phi_0, \phi_0)$$

or, alternatively, $\quad \dfrac{d}{d\varepsilon}\left(\dfrac{1}{\lambda(\varepsilon)}\right)\bigg|_{\varepsilon=0} = \left(\dfrac{\partial K}{\partial\varepsilon}\phi,\phi\right)\bigg|_{\varepsilon=0}$ $\tag{9}$

where ϕ is the normalized cf of K corresponding to the cv λ. Actually, the relation (9) is valid for arbitrary ε, as can easily be derived from the expression $\phi = \lambda K\phi$ with $(\phi,\phi) = 1$. Although the higher-order expressions are generally more cumbersome, their derivation is in all cases straightforward.

On the other hand, it should also be clear—if not from the above exercise, certainly from the several other examples discussed earlier in this chapter—that linearity in the perturbation parameter does not significantly simplify matters. Indeed, as we have observed, the most general behavior can be shown to occur already in the linear case.

PROBLEMS

1. [Rellich (1942, p. 462; 1950, p. 608) and Kato (1966, p. 371).] Consider the kernel

$$K(x,y;\varepsilon) = \min(x,y) - (1+\varepsilon)xy \qquad 0 \le x,y \le 1$$

a. Determine that the characteristic functions of K are solutions of the boundary-value problem

$$\phi'' + \lambda\phi = 0 \qquad 0 \le x \le 1$$
$$\phi(0) = 0$$
$$\varepsilon\phi'(1) - (1+\varepsilon)\phi(1) = 0$$

b. Verify that, for $\varepsilon > 0$, there is a single negative characteristic value and associated cf given by

$$\lambda = -\mu^2$$
$$\phi = \text{const } \sinh(\mu x)$$

where μ is the unique positive root of

$$\tanh \mu = \frac{\varepsilon\mu}{1+\varepsilon}$$

2. Let the $\mathfrak{L}^2$ kernel K have the form

$$K(x,y;\varepsilon) = K_1(x,y) + \varepsilon K_2(x,y)$$

where K_1 and K_2 are individually square integrable and mutually orthogonal. Assume, furthermore, that all of the higher-order iterates of K_1 vanish.
a. Satisfy yourself that any characteristic values which K possesses come entirely from K_2.
b. Give an example of such a kernel K for which the number of cv's is countably infinite.

3. Consider the 2×2 Hermitian matrix given by

$$A(\varepsilon) \equiv e^{-1/\varepsilon^2}\begin{pmatrix} \cos\dfrac{2}{\varepsilon} & \sin\dfrac{2}{\varepsilon} \\ \sin\dfrac{2}{\varepsilon} & -\cos\dfrac{2}{\varepsilon} \end{pmatrix}$$
$$A(0) = 0$$

a. Determine the cv's and cf's of A.
b. Satisfy yourself that, although the cf's of A are infinitely differentiable for all $\varepsilon \ne 0$, they cannot be continued to the origin as even continuous functions. [See Kato (1966, p. 111).]

4. a. If A is an $n \times n$ matrix with complex-valued entries $a_{\mu\nu}$ and we define the norm of A by

$$\|A\| = \max_{\mu,\nu} |a_{\mu\nu}|$$

verify that the norm of the kth iterate of A satisfies the following inequality:

$$\|A^k\| \le n^{k-1}\|A\|^k$$

b. Let I be the identity matrix in n space. Satisfy yourself that

$$[I+A]^{-\frac{1}{2}} \equiv \sum_{k=0}^{\infty} \binom{-\frac{1}{2}}{k} A^k \qquad (*)$$

is a meaningful definition for the negative square root of $I+A$ if the series on the right converges.

c. Show that

$$\sum_{k=0}^{\infty} (-1)^k \binom{-\frac{1}{2}}{k} \|A^k\|$$

is majorized by the series for

$$(1 - n\|A\|)^{-\frac{1}{2}}$$

and hence $[I+A]^{-\frac{1}{2}}$ is well defined by (*), at least whenever $n\|A\| < 1$.

d. Deduce from the result of part c that $Q(\varepsilon) \equiv P^{-\frac{1}{2}}(\varepsilon)$, where $P(\varepsilon)$ is given by (12.5-5), is well defined and analytic in ε for all small enough values of this perturbation parameter.

5. Let K be an $\mathfrak{L}^2$ kernel and R its resolvent. Assume that ϕ_ν is a (not generalized) cf of K associated with the cv λ_ν.

a. Show that

$$(\lambda_\nu - \lambda) R \phi_\nu = \phi_\nu$$

[*Hint:* See Sec. 6.4 or use the Fredholm identity (3.2-5).]

b. If C is a simple closed contour in the λ plane, define Π as in Sec. 12.4 by

$$\Pi \equiv -\frac{1}{2\pi i} \oint_C R \, d\lambda$$

Then verify that

$$\Pi \phi_\nu = \phi_\nu$$

whenever C encloses λ_ν. [The role of Π as a *projection* is thoroughly discussed by Kato (1966, pp. 67ff).]

c. Define H as in Sec. 12.4 by

$$H \equiv \frac{1}{\varepsilon}\left[\Pi K - \frac{1}{\lambda_0}\, \Pi \right]$$

Demonstrate that, under the assumptions of part b,

$$\frac{\lambda_0 \lambda_\nu \varepsilon}{\lambda_0 - \lambda_\nu}\, H\phi_\nu = \phi_\nu$$

6. Assume that the continuous functions $h_\nu(\varepsilon)$, $\nu = 1, 2, \ldots, N$, satisfy

$$\sum_{\nu=1}^{N}\left[h_\nu{}^k(\varepsilon) - h_\nu{}^k(0) \right] = O(\varepsilon^\beta) \qquad 0 < \beta < 1$$

for integer $k = 1, 2, \ldots, N$. If the $h_\nu(0)$ are all distinct, show that the $h_\nu(\varepsilon)$ themselves must be given by

$$h_\nu(\varepsilon) = h_\nu(0) + O(\varepsilon^\beta)$$

7. Let I be the $M \times M$ identity matrix and assume that the matrix A is given by

$$A = I + O(\varepsilon^\beta) \qquad 0 < \beta < 1$$

Verify that each of the characteristic values of A also satisfies

$$\lambda = 1 + O(\varepsilon^\beta)$$

8. For the example of Sec. 12.7:
a. Verify the expression (12.7-8).
b. Show that, in the notation of that section,

$$(\phi_1, \phi_0) = 0$$
$$\lambda_2 = \lambda_0(\rho\phi_0, \phi_0)^2 - \lambda_0(\rho\phi_1, \phi_0)$$

c. Demonstrate that $1/\lambda(\varepsilon)$ is linear in ε whenever $\rho(x) = $ constant.
9. Assume that the $\mathfrak{L}^2$ kernel $K(x,y;\varepsilon) = K_0(x,y) + \varepsilon K_1(x,y)$ is Hermitian for real ε. For arbitrary real ε_0 let $K' = K_0 + \varepsilon_0 K_1$, $K'' = K_0$. Associate with K, K', and K'' the cv's λ_ν, λ'_ν, and λ''_ν, respectively.
a. Deduce from the Weyl inequalities (9.5-1) that

$$\left| \frac{1}{\lambda_{\nu+\mu-1}} \right| \leq \left| \frac{\varepsilon/\varepsilon_0}{\lambda'_\nu} \right| + \left| \frac{1 - \varepsilon/\varepsilon_0}{\lambda''_\mu} \right| \qquad \mu, \nu \geq 1$$

b. Show that, whenever there exists a finite least negative (positive) cv, the result in part a implies that its reciprocal $1/\lambda_1^-$ $(1/\lambda_1^+)$ is a *convex* (*concave*) function of ε.

c. If K is a nonnegative definite degenerate kernel of rank N, devise a proof that demonstrates that $1/\lambda_N$ is also a *convex* function of ε.

[An elegant proof of essentially b and c above is to be found in Rellich (1953b, p. 55).]

chapter thirteen

Series Expansions and Their Convergence

13.1 GENERALIZED FOURIER SERIES AND BESSEL'S INEQUALITY

For a number of chapters we have been concerned with methods for the calculation of the characteristic values and characteristic functions of various kernels, with special emphasis on those of Hermitian type. It is reasonable to inquire now whether, having determined, say, the full characteristic system of a given Hermitian $\mathfrak{L}^2$ kernel K, we can use these cv's and cf's in some additional constructive manner, for instance, as the basis elements for generalized Fourier series developments of other functions of one (or more) independent variables. In this and sub-sequent chapters we shall do just that. To be specific, if the cv's and cf's of K are denoted by λ_ν and ϕ_ν, respectively, we shall study the nature of expansions of the form

$$f(x) \approx \sum_{\nu=1}^{\infty} \alpha_\nu \phi_\nu(x) \tag{1}$$

and

$$f(x,y) \approx \sum_{\mu,\nu=1}^{\infty} \beta_{\mu\nu} \phi_\mu(x) \overline{\phi}_\nu(y) \tag{2}$$

In view of our earlier results on the characteristic decomposition of Hermitian kernels themselves (see Secs. 6.2 and 6.4), the special cases of (2) wherein $\beta_{\mu\nu} = \delta_{\mu\nu}/\lambda_\nu{}^p$, $p = 1, 2, \ldots$, will be of some interest.

As we pursue our investigations, the following result [see e.g., Courant and Hilbert (1953, pp. 51ff) and Smithies (1962, pp. 54ff)] will be of fundamental importance.†

THEOREM 13.1-1

Given an orthonormalized set of square-integrable functions $\{\phi_\nu\}$, the *best* mean-square approximation of the form (1) to an arbitrary $\mathfrak{L}^2$ function f occurs when the coefficients α_ν are given by

$$\alpha_\nu = (f, \phi_\nu)$$

We have then the so-called *Bessel's inequality*‡

$$\sum_{\nu=1}^{\infty} |(f, \phi_\nu)|^2 \leq \|f\|^2 \tag{3}$$

Proof   Expansion and completion of the square shows that

$$\left\| f - \sum_{\nu=1}^{n} \alpha_\nu \phi_\nu \right\|^2 = \|f\|^2 - \sum_{\nu=1}^{n} |(f, \phi_\nu)|^2 + \sum_{\nu=1}^{n} |(f, \phi_\nu) - \alpha_\nu|^2$$

The right-hand side of this expression clearly attains its minimum value when (and only when) $\alpha_\nu = (f, \phi_\nu)$ for each ν. The arbitrariness of n and the nonnegative character of the minimum value itself immediately permit the desired conclusions.   Q.E.D.

In analogy with the classical situation, the numbers $\alpha_\nu = (f, \phi_\nu)$ may be termed the *Fourier coefficients* and a series of the form (1) with these coefficients called a *generalized Fourier series*. The above theorem also has obvious analogues for functions of two (or more) variables, and thus similar remarks apply to series of the form (2) if $\beta_{\mu\nu} = (f\phi_\nu, \phi_\mu)$. Although it is not necessary, for convenience we shall assume here that the ϕ_ν have arisen as the cf's of a given Hermitian $\mathfrak{L}^2$ kernel K.§ If K is a

---

† Recall that we already had need for this result in our study of error bounds occurring with iterative approximations.

‡ Bessel (1828) first discussed these matters with reference to ordinary Fourier series.

§ We shall see later in this chapter how, given a priori the ϕ_ν and appropriate λ_ν, a Hermitian kernel K can be constructed that has the λ_ν and ϕ_ν as cv's and cf's, respectively.

periodic difference kernel, the ϕ_ν are the familiar trigonometric functions (see Prob. 1).

In the remainder of this chapter we shall often need to use the fact that $\mathfrak{L}^2$ is a *complete space*, i.e., one in which Cauchy sequences are *convergent*. Proofs of this result, which is commonly referred to as the Riesz-Fischer theorem, may be found in a number of the references already cited [see, for example, Riesz and Sz.-Nagy (1955, pp. 57ff) and Smithies (1962, pp. 56ff)]. It is worth noting, however, that Riesz's original form of the result was essentially the following [see F. Riesz (1907a, b, and c); compare Fischer (1907)]:

THEOREM 13.1-2

Given an orthonormalized set of square-integrable functions $\{\phi_\nu\}$ and an arbitrary sequence of complex numbers $\{\alpha_\nu\}$, the series

$$\sum_{\nu=1}^{\infty} \alpha_\nu \phi_\nu$$

is mean convergent to $f \in \mathfrak{L}^2$ iff

$$\sum_{\nu=1}^{\infty} |\alpha_\nu|^2 < \infty$$

If the series is convergent, then f is such that

$$(f, \phi_\nu) = \alpha_\nu \qquad \nu = 1, 2, \ldots$$

In view of this result, therefore, we know, using Bessel's inequality, that generalized Fourier series of the form

$$\sum_{\nu=1}^{\infty} (f, \phi_\nu) \phi_\nu$$

are mean convergent whatever the function f in $\mathfrak{L}^2$. That the limit function is equal (almost everywhere) to f itself if and only if the set $\{\phi_\nu\}$ is complete will be shown in the next section.

13.2   HILBERT'S THEOREM AND A COMPLETENESS CRITERION

The principal result (of the type in which we are interested) concerning series representations of functions of a single independent variable was first essentially obtained by Hilbert (1904) and a year later also appeared in Schmidt's Göttingen dissertation [see Schmidt (1907)]. Although the original proof concerned continuous functions, the argument retains its validity even in more general settings.

THEOREM (HILBERT, $\mathfrak{L}^2$)

Let K be a given Hermitian† $\mathfrak{L}^2$ kernel with cv's λ_ν and cf's ϕ_ν, respectively.

a. Every function $f \in \mathfrak{L}^2$ admits the development, convergent in the mean,

$$f = h + \sum_{\nu=1}^{\infty} (f, \phi_\nu) \phi_\nu \tag{1}$$

where h is a function (depending on f) satisfying $Kh = 0$.

b. Thus, if g can be represented as

$$g = Kf \tag{2}$$

then it has a mean-convergent generalized Fourier series

$$\sum_{\nu=1}^{\infty} \frac{(f, \phi_\nu)}{\lambda_\nu} \phi_\nu \tag{3}$$

Proof   We treat first the second part of the theorem. For arbitrary f in $\mathfrak{L}^2$, let

$$f_n = f - \sum_{\nu=1}^{n} (f, \phi_\nu) \phi_\nu$$

Then

$$\|f_n\|^2 = \|f\|^2 - \sum_{\nu=1}^{n} |(f, \phi_\nu)|^2 \leq \|f\|^2$$

Moreover, since f_n belongs to the subspace $c_n(\phi_n)$ of functions orthogonal to the first n of the ϕ_ν, we also have

$$\| Kf_n \| \leq \frac{\|f_n\|}{|\lambda_{n+1}|} \leq \frac{\|f\|}{|\lambda_{n+1}|}$$

---

† If K is not Hermitian, the situation is considerably more complicated, of course. Certain results which retain some of the flavor of Hilbert's theorem do exist, however, for kernels that are either only *slightly* non-Hermitian or are *almost* degenerate. For a quantitative description of these matters, see Lidskii (1962). Of interest here also are Livsic (1954), Brodskii and Livsic (1958), Lidskii (1959a and b), and Gohberg and Krein (1969).

But $|\lambda_{n+1}| \to \infty$ as $n \to \infty$, and thus $\|Kf_n\| \to 0$, which implies that the series

$$\sum_{\nu=1}^{\infty} (f, \phi_\nu) K\phi_\nu = \sum_{\nu=1}^{\infty} \frac{(f, \phi_\nu)}{\lambda_\nu} \phi_\nu$$

$$= \sum_{\nu=1}^{\infty} (f, K\phi_\nu) \phi_\nu \qquad (4)$$

$$= \sum_{\nu=1}^{\infty} (Kf, \phi_\nu) \phi_\nu$$

converges in the mean to Kf.

Returning now to the first portion of the theorem, as we previously noted, for the complete space $\mathcal{L}^2$ the generalized Fourier series

$$\sum_{\nu=1}^{\infty} (f, \phi_\nu) \phi_\nu$$

is mean convergent (for arbitrary $f \epsilon \mathcal{L}^2$) to some function $\tilde{f} \epsilon \mathcal{L}^2$. It follows, then, from the portion of the theorem that has already been established that

$$K\tilde{f} = \sum_{\nu=1}^{\infty} (f, \phi_\nu) K\phi_\nu = Kf$$

Setting $h \equiv f - \tilde{f}$, we have $Kh = 0$ and

$$f = h + \sum_{\nu=1}^{\infty} (f, \phi_\nu) \phi_\nu \qquad \text{Q.E.D.}$$

A few remarks are in order. The full Hilbert theorem is equally valid for completely continuous symmetric transformations in arbitrary separable Hilbert spaces. Moreover, as the above argument shows, for the second part of the theorem, even *incomplete* or *pre*-Hilbert spaces will suffice. If the kernel K enjoys additional structure, the expected mean convergence is improved upon. In fact, already for the theorem in its given form, the convergence of the series (3) to the function Kf is *relatively uniform and absolute* [see Smithies (1962, p. 117)]. Verification of this fact necessitates only the recognition that, by Schwarz's and Bessel's inequalities as applied to the kernel K considered as a function of its second variable (see Property 6 of Sec. 7.1), we have

$$\left[\sum_{\nu=n}^{m} |(f, \phi_\nu)| \frac{|\phi_\nu|}{|\lambda_\nu|} \right]^2 \le \sum_{\nu=n}^{m} |(f, \phi_\nu)|^2 \sum_{\nu=n}^{m} \left| \frac{\phi_\nu}{\lambda_\nu} \right|^2$$

$$\le \left[\sum_{\nu=n}^{m} |(f, \phi_\nu)|^2 \right] \|K\|_y^2$$

If $\|K\|_y$ is uniformly bounded, as is the case when K is a *continuous* kernel, then the convergence in the second part of Hilbert's theorem is actually *uniform and absolute*.†

Hilbert's theorem has an immediate and obvious corollary, namely, the following:

COMPLETENESS THEOREM

The set $\{\phi_\nu\}$ is *complete* in $\mathcal{L}^2$ iff K is *closed*, that is, iff $Kh = 0$ implies $\|h\| = 0$.

(See Sec. 7.1 and the Auxiliary Result of Sec. 8.5.) For such sets of $\{\phi_\nu\}$ we have

$$\left\| f - \sum_{\nu=1}^{n} (f, \phi_\nu) \phi_\nu \right\| \to 0$$

as $n \to \infty$ for all $f \in \mathcal{L}^2$, in which case Bessel's inequality becomes *Parseval's equality*‡

$$\sum_{\nu=1}^{\infty} |(f, \phi_\nu)|^2 = \|f\|^2 \tag{5}$$

If the set $\{\phi_\nu\}$ is not a priori complete, i.e., if the kernel K is not closed, additional linearly independent functions may be included in order to render the augmented set complete. These additional functions, of

---

† Even for continuous K, however, the convergence in the first part of Hilbert's theorem cannot generally be uniform and absolute. Why?

‡ The apparently more general [although actually equivalent (see Prob. 2)] result

$$\sum_{\nu=1}^{\infty} (f, \phi_\nu) (\phi_\nu, g) = (f, g)$$

valid for arbitrary $f, g \in \mathcal{L}^2$ also goes by the same name. It was this formula that Parseval (1805) actually established for trigonometric series under very restrictive assumptions [see Hobson (1957, vol. II, pp. 575–576) for further historical discussion].

course, are "cf's" of K associated with the "cv" $\lambda = \infty$, and, taken together, they span the complement of the range of K in $\mathfrak{L}^2$.

Given an orthonormalized set of $\{\phi_\nu\}$ that is believed to be complete, if an antecedent kernel K is known, the above corollary of Hilbert's theorem provides a convenient means of verifying this property. For example, the set $\{\sin \nu\pi x\}$ for $0 \le x \le 1$ has the familiar

$$K(x,y) = \min\,(x,y) - xy \equiv \begin{cases} x(1-y) & x \le y \\ y(1-x) & y \le x \end{cases} \tag{6}$$

as an antecedent kernel, while the kernel

$$K(x,y) = \tfrac{1}{3} + \tfrac{1}{2}(x^2 + y^2) - \max\,(x,y) \tag{7}$$

gives rise to the set $\{\cos \nu\pi x\}$, $\nu = 1, 2, \ldots$, on the same interval. It is easily established that the first of the above kernels is closed in $\mathfrak{L}^2$ and that the only functions orthogonal to the second must be constant (almost everywhere) in the fundamental interval. From the combination of these two results, the classical completeness properties of the trigonometric functions on the interval $[0,1]$ (and on $[-1,1]$ also) follow immediately. Hochstadt (1967) has used this same line of reasoning to show the completeness of various sets of Bessel functions, thereby providing an alternate proof of the mean convergence of Fourier-Bessel and Fourier-Dini expansions [also see Hochstadt (1970) and Kneser (1907)]. Various other examples are suggested in the problems at the end of this chapter.

13.3   SERIES DEVELOPMENTS FOR HERMITIAN KERNELS AND THEIR RESOLVENTS

For a square-integrable function K of two variables, the analogue of Theorem 13.1-1 affirms that the best mean-square approximation of the form (13.1-2) in terms of the orthonormalized functions $\{\phi_\nu\}$ is given by

$$\sum_{\mu,\nu=1}^{\infty} (K\phi_\nu, \phi_\mu)\phi_\mu(x)\,\overline{\phi}_\nu(y) \tag{1}$$

If, in addition, K is a Hermitian kernel that has as its cf's precisely the set $\{\phi_\nu\}$, then (1) becomes (as expected)

$$\sum_{\nu=1}^{\infty} \frac{\phi_\nu(x)\,\overline{\phi}_\nu(y)}{\lambda_\nu} \tag{2}$$

where the λ_ν are the real cv's of K associated with these cf's ϕ_ν. In this case the deviation from K (in norm) of truncated versions of the expansion (2) is given by

$$\Delta_n \equiv \left\| K(x,y) - \sum_{\nu=1}^{n} \frac{\phi_\nu(x)\overline{\phi}_\nu(y)}{\lambda_\nu} \right\|^2$$

$$= \|K\|^2 - \sum_{\nu=1}^{n} \left(\frac{1}{\lambda_\nu}\right)^2$$

It follows, however, from (7.1-5) that $\Delta_n \to 0$ as $n \to \infty$, and thus we have the following theorem.

THEOREM 13.3-1

Given a Hermitian $\mathfrak{L}^2$ kernel K with cv's λ_ν and cf's ϕ_ν, it can be expanded into the *mean-convergent* generalized Fourier series†

$$\sum_{\nu=1}^{\infty} \frac{\phi_\nu(x)\overline{\phi}_\nu(y)}{\lambda_\nu} \tag{3}$$

A converse to this result is also true, namely:

THEOREM 13.3-2

Given an orthonormal set of $\mathfrak{L}^2$ functions $\{\phi_\nu\}$ and a sequence of real numbers λ_ν satisfying $|\lambda_\nu| \leq |\lambda_{\nu+1}|$ $(\nu \geq 1)$ and

$$\sum_{\nu=1}^{\infty} \left(\frac{1}{\lambda_\nu}\right)^2 < \infty$$

then the series

$$\sum_{\nu=1}^{\infty} \frac{\phi_\nu(x)\overline{\phi}_\nu(y)}{\lambda_\nu}$$

is mean convergent to a Hermitian $\mathfrak{L}^2$ kernel K with the λ_ν and ϕ_ν as cv's and cf's, respectively.

---

† It is important to note that this mean convergence holds whether or not the set $\{\phi_\nu\}$ of cf's is complete.

The proof of this theorem is accomplished in three quick steps: (1) direct application of Riesz's form of the Riesz-Fischer theorem (Theorem 13.1-2) for functions of two variables; (2) modification of the resulting $\mathfrak{L}^2$ limit function K on a set of measure zero, if necessary, to ensure its Hermitian character; (3) trivial verification of the characteristic properties of ϕ_ν, λ_ν vis-à-vis this Hermitian kernel K.

Earlier we have observed that for square-integrable functions f of one variable,

$$Kf = \sum_{\nu=1}^{\infty} \frac{(f, \phi_\nu)}{\lambda_\nu} \phi_\nu \qquad (13.2\text{-}3')$$

with relatively uniform and absolute convergence of the series, from which the *Hilbert formula*

$$(Kf, g) = \sum_{\nu=1}^{\infty} \frac{(f, \phi_\nu)(\phi_\nu, g)}{\lambda_\nu} \qquad (4)$$

easily follows. It is natural now to inquire concerning analogues of these results for functions f, say, of two independent variables and, in particular, for the special case wherein f is an iterate of K. In this last situation, merely substituting above first K^{p-1} and then $K^{p-1} f$ with f in $\mathfrak{L}^2$, we clearly have the following:

THEOREM 13.3-3

For the given Hermitian $\mathfrak{L}^2$ kernel K with cv's λ_ν and cf's ϕ_ν, the expansion

$$K^p(x, y) = \sum_{\nu=1}^{\infty} \frac{\phi_\nu(x)\overline{\phi}_\nu(y)}{\lambda_\nu^p} \qquad p \geq 2 \qquad (5)$$

is relatively uniformly and absolutely convergent in x for fixed y and y for fixed x.† [Actually for $p \geq 3$, the series is relatively uniformly and absolutely convergent in (x, y) (see Prob. 7).] More-

---

† For completeness we should show that a *full characteristic system* of the iterate K^p is given by the cf's $\{\phi_\nu\}$ of K, with the associated cv's $\{\lambda_\nu^p\}$. This result, however, is an immediate consequence of our earlier work. See Smithies (1962, pp. 121–122) for a version of the classical proof based upon the pth roots of unity. Also see Prob. 13 of Chap. 3 and Prob. 5 of Chap. 7.

over, if f is an $\mathfrak{L}^2$ function, then

$$K^p f = \sum_{\nu=1}^{\infty} (K^p f, \phi_\nu) \phi_\nu = \sum_{\nu=1}^{\infty} \frac{(f, \phi_\nu)}{\lambda_\nu^{\,p}} \, \phi_\nu \tag{6}$$

these series also being relatively uniformly and absolutely convergent.

With the above results in hand, we are in a position to discuss the nature of the generalized Fourier series expansion for the resolvent R_K of the given kernel K. We find (as expected) the following result.

THEOREM 13.3-4

Given a Hermitian $\mathfrak{L}^2$ kernel K with cv's λ_ν and cf's ϕ_ν, the resolvent R_K has the mean-convergent expansion

$$\sum_{\nu=1}^{\infty} \frac{\phi_\nu(x) \overline{\phi}_\nu(y)}{\lambda_\nu - \lambda} \tag{7}$$

valid for all $\lambda \neq \lambda_\nu$. Alternatively, R_K may be expressed, for such λ, as

$$R_K = K + \lambda K^2 + \lambda^2 \sum_{\nu=1}^{\infty} \frac{\phi_\nu(x) \overline{\phi}_\nu(y)}{\lambda_\nu^2 (\lambda_\nu - \lambda)} \tag{8}$$

this last series being relatively uniformly and absolutely convergent.

Proof   Since $(\lambda_\nu - \lambda) R_K \phi_\nu = \phi_\nu$, from which it follows that for all $\lambda \neq \lambda_\nu$

$$(R_K \phi_\nu, \phi_\mu) = \frac{\delta_{\mu\nu}}{\lambda_\nu - \lambda}$$

we know that (7) is the best mean-square approximation of the form (13.1-2) in terms of the ϕ_ν. Exactly as for the expansion (3) for K, however, it is a consequence of (7.1-6) that the series (7) is actually mean convergent to R_K. The expression (8) is then obtained using relations (3), (5), and (7); the relatively uniform and absolute convergence of the series therein may be verified straightforwardly once (8) has been rewritten as

$$\lambda^2 K R_K K = \lambda^2 K \sum_{\nu=1}^{\infty} \frac{\phi_\nu(x)\overline{\phi}_\nu(y)}{\lambda_\nu - \lambda} K \tag{8'}$$

(see Prob. 7).

The expression (7) allows us to deduce the analogous expansions for the solution of the general nonhomogeneous equation

$$\phi = f + \lambda_0 K \phi \tag{9}$$

In particular, if f is an arbitrary $\mathfrak{L}^2$ function and λ_0 is a regular value of K, then (9) has the unique solution

$$\phi = f + \lambda_0 \sum_{\nu=1}^{\infty} \frac{(f, \phi_\nu)}{\lambda_\nu - \lambda_0} \phi_\nu \tag{10}$$

On the other hand, if λ_0 is a cv of K and has associated with it the cf's $\Phi_1, \ldots, \Phi_p$, we know that $\mathfrak{L}^2$ solutions of (9) exist iff

$$(f, \Phi_\nu) = 0 \qquad 1 \le \nu \le p$$

When this condition is satisfied, the (nonunique) solutions have the form

$$\phi = f + \sum_{\nu=1}^{p} \alpha_\nu \Phi_\nu + \lambda_0 \sum_{\lambda_\nu \neq \lambda_0} \frac{(f, \phi_\nu)}{\lambda_\nu - \lambda_0} \phi_\nu \tag{11}$$

where the α_ν are arbitrary constants (compare 11.3-3). We leave it to the interested reader to verify that the series appearing in (10) and (11) have the usual relatively uniform and absolute convergence.

13.4   CONTINUOUS KERNELS AND MERCER'S THEOREM

In the same manner that the expected relatively uniform convergence in the second part of Hilbert's theorem is improved upon whenever the given Hermitian kernel K is significantly smoother than $\mathfrak{L}^2$, so also is the convergence of the various series developments of Sec. 13.3. For instance, in Theorem 13.3-3, if K is continuous, then the convergence is uniform and absolute for either fixed x or fixed y.

In order to explore such improvements further, we assume for the remainder of this section that K is a *continuous* function of its independent variables x and y. The following result due to Dini (1892) [see also Courant and Hilbert (1953, p. 57), for example] will then be valuable in the sequel.

DINI'S THEOREM

If a series of *positive, continuous* functions converges in a closed bounded domain to a *continuous* function, then the convergence is *uniform.*

Proof   See Prob. 8.

The Dini theorem shows, for example, that in the case of continuous kernels, the convergence of the series (13.3-5) to the various iterates is uniform (and absolute) in x and y independently. To be sure, since the cf's of continuous kernels, by convention, are taken to be continuous, the convergence of

$$\sum_{\nu=1}^{\infty} \frac{|\phi_\nu(x)|^2}{\lambda_\nu^2}$$

to $K^2(x,x)$ is then uniform (and trivially absolute). From this it follows, using Schwarz's inequality, that

$$\left[\sum_{\nu=n}^{m} \left| \frac{\phi_\nu(x)\overline{\phi}_\nu(y)}{\lambda_\nu^p} \right| \right]^2 \leq \sum_{\nu=n}^{m} \frac{|\phi_\nu(x)|^2}{|\lambda_\nu|^p} \sum_{\nu=n}^{m} \frac{|\phi_\nu(y)|^2}{|\lambda_\nu|^p} \to 0 \qquad \text{as } n,m \to \infty$$

for all $p \geq 2$.

It is natural now to inquire relative to the case $p = 1$. In other words, what analogous improvements can be made in Theorem 13.3-1? A first result is the following:

LEMMA   If the series

$$\sum_{\nu=1}^{\infty} \frac{\phi_\nu(x)\overline{\phi}_\nu(y)}{\lambda_\nu} \tag{1}$$

where the λ_ν and ϕ_ν are the cv's and cf's of a given continuous Hermitian kernel K, converges to a limit function *uniformly* in either variable alone, then that limit function must be the kernel K.

Proof   Call the original limit function $L(x,y)$ and assume that the uniform convergence is in the second variable y. We know that by Dini's theorem

$$\int_a^b \left| K(x,y) - \sum_{\nu=1}^n \frac{\phi_\nu(x)\overline{\phi_\nu(y)}}{\lambda_\nu} \right|^2 dy = K^2(x,x) - \sum_{\nu=1}^n \frac{|\phi_\nu(x)|^2}{\lambda_\nu^2}$$

tends to zero uniformly in x as $n \to \infty$. As a consequence

$$\int_a^b |K(x,y) - L(x,y)|^2 \, dy = 0$$

for all x, and thus $K(x,y) = L(x,y)$ for almost all y. But L, being the uniform limit (in y) of a set of continuous functions, is continuous in y, from which it follows immediately that $K(x,y) \equiv L(x,y)$.

Notwithstanding this result, the series (1) need not be convergent for all (x,y). Examples of continuous kernels having such representations may be developed using known counterexamples from the theory of Fourier series [see Hellinger and Toeplitz (1927, pp. 1521–1522) and Smithies (1962, pp. 126–127)]. In particular, one makes use of the property, which we have observed earlier, that the cf's of periodic difference kernels are just the familiar trigonometric functions. Salem (1954), in fact, has then shown that there are continuous such kernels (with cv's ordered in the natural way), which have series expansions (1) *diverging* along various lines $x - y =$ constant.

   If some additional assumption is made, over and above the continuity of the given kernel, the improved convergence of (1) often follows readily. In 1909 Mercer verified such a result for kernels all of whose cv's are of one sign. To be precise, he established the following:

MERCER'S THEOREM

   Let K be a *continuous, nonnegative (nonpositive) definite*, Hermitian kernel with cv's λ_ν and cf's ϕ_ν. Then

$$K(x,y) = \sum_{\nu=1}^\infty \frac{\phi_\nu(x)\overline{\phi}_\nu(y)}{\lambda_\nu} \tag{1$'$}$$

the series being uniformly and absolutely convergent in (x,y).

Proof   We assume that K is nonnegative definite, in which case all of the λ_ν are positive and $(K\phi,\phi) \geq 0$ for arbitrary ϕ. It follows, moreover, that†

$$K(x,x) \geq 0 \qquad \text{for all } x$$

For if $K(x_0,x_0) < 0$ for some x_0, then by continuity there exists a positive δ such that Re $[K(x,y)] < 0$, at least within the neighborhood $|x - x_0| < \delta$, $|y - x_0| < \delta$. For the continuous, positive function

$$\psi(x) = \begin{cases} 0 & |x - x_0| \geq \delta \\ \\ \delta^2 - (x - x_0)^2 & |x - x_0| < \delta \end{cases}$$

we then have $(K\psi, \psi) < 0$ in contradiction of our hypotheses.

From previous work, (Chap. 8, Prob. 8), we know that if K is nonnegative definite, so also is

$$K_n(x,y) \equiv K(x,y) - \sum_{\nu=1}^{n} \frac{\phi_\nu(x)\overline{\phi}_\nu(y)}{\lambda_\nu}$$

In view of the above result, then, $K_n(x,x) \geq 0$ or

$$\sum_{\nu=1}^{n} \frac{|\phi_\nu(x)|^2}{\lambda_\nu} \leq K(x,x) \qquad \text{for all } x,n \qquad (2)$$

and thus, letting n tend to ∞, the series on the left-hand side of this inequality converges (absolutely) for all x. We note in passing that integration of (2) with respect to x shows that the series $\sum_\nu (1/\lambda_\nu)$ is convergent also.

---

† Actually this is a special case of the more general result [Mercer (1909, p. 426)] that a continuous Hermitian kernel $K(x,y)$ is nonnegative definite *iff* the determinants

$$\begin{vmatrix} K(x_1,x_1) & K(x_1,x_2) & \cdots & K(x_1,x_n) \\ K(x_2,x_1) & K(x_2,x_2) & \cdots & K(x_2,x_n) \\ \cdots\cdots\cdots\cdots\cdots\cdots\cdots\cdots\cdots\cdots\cdots\cdots \\ K(x_n,x_1) & K(x_n,x_2) & \cdots & K(x_n,x_n) \end{vmatrix}$$

with arbitrary integral n, are nonnegative for all $a \leq x_1, x_2, \ldots, x_n \leq b$ (see Sec. 4.1).

Returning now to (1'), we have the following for fixed x and given positive ε:

$$\left(\sum_{\nu=n}^{m} \left| \frac{\phi_\nu(x)\overline{\phi_\nu(y)}}{\lambda_\nu} \right| \right)^2 \leq \sum_{\nu=n}^{m} \frac{|\phi_\nu(x)|^2}{\lambda_\nu} \sum_{\nu=n}^{m} \frac{|\phi_\nu(y)|^2}{\lambda_\nu}$$

$$\leq \sum_{\nu=n}^{m} \frac{|\phi_\nu(x)|^2}{\lambda_\nu} K(y,y)$$

$$\leq \varepsilon M \qquad \text{for } n,m \geq N(x,\varepsilon) \qquad (3)$$

where $K(y,y) \leq M$ for all y. Here we have made explicit use of the inequality (2) and the fact that the series therein converges. It follows from (3) that the series in (1') converges uniformly *in y* for each x; an analogous argument would show the uniform convergence *in x* for each y. In either case, by our lemma above we know now that the limit function is the given kernel K itself. As a consequence of this, the series of positive continuous functions in (2) must have as its limit the continuous function $K(x,x)$, and therefore, by Dini's theorem, the convergence to this limit function must be uniform. The dependence upon x of $N(x,\varepsilon)$ in (3) is thus actually extrinsic, from which fact the desired uniform (and absolute) convergence of the series in (1') to $K(x,y)$ immediately follows.

In view of considerations made earlier (see Sec. 8.1), Mercer's theorem obviously has the following corollary.

COROLLARY

The conclusions of Mercer's theorem remain valid as long as all but a *finite* number of the cv's of the given continuous Hermitian kernel are of the *same* sign.

As a related result, we have then the following theorem.

THEOREM 13.4-1

The equation

$$\text{Tr} (K) = \sum_{\nu=1}^{\infty} \frac{1}{\lambda_\nu} \qquad (4)$$

holds for continuous Hermitian kernels K with all but a finite number of cv's λ_ν of one sign (or the other).†

Such continuous and Hermitian functions are thus among the class of kernels for which the important relationship (4.3-3) mentioned in Chap. 4 is valid in the special case $n = 1$. Other members of this class (which includes both Hölder continuous and composite kernels) will be encountered in subsequent chapters.

If the given kernel K enjoys still additional smoothness, some further generalizations ensue. For instance, Kadota (1967) has established the following natural extension of Mercer's theorem valuable in certain noise-theoretic investigations.

THEOREM 13.4-2

If in addition to the assumptions of Mercer's theorem, the symmetric derivative function

$$K_n(x,y) \equiv \frac{\partial^{2n}}{\partial x^n \partial y^n} K(x,y)$$

exists and is continuous in the fundamental domain, then $\phi_\nu^{(n)}$, the nth derivative of ϕ_ν, exists and is continuous for each value of ν and

$$K_n(x,y) = \sum_{\nu=1}^{\infty} \frac{\phi_\nu^{(n)}(x)\overline{\phi}_\nu^{(n)}(y)}{\lambda_\nu}$$

the series converging uniformly and absolutely in (x,y).

The proof of this result by and large mirrors the argument presented above for Mercer's theorem itself.

13.5 THE WEINSTEIN-KATO METHOD OF CHARACTERISTIC-VALUE APPROXIMATION

We want to close this chapter with discussion of a relatively simple, yet often reasonably accurate, characteristic-value-approximation technique that is an easy consequence of the series expansions considered recently

---

† Recent work of Weidmann (1966) has shown that (4) holds under much weaker conditions, notably K merely bounded in the fundamental domain, provided Tr (K) is properly interpreted as the so-called *matrical trace* [see (15.2-5)].

above. The procedure provides two-sided bounds and, similar to the more sophisticated approaches of Lehmann (1949 and 1950) and Maehly (1952), it is often used to obtain further improvements to rough approximations provided by other techniques.

Assume, as usual, that we have a given Hermitian $\mathfrak{L}^2$ kernel K in whose cv's λ_ν (and cf's ϕ_ν) we are interested. Let ϕ be an arbitrary function of unit norm in the range of K so that Parseval's relation (13.2-5)

$$\sum_{\nu=1}^{\infty} |(\phi, \phi_\nu)|^2 = \|\phi\|^2 = 1$$

is valid, and define the real numbers η and ε by $\eta \equiv (K\phi, \phi)$ and $\varepsilon \equiv \|K\phi - \eta\phi\|$. For this *trial function* ϕ, it then follows from the Hilbert formula (13.3-4) that

$$\sum_{\nu=1}^{\infty} \frac{|(\phi, \phi_\nu)|^2}{\lambda_\nu} = \eta$$

$$\sum_{\nu=1}^{\infty} \frac{|(\phi, \phi_\nu)|^2}{\lambda_\nu{}^2} = \varepsilon^2 + \eta^2$$

Thus, if we consider the series $\Sigma |z - (1/\lambda_\nu)|^2 |(\phi, \phi_\nu)|^2$ with z an arbitrary complex number, we find

$$|z - \eta|^2 + \varepsilon^2 = \sum_{\nu=1}^{\infty} \left| z - \frac{1}{\lambda_\nu} \right|^2 |(\phi, \phi_\nu)|^2$$

$$\geq \left| z - \frac{1}{\lambda_\mu} \right|^2 \tag{1}$$

where λ_μ is that cv (or among those cv's, if more than one) for which $|z - (1/\lambda_\nu)|$ assumes its *minimum* value. Since $|z - \eta|^2 + \varepsilon^2$ is by definition equivalent to $\|K\phi - z\phi\|^2$, we have, up to this point, actually derived the following analogue of a familiar matrix-theoretic result [Isaacson and Keller (1966, p. 141)], namely:

THEOREM 13.5-1

If ϕ is an arbitrary square-integrable function of unit norm in the range of a given Hermitian $\mathfrak{L}^2$ kernel K, then for every complex number z there is at least one cv λ_μ of K whose reciprocal lies in the circular disk of radius $\|K\phi - z\phi\|$ centered at z.

For our further purposes it is convenient to restrict attention to real z. Taking the square root in (1) then leads to

$$z - \sqrt{(z-\eta)^2 + \varepsilon^2} \le \frac{1}{\lambda_\mu} \le z + \sqrt{(z-\eta)^2 + \varepsilon^2} \tag{2}$$

We note that the left-hand side of this expression is a monotonically *increasing* function of z that approaches the limiting value η (from below) as $z \to \infty$. If we set

$$z - \sqrt{(z-\eta)^2 + \varepsilon^2} \equiv \eta - p\varepsilon$$

therefore, where p is a *positive* parameter, we obtain

$$z - \eta = \frac{\varepsilon}{2}\left(\frac{1}{p} - p\right)$$

and (2) becomes

$$\eta - p\varepsilon \le \frac{1}{\lambda_\mu} \le \eta + \frac{\varepsilon}{p} \tag{3}$$

We have thus established the following result†

THEOREM 13.5-2

Let ϕ be an arbitrary square-integrable function of unit norm in the range of a given Hermitian $\mathcal{L}^2$ kernel K, and define η and ε by

$$\eta \equiv (K\phi, \phi)$$
$$\varepsilon \equiv \|K\phi - \eta\phi\|$$

Then for every $p \ge 0$ there is *at least one* cv λ_μ of K whose reciprocal lies within the interval $[\eta - p\varepsilon, \eta + \varepsilon/p]$.

The smallest interval occurs when $p = 1$, in which case we obtain the following corollary.

---

† This and other inclusion theorems for cv's, as well as various error bounds for estimates of cf's, can be alternatively derived in a clever and noteworthy fashion for kernels of finite rank from certain "analogous" results in probability theory; see Marshall and Olkin (1964).

COROLLARY

Under the above assumptions, there is *at least one* cv λ_μ of K for which

$$\eta - \varepsilon \leq \frac{1}{\lambda_\mu} \leq \eta + \varepsilon \tag{4}$$

The above results appear to have been first suggested by Krylov and Bogoliubov (1929). They were subsequently rediscovered by D. H. Weinstein (1934), whose work became the source for the later extensions of Kohn (1947) and, to some extent, for the substantive generalizations of Kato (1949a). The conclusions are *sharp* in the sense that, given η, ε, and p, the interval specified in the theorem cannot be generally replaced by a smaller one $[\alpha, \beta]$ with $\eta - p\varepsilon < \alpha < \beta < \eta + \varepsilon/p$ (see Prob. 9). On the other hand, the results are less than complete in that they do not delineate precisely which cv (or cv's) it is whose reciprocal(s) is(are) contained in the given interval.

In many situations, however, this last potentially troublesome feature is of little consequence. For instance, if by alternative estimation procedures we have ascertained that the reciprocal of the μth cv (and only the μth) is contained in the interval (α, β), then any trial function ϕ for which

$$\varepsilon^2 < (\eta - \alpha)(\beta - \eta)$$

suffices to sharpen the inequalities, viz.,†

$$\alpha < \eta - \frac{\varepsilon^2}{\beta - \eta} \leq \frac{1}{\lambda_\mu} \leq \eta + \frac{\varepsilon^2}{\eta - \alpha} < \beta \tag{5}$$

Used in this fashion, the so-called Weinstein-Kato method often serves very well as an *estimation-improvement* technique.

If, for the moment, we also assume that K is nonnegative definite, an interesting special case of (5) then occurs when $\alpha = 1/l_{\mu+1}$, where $l_{\mu+1}$ is a *lower bound* for $\lambda_{\mu+1}$, and $\beta = 1/L_{\mu-1}$, where $L_{\mu-1}$ is an *upper bound* for $\lambda_{\mu-1}$. [This is the situation discussed by Kohn (1947).] The relation (5) then shows how, in systematic fashion, additional lower bounds $l_\nu (\nu \leq \mu)$ and upper bounds $L_\nu (\nu \geq \mu)$ for other cv's can be de-

---

† Alternately set $\eta - p\varepsilon = \alpha$, $\eta + \varepsilon/p = \beta$ in (3). This is the best possible general bound using only the four numbers α, β, η, and ε and thus is stronger, for example, than the considerably more complicated result of Trefftz (1933, p. 601).

termined. When $\mu = 1$, we may take $\beta = \infty$, thus obtaining

$$\eta \leq \frac{1}{\lambda_1} \leq \eta + \frac{\varepsilon^2}{\eta - (1/l_2)}$$

We recognize the left-hand inequality as the familiar Ritz relation (9.1-1). The right-hand inequality we have also encountered previously; it is simply the Temple-Collatz result (10.4-6) specialized by setting $n = k = 1$.

As might be expected, there are simple characteristic-function error bounds that accompany the characteristic-value inequalities (5). To be sure, under the same assumptions as above, we have

$$\varepsilon^2 + \eta^2 - (\alpha + \beta)\eta + \alpha\beta = \sum_{\nu=1}^{\infty} \left(\frac{1}{\lambda_\nu} - \alpha\right)\left(\frac{1}{\lambda_\nu} - \beta\right) |(\phi, \phi_\nu)|^2$$

$$\geq \left(\frac{1}{\lambda_\mu} - \alpha\right)\left(\frac{1}{\lambda_\mu} - \beta\right) |(\phi, \phi_\mu)|^2$$

from which it follows that

$$1 - |(\phi, \phi_\mu)|^2 \leq \frac{\varepsilon^2 + (1/\lambda_\mu - \eta)(\alpha + \beta - \eta - 1/\lambda_\mu)}{(1/\lambda_\mu - \alpha)(\beta - 1/\lambda_\mu)}$$

In the special case $\mu = 1$ we may again take $\beta = \infty$, thence obtaining

$$1 - |(\phi, \phi_1)|^2 \leq \frac{1/\lambda_1 - \eta}{1/\lambda_1 - \alpha}$$

a result that is very reminiscent of (10.5-6).

The case of closely spaced or (possibly) degenerate cv's is, of course, more difficult to analyze. Using a set of m orthonormalized trial functions $\psi_1, \ldots, \psi_m$, which also satisfy

$$(K\psi_\mu, \psi_\nu) = \eta_\nu \delta_{\mu\nu}$$

with $\alpha < \eta_1 \leq \eta_2 \leq \cdots \leq \eta_m < \beta$ and $\varepsilon_\nu = \|K\psi_\nu - \eta_\nu \psi_\nu\|$,   Kato (1949a) has shown that if the interval (α, β) contains *at most m* reciprocal cv's of K, then the intervals

$$[\min(\eta_\nu, \sigma_\nu), \max(\eta_\nu, \rho_\nu)] \qquad 1 \leq \nu \leq m \qquad (6)$$

each contain *at least* one of the reciprocal cv's. In this expression ρ_ν is the *largest* root of the equation

$$\sum_{\mu=1}^{\nu} \frac{\varepsilon_\mu{}^2}{(\rho - \eta_\mu)(\eta_\mu - \alpha)} = 1 \tag{7}$$

while σ_ν is the *smallest* root of

$$\sum_{\mu=\nu}^{m} \frac{\varepsilon_\mu{}^2}{(\beta - \eta_\mu)(\eta_\mu - \sigma)} = 1 \tag{8}$$

An occasionally useful (but weaker) form of (6) may be obtained by recognizing from (7) and (8) that

$$\sigma_\nu \geq \eta_\nu - \sum_{\mu=\nu}^{m} \frac{\varepsilon_\mu{}^2}{\beta - \eta_\mu} \quad \text{and} \quad \rho_\nu \leq \eta_\nu + \sum_{\mu=1}^{\nu} \frac{\varepsilon_\mu{}^2}{\eta_\mu - \alpha}$$

The interested reader should consult the relevant literature for further discussion of these matters.

In order to provide at least the flavor of cv estimation using the approach of D. H. Weinstein and Kato, let us briefly analyze the familiar nonnegative definite kernel

$$K(x,y) = \min{(x,y)} - xy \qquad 0 \leq x,y \leq 1$$

For the simple trial function $\phi \equiv 1$, we find $\eta = \frac{1}{12}, \varepsilon = \sqrt{5}/60$. It follows then from (4) that

$$8.3 \doteq 15 - 3\sqrt{5} \leq \lambda_\mu \leq 15 + 3\sqrt{5} \doteq 21.7$$

Sharper bounds (for what is obviously the smallest cv $\lambda_1 = \pi^2$) are obtained using the trial function $\phi = \sqrt{30}\,(x - x^2)$. In this case we have $\eta = \frac{17}{168}$, $\varepsilon = \sqrt{3}/504$, and thus

$$9.6 \doteq \frac{84}{433}(51 - \sqrt{3}) \leq \lambda_\mu \leq \frac{84}{433}(51 + \sqrt{3}) \doteq 10.2$$

(Compare these results with those of the example at the end of Sec. 7.5 and Prob. 1 of Chap. 9 and Prob. 2 of Chap. 10.)

PROBLEMS

1. Let $K(x,y) = k(x - y)$ for $0 \leq x,y \leq 2\pi$ where k is square integrable and periodic with period 2π.
   a. Verify that the cf's of K are just the familiar trigonometric functions $e^{(\pm inx)}$ appropriately normalized.

b. Show that the associated cv's λ_n are real if K is Hermitian.

2. a. Verify the useful identity

$$4(f,g) = \|f+g\|^2 - \|f-g\|^2 + i\|f+ig\|^2 - i\|f-ig\|^2$$

b. By successively applying Parseval's equality (13.2-5) to $f+g$, $f-g$, $f+ig$, and $f-ig$ in place of f and using the above identity, show that

$$\sum_{\nu=1}^{\infty} (f,\phi_\nu)(\phi_\nu,g) = (f,g)$$

is valid for arbitrary $f, g \in \mathfrak{L}^2$ whenever the orthonormalized set $\{\phi_\nu\}$ is complete.

3. Use Hilbert's theorem and the kernels (13.2-6) and (13.2-7) to show that functions with square-integrable second derivatives can be expanded in uniformly and absolutely convergent ordinary Fourier series. [Compare the classical results of Dirichlet (1829) and others; see Hobson (1957, vol. II, chap. 8), for example.]

4. Given an orthonormalized set $\{\phi_\nu\}$ and a sequence of real numbers λ_ν tending to infinity, define A to be the linear transformation such that

$$Af = \sum_{\nu=1}^{\infty} \frac{(f,\phi_\nu)}{\lambda_\nu}\,\phi_\nu$$

is valid for arbitrary $f \in \mathfrak{L}^2$. Show that A transforms bounded infinite sets into (conditionally sequentially) compact sets (see Appendix B) and hence is *completely continuous*. (In a very real sense, then, the above relation characterizes such symmetric transformations.)

5. The completeness properties of various sets of polynomials orthogonal on a bounded interval, such as the Legendre or Chebyshev polynomials, follows readily from the classical Weierstrass approximation theorem. A different technique is needed, however, if the interval is infinite.

a. Using the ideas developed in Sec. 13.2, formally establish that the Hermite functions

$$e^{-x^2/2}\,H_\nu(x) \qquad \nu = 0, 1, 2, \ldots$$

which arise from the Hermitian kernel

$$K(x,y) = e^{\frac{1}{2}(x^2+y^2)} \int_{-\infty}^{\min(x,y)} e^{-t^2}\,dt \int_{\max(x,y)}^{\infty} e^{-t^2}\,dt$$

are complete in $\mathfrak{L}^2(-\infty,\infty)$.

b. In similar fashion show that the kernel

$$K(x,y) = e^{\frac{1}{2}(x+y)} \int_{\max(x,y)}^{\infty} \frac{e^{-t}}{t} \, dt \qquad 0 \le x, y < \infty$$

whose cf's are the Laguerre functions $e^{-x/2} L_\nu(x)$, is closed.

6. a. Reformulate the classical *Sturm-Liouville problem*

$$(p(x)u')' + q(x)u + \lambda r(x)u = 0$$

$$u(0) = u(1) = 0$$

as a Fredholm integral equation (see Appendix A), and then use the completeness theorem to show that the resulting characteristic functions are complete whenever $r(x)$ is uniformly of one sign in $[0,1]$.

b. Show that analogous results are valid for the cf's associated with self-adjoint boundary-value problems in two (and more) dimensions [see Garabedian (1964, pp. 382ff)].

7. a. [Smithies (1962, p. 123).] Show that for $p \ge 3$ the series expansion (13.3-5) is relatively uniformly and absolutely convergent in (x,y). *Hint:* Write K^p as $KK^{p-2}K$.

b. Verify the same convergence for the series (13.3-8).

8. Establish Dini's theorem (Sec. 13.4) by demonstrating that the assumption that the series of positive continuous functions is *not* uniformly convergent leads to a *contradiction* of the fact that the limit function is continuous.

9. Given real numbers $\eta, \varepsilon \ge 0$, and p positive (with η and ε not both vanishing), show that a degenerate kernel K of (at most) second order can be constructed which has the cv's

$$\lambda_1 = \frac{1}{\eta - p\varepsilon}$$

$$\lambda_2 = \frac{p}{\eta p + \varepsilon}$$

and is such that $(K\phi, \phi) = \eta$, $\|K\phi - \eta\phi\| = \varepsilon$ for an appropriately chosen ϕ.

10. The boundary-value problem

$$\frac{d^2\phi}{dx^2} + \lambda\phi(x) = 0 \qquad -1 \le x \le 1$$

$$\phi(-x) = \phi(x)$$

$$\phi(1) = 0$$

gives rise to the characteristic values and characteristic functions

$$\lambda_\nu = \left[\frac{(2\nu - 1)\pi}{2}\right]^2$$

$$\phi_\nu = \cos\left[\frac{(2\nu - 1)\pi}{2}\right]x \qquad \nu = 1, 2, \ldots$$

respectively.

a. Determine the Fredholm integral equation equivalent to this boundary-value problem.

b. Show that for the trial function

$$\phi = \frac{\sqrt{15}}{4}(1 - x^2)$$

we have

$$\eta \equiv (K\phi, \phi) = \frac{17}{42}$$

$$\varepsilon \equiv \|K\phi - \eta\phi\| = \frac{\sqrt{3}}{126}$$

c. Using the fact that only the reciprocal of λ_1 is contained in the interval (α, β) where $\alpha = 4/(9\pi^2)$ and $\beta = \infty$, employ (13.5-5) to verify the bounds

$$2.4674 \le \lambda_1 \le 2.4706$$

[These results may be compared with those of Stadter (1966, sec. 4).]

11. [Maehly (1952)]. Given the nonnegative definite kernel K, assume that the normalized trial function ϕ is such that

$$\frac{1}{\lambda_{\mu+1}} \le \eta \le \frac{1}{\lambda_\mu}$$

where $\eta \equiv (K\phi, \phi)$. Define

$$\varepsilon \equiv \|K\phi - \eta\phi\|$$

$$d_0 \equiv \frac{1}{2}\left(\frac{1}{\lambda_\mu} - \frac{1}{\lambda_{\mu+1}}\right)$$

$$\Delta \equiv \eta - \frac{1}{2}\left(\frac{1}{\lambda_\mu} + \frac{1}{\lambda_{\mu+1}}\right)$$

a. Verify that

$$d_0{}^2 - \Delta^2 = \left(\frac{1}{\lambda_\mu} - \eta\right)\left(\eta - \frac{1}{\lambda_{\mu+1}}\right) \le \varepsilon^2$$

Hint: See (13.5-5).

b. Using this result, deduce that for every d satisfying $\varepsilon \le d \le d_0$,

$$\frac{1}{\lambda_\mu} \le \eta + d - \sqrt{d^2 - \varepsilon^2}$$

whenever $\Delta \ge 0$, and

$$\eta - d + \sqrt{d^2 - \varepsilon^2} \le \frac{1}{\lambda_{\mu+1}}$$

whenever $\Delta < 0$,

c. Satisfy yourself, therefore, that for every d for which $\varepsilon \le d \le d_0$, the reciprocal of either λ_μ or $\lambda_{\mu+1}$ (but generally not both) lies within the interval

$$[\eta - d + \sqrt{d^2 - \varepsilon^2}, \ \eta + d - \sqrt{d^2 - \varepsilon^2}]$$

(This result can be of value whenever lower bounds are known for the *difference* between *successive* reciprocal cv's.)

chapter fourteen

Expansion Theorems for General Kernels

14.1 HILBERT-SCHMIDT KERNELS

The analysis of the last chapter is generally not applicable in the case of an arbitrary square-integrable kernel K. Indeed, as we have already observed, such kernels need not possess cv's and cf's, and thus the various characteristic expansions presented previously have no meaning. Since $\mathfrak{L}^2$ is a separable Hilbert space, however, there do exist complete orthonormalized sets of $\mathfrak{L}^2$ functions $\{\phi_\nu\}$ † that are countable and hence can be used, in place of the cf's, as the basis elements for generalized Fourier series developments. In this chapter we begin to explore the implications of such expansions for the kernel K. ‡

---

† No confusion should result from this new notation.

‡ This was essentially the starting point for the original investigations of Hilbert (1904) who exploited (for a restricted class of $\mathfrak{L}^2$ kernels) the equivalence between Fredholm integral equations and related sets of linear equations in infinitely many unknowns. See Hellinger and Toeplitz (1927) for an illuminating account of this approach. Also consult the comprehensive paper of Hille and Tamarkin (1931) where matrix-theoretic procedures are used to establish growth estimates for the cv's (if they exist) of general kernels belonging to various broad "smoothness" classes. Some of these latter results are explored in Chap. 16.

We know from the two-dimensional analogue of Theorem 13.1-1 that

$$\sum_{\mu,\nu=1}^{\infty} (K\phi_\nu,\phi_\mu)\phi_\mu(x)\,\overline{\phi}_\nu(y) \qquad (13.3\text{-}1')$$

gives the best mean-square approximation to K of the form (13.1-2). In view of the completeness of the $\{\phi_\nu\}$, moreover, Parseval's relation

$$\sum_{\mu,\nu=1}^{\infty} |(K\phi_\nu,\phi_\mu)|^2 = \|K\|^2 \qquad (2)$$

is valid. In fact, if $\{\psi_\nu\}$ is any other complete orthonormalized set, we have [using (13.2-5) repeatedly] †

$$\sum_{\nu} \|K\psi_\nu\|^2 = \sum_{\mu,\nu} |(K\psi_\nu,\phi_\mu)|^2$$

$$= \|K\|^2 \qquad (3)$$

$$= \sum_{\mu,\nu} |(\psi_\nu,K^*\phi_\mu)|^2 = \sum_{\mu} \|K^*\phi_\mu\|^2$$

Transformations for which

$$\sigma(K) \equiv \left[\sum_\nu \|K\psi_\nu\|^2\right]^{\frac{1}{2}} \qquad (4)$$

is finite for some (and hence every) complete orthonormalized set $\{\psi_\nu\}$ are said to be of *Hilbert-Schmidt* type. On the Hilbert space $\mathfrak{L}^2$, then, these are just the transformations generated by square-integrable kernels. In a more general setting, the transformations of Hilbert-Schmidt type comprise that particular subclass of completely continuous transformations which plays the role of *integral operators* on a given abstract separable Hilbert space [see Dunford and Schwartz (1963, sec. 11.6) and Schatten (1960, chap. 2)].‡ In either case, the Hilbert-Schmidt transformations form a Banach space, in fact, a Hilbert space, with norm $\sigma(K)$.

---

† Where explicit limits for sums are omitted, it is to be understood that the summation extends over all positive integers.

‡ Nothing new actually occurs in the general situation since, by utilizing the natural isomorphisms provided by the Fourier coefficients in terms of given complete orthonormalized sets, every transformation of Hilbert-Schmidt type, acting in an abstract separable Hilbert space, can easily be shown to be unitarily equivalent to an integral operator in $\mathfrak{L}^2$ with a square-integrable kernel.

For our further study of $\mathfrak{L}^2$ kernels K, it will be convenient to generalize the representation (13.3-1') by introducing a second complete orthonormalized set of square-integrable functions. Then for given $\{\phi_\mu\}$, $\{\psi_\nu\}$ we find

$$K(x,y) = \sum_{\mu,\nu} (K\psi_\nu, \phi_\mu)\phi_\mu(x)\overline{\psi}_\nu(y) \tag{5}$$

where, in view of (3), the generalized double Fourier series is *mean convergent*.

The infinite matrix formed from the coefficients $(K\psi_\nu, \phi_\mu)$ of (5) is often termed the *kernel matrix*, associated with K, relative to the sets $\{\phi_\mu\}$ and $\{\psi_\nu\}$ [see Schmeidler (1950 and 1965, sec. 6), for example]. Moreover, since the choice of these orthonormal sets is completely arbitrary, the kernel matrix of a given K can have many distinct forms. In the next section, we consider a particular choice of the $\{\phi_\mu\}$ and $\{\psi_\nu\}$ and show that we obtain thereby an extremely simple matrix, i.e., one with entries *only* on the main diagonal.†

14.2 SINGULAR FUNCTIONS AND SINGULAR VALUES

Given an arbitrary $\mathfrak{L}^2$ kernel K, let us define a new kernel $[K]^2$ by the composition

$$[K]^2 \equiv KK^* \tag{1}$$

Here we choose to regard the superscript 2 as merely part of the notation; it will be clear in due course, however, that the kernel $[K]^2$ is actually (as expected) the second iterate of a well-defined square-summable and nonnegative definite kernel $[K] = (KK^*)^{\frac{1}{2}}$.‡ In an analogous fashion we also define

$$[K^*]^2 \equiv K^*K \tag{2}$$

Now both $[K]^2$ and $[K^*]^2$ are nonnegative definite Hermitian kernels. As a consequence, both kernels possess full sets of cf's and associated cv's. In fact, if we denote the cf's and (positive) cv's of the

---

† Such a simplified form would not, in general, exist using only *one* complete orthonormalized set and the representation (13.3-1'), of course, since this would imply the existence of cv's.

‡ The notation $[K]$ for $(KK^*)^{\frac{1}{2}}$ is essentially due to Schatten (1960).

former by ϕ_ν and μ_ν^2, respectively, then it follows readily that the functions

$$\psi_\nu \equiv \mu_\nu K^* \phi_\nu \tag{3}$$

are cf's of the latter corresponding to the *same* cv's. An analogous result is valid if we begin instead with the cf's ψ_ν of $[K^*]^2$ and investigate the nature of the functions ϕ_ν given by

$$\phi_\nu \equiv \mu_\nu K \psi_\nu \tag{4}$$

As might be anticipated, even more is true.

THEOREM 14.2

Let the cf's and associated cv's of the kernel $[K]^2$, formed from a given arbitrary $\mathfrak{L}^2$ kernel K, be denoted by ϕ_ν and μ_ν^2, respectively. Then the μ_ν^2 are also the only cv's of $[K^*]^2$, and a full set of cf's of this kernel is given by the totality of ψ_ν as determined from the ϕ_ν by (3). A similar result holds if we start first with the ψ_ν.

Proof   Since $\mu_\nu^2 > 0$ for all ν, we have

$$
\begin{aligned}
(\psi_\nu, \psi_{\nu'}) &= (\mu_\nu K^* \phi_\nu, \mu_{\nu'} K^* \phi_{\nu'}) \\
&= \mu_\nu \mu_{\nu'} (\phi_\nu, KK^* \phi_{\nu'}) \\
&= \frac{\mu_\nu}{\mu_{\nu'}} (\phi_\nu, \phi_{\nu'}) \\
&= \delta_{\nu\nu'}
\end{aligned}
$$

for arbitrary ν and ν', and thus the $\{\psi_\nu\}$ form an orthonormalized set. Moreover, for any cf ψ of $[K^*]^2$ with cv μ^2, we know from above that $\phi = \mu K \psi$ defines a cf of $[K]^2$ associated with the same cv. It follows from our assumptions, then, that ϕ must be representable as

$$\phi = \sum_{\nu=1}^n \alpha_\nu \phi_\nu$$

and hence $\psi = \mu K^* \phi = \sum_{\nu=1}^n \mu \alpha_\nu K^* \phi_\nu = \sum_{\nu=1}^n \frac{\mu \alpha_\nu}{\mu_\nu} \psi_\nu$

We conclude from this last relation and the orthonormal nature of the set $\{\psi_\nu\}$ that the ψ_ν, with their associated cv's μ_ν^2, comprise a full characteristic system. The converse part of the theorem is verified in analogous fashion.

Functions ϕ_ν and ψ_ν as given above were first introduced by Schmidt (1907)[†] and are called *singular functions* (sf's) of K corresponding to the *singular value* (sv) μ_ν. In view of (3) and (4) they are biorthogonal with respect to the kernel K, and thus the collections $\{\phi_\nu\}$ and $\{\psi_\nu\}$ appear to have the special property mentioned at the close of Sec. 14.1. That these sets may not be complete in $\mathfrak{L}^2$ is actually of no consequence since they do at least span the ranges of K and K^*, respectively. Substituting, then, in (14.1-5), we see that the general expansion assumes the simplified form[‡]

$$K(x,y) = \sum_\nu \frac{\phi_\nu(x)\,\overline{\psi_\nu(y)}}{\mu_\nu} \tag{5}$$

with the kernel matrix of K relative to the $\{\phi_\nu\}$ and $\{\psi_\nu\}$ being *diagonal*. Moreover, we have from (14.1-3) that

$$\sum_\nu \left(\frac{1}{\mu_\nu}\right)^2 = \|K\|^2 \tag{6}$$

14.3   SERIES EXPANSIONS IN TERMS OF SINGULAR FUNCTIONS

The mean-convergent representation (14.2-5), as the analogue in the general case of the expansion (13.3-3) for Hermitian kernels, can be the starting point for a number of investigations.[§]   For instance, we have the following analogue of Hilbert's theorem (Sec. 13.2).

THEOREM 14.3

Let K be an arbitrary $\mathfrak{L}^2$ kernel with sv's μ_ν and sf's ϕ_ν and ψ_ν, respectively.

(a)   Every function $f \in \mathfrak{L}^2$ admits the developments, convergent in the mean,

$$f = h + \sum_\nu (f,\phi_\nu)\phi_\nu \tag{1}$$

[†] Schmidt's original theory was subsequently extended by Smithies (1937 and 1962). Also see Vergerio (1917), Mollerup (1923), and Lewis (1950) for alternate approaches.

[‡] The custom is to associate the positive sign with the real number μ_ν; we shall adhere to that practice since there is no loss of generality in so doing. We then make the further assumption that the sv's are ordered in the natural manner $0 < \mu_1 \leq \mu_2 \leq \mu_3 \leq \cdots$.

[§] Truncated versions of the series (14.2-5) can play the role of the arbitrarily accurate degenerate kernels used in the Schmidt approach to the solution of integral equations (Chap. 3). See Schmeidler (1965, sec. 7), for a proof of the Fredholm alternative (Chap. 2) employing this procedure and therefore making explicit use of the diagonal form of the kernel matrix.

and
$$f = \widetilde{h} + \sum_{\nu} (f, \psi_\nu)\psi_\nu \qquad (2)$$

where h and $\widetilde{h}$ are functions (depending on f) satisfying $K^*h = 0$ and $K\widetilde{h} = 0$, respectively.

(b) If g can be represented as

$$g = Kf \qquad (3)$$

then it has a mean-convergent generalized Fourier series

$$\sum_{\nu} \frac{(f, \psi_\nu)}{\mu_\nu}\, \phi_\nu \qquad (4)$$

Similarly, for $\widetilde{g} = K^*f$ we have

$$\sum_{\nu} \frac{(f, \phi_\nu)}{\mu_\nu}\, \psi_\nu \qquad (5)$$

Proof Since from Sec. 14.2 the $\{\phi_\nu\}$ and $\{\psi_\nu\}$ constitute full characteristic systems for the nonnegative definite Hermitian kernels $[K]^2$ and $[K^*]^2$, respectively, the expansions (1) and (2) follow directly from the classical Hilbert theorem. Note particularly that $KK^*h = 0$ implies

$$0 = (KK^*h, h) = \|K^*h\|^2$$

and so h has the desired behavior. A similar argument suffices to show $K\widetilde{h} = 0$. The representations (4) and (5) then follow by operating on (2) with K and (1) with K^* and using the relations (14.2-4) and (14.2-3).

As was the case with the classical result, the convergence of the series (4) and (5) is actually *relatively* uniform and absolute. The argument proceeds in precisely the same fashion as before, only this time we note that the necessary inequalities

$$\sum_{\nu=n}^{m} \left|\frac{\phi_\nu}{\mu_\nu}\right|^2 \le \|K\|_y^2 \qquad \text{and} \qquad \sum_{\nu=n}^{m} \left|\frac{\psi_\nu}{\mu_\nu}\right|^2 \le \|K^*\|_y^2$$

are consequences of the fact that, by (14.2-4) and (14.2-3), the quotients $\phi_\nu(x)/\mu_\nu$ and $\psi_\nu(x)/\mu_\nu$ are, for fixed x, the Fourier coefficients of $K(x, y)$ and $K^*(x, y)$ with respect to the $\{\psi_\nu\}$ and $\{\phi_\nu\}$, respectively. Of course, if either $\|K\|_y$ or $\|K^*\|_y$ is uniformly bounded, the convergence of the appropriate series becomes *uniform and absolute*.

The above analogue of the Hilbert result has the following corollary, which extends, in a sense, the completeness theorem of Sec. 13.2.

COROLLARY

The full set $\{\phi_\nu\}$ of sf's is *complete* in $\mathfrak{L}^2$ *iff* K^* is *closed*. Similarly, the full set $\{\psi_\nu\}$ is *complete iff* K is *closed*.

As another immediate and obvious corollary, we have the following.

GENERALIZED HILBERT FORMULA

For arbitrary square-integrable functions f and g,

$$(Kf, g) = \sum_\nu \frac{(f, \psi_\nu)(\phi_\nu, g)}{\mu_\nu} \tag{6}$$

Returning now to the representation (14.2-5) and its derivation, it is clear that replacement of K by K^* merely leads to an interchange of the roles of the ϕ_ν and ψ_ν. We have, then, the companion expansion

$$K^*(x, y) = \sum_\nu \frac{\psi_\nu(x)\overline{\phi}_\nu(y)}{\mu_\nu} \tag{7}$$

from which it follows naturally that

$$[K]^2 \equiv KK^* = \sum_\nu \frac{\phi_\nu(x)\overline{\phi}_\nu(y)}{\mu_\nu^2} \tag{8}$$

and

$$[K^*]^2 \equiv K^*K = \sum_\nu \frac{\psi_\nu(x)\overline{\psi}_\nu(y)}{\mu_\nu^2}$$

Alternatively, of course, these expressions are a direct consequence of (13.3-3), owing to the nature of the ϕ_ν, ψ_ν, and μ_ν.

The representations (8) suggest the definitions

$$[K] \equiv \sum_\nu \frac{\phi_\nu(x)\overline{\phi}_\nu(y)}{\mu_\nu} \tag{9}$$

$$[K^*] \equiv \sum_\nu \frac{\psi_\nu(x)\overline{\psi}_\nu(y)}{\mu_\nu}$$

for the unique, nonnegative definite kernels $(KK^*)^{\frac{1}{2}}$ and $(K^*K)^{\frac{1}{2}}$, respectively. In turn, the kernels $[K]^2$ and $[K^*]^2$ then appear as the second iterates of these square-summable functions. Moreover, if we denote by U the (not generally $\mathfrak{L}^2$) kernel

$$U = \sum_\nu \phi_\nu(x)\,\overline{\psi}_\nu(y)$$

which gives rise to the *partially isometric*† transformation of the space spanned by the $\{\psi_\nu\}$ into the space spanned by the $\{\phi_\nu\}$, the fundamental expansion (14.2-5) and hence the given kernel K can be seen to possess the interesting decompositions

$$K = [K]\,U = U[K^*] \tag{10}$$

These *polar* representations are the analogues of the modulus and phase form of expression in complex variables and are well known in finite-dimensional operator theory [see Halmos (1958, pp. 169–171); also see Schatten (1960, pp. 4ff) and Riesz and Sz.-Nagy (1955, pp. 284–286) for the general case].

14.4 NORMAL KERNELS

In various special cases, the results of the preceding sections simplify considerably. For instance, if K is Hermitian, $[K]^2 = K^2$ and the ϕ_ν are thus just the cf's of the second iterate of K and hence of K itself. In view of our convention of always choosing μ_ν positive, we have, then, from (14.2-3) that

$$\psi_\nu = \mu_\nu K^* \phi_\nu$$
$$= |\lambda_\nu|\,K\phi_\nu$$
$$= \frac{|\lambda_\nu|}{\lambda_\nu}\,\phi_\nu$$

The expansion (14.2-5), therefore, assumes the familiar form

$$K(x,y) = \sum_\nu \frac{\phi_\nu(x)\,\overline{\phi}_\nu(y)}{\lambda_\nu}$$

---

† A transformation that is *isometric* on a proper subspace and vanishes on the remainder of the Hilbert space in question is said to be *partially isometric*. If both the transformation and its adjoint are isometric throughout the entire Hilbert space, the transformation is *unitary*.

and the several other relations become identical with those derived in earlier chapters specifically for Hermitian kernels.

Similar simplifications occur even when the kernel is known merely to be *normal*. In order to show this, however, we will need to make use of two well-known results from matrix theory [see Halmos (1958, pp. 144ff), for instance, and Prob. 5 at the end of this chapter].

LEMMA 1   Let A be an arbitrary matrix having complex elements. Then there exists a *unitary* matrix V such that

$$A = V^*TV \qquad V^* = V^{-1} \tag{1}$$

where T is *triangular*.† If A is *normal*, then T is diagonal and its entries are merely the (not necessarily distinct) eigenvalues of A.

LEMMA 2   The eigenvalues of unitary matrices have unit magnitude.

In our analysis‡ we shall also have need of the following elementary, yet fundamental, result.

THEOREM 14.4-1

Let K be a normal $\mathfrak{L}^2$ kernel, i.e., one that satisfies $KK^* = K^*K$.§ If ϕ is a cf of K with cv λ, then ϕ is also a cf of K^* associated with the cv $\bar{\lambda}$.

Proof   Note that

$$
\begin{aligned}
\|\phi - \bar{\lambda}K^*\phi\|^2 &= (\phi,\phi) - \bar{\lambda}(K^*\phi,\phi) - \lambda(\phi,K^*\phi) + \lambda\bar{\lambda}(K^*\phi,K^*\phi)\\
&= (\phi,\phi) - \bar{\lambda}(\phi,K\phi) - \lambda(K\phi,\phi) + \lambda\bar{\lambda}(KK^*\phi,\phi)\\
&= (\phi,\phi) - \lambda(K\phi,\phi) - \bar{\lambda}(\phi,K\phi) + \lambda\bar{\lambda}(K^*K\phi,\phi)\\
&= \|\phi - \lambda K\phi\|^2\\
&= 0 \qquad\qquad\qquad\qquad\qquad\qquad\qquad\qquad \text{Q.E.D.}
\end{aligned}
$$

---

† For analogues of this result in the case of integral operators acting on arbitrary separable Hilbert spaces see Livsic (1954), J. Schwartz (1962), and Gohberg and Krein (1969).

‡ This analysis owes much to the concise account of Smithies (1962, pp. 150–163).

§ In keeping with our usual practice, we assume that this equality need only hold "almost everywhere." Some authors choose to term such kernels *almost normal*, reserving the designation *normal* for kernels satisfying $KK^* = K^*K$ for all (x,y). Nothing new would be added to our analysis, however, by following this convention since every almost normal kernel is known to be equivalent to a normal kernel [Smithies (1962, pp. 160–161)].

It is worth observing that the following corollary, then, is an immediate consequence (see the general proof of Property 4, Sec. 7.1).

COROLLARY

> Characteristic functions of normal kernels belonging to distinct characteristic values are *orthogonal*.

Other useful properties of normal kernels are to be found among the problems at the end of this chapter.

Returning now to our main considerations, we know that for a normal $\mathcal{L}^2$ kernel K, the auxiliary kernels $[K]^2$ and $[K^*]^2$ are identical (equivalent), and thus full sets of their respective cf's ϕ_ν and ψ_ν can be generally selected so that

$$\phi_\nu = a_\nu \psi_\nu \qquad \text{for all } \nu \tag{2}$$

with the complex constants of proportionality satisfying $|a_\nu| = 1$. Unfortunately, pairs of functions $\{\phi_\nu, \psi_\nu\}$ chosen arbitrarily in this manner need not be singular functions since the fundamental relations (14.2-3) and (14.2-4) may not hold between them. We want to show, however, that given *any* full sets of sf's, other full sets can be simply constructed that do have the property (2). Moreover, it will turn out that each ϕ_ν so created is actually a characteristic function of the given kernel K associated with a well-defined cv λ_ν. In fact, the totality of the ϕ_ν, with their corresponding λ_ν, constitute a *full characteristic system* for K.

Consider, then, that we are given arbitrary full sets $\{\tilde{\phi}_\nu\}$ and $\{\tilde{\psi}_\nu\}$ of sf's of a normal kernel K. Let μ be any sv of K and, for simplicity of notation, assume that the first n pairs of sf's are associated with this sv. In view of the normality of K, each $\tilde{\phi}_\nu (1 \leq \nu \leq n)$ is not only a cf of $[K]^2$ with corresponding cv μ^2, but also is a cf of $[K^*]^2$ belonging to the same cv. It follows then that each $\tilde{\phi}_\nu$ must be representable as

$$\tilde{\phi}_\nu = \sum_{\alpha=1}^{n} a_{\nu\alpha} \tilde{\psi}_\alpha \tag{3}$$

Moreover, the $n \times n$ coefficient matrix $A = (a_{\nu\mu})$ is *unitary* since

$$\delta_{\nu\mu} = (\tilde{\phi}_\nu, \tilde{\phi}_\mu) = \sum_{\alpha,\beta=1}^{n} a_{\nu\alpha} \bar{a}_{\mu\beta} (\tilde{\psi}_\alpha, \tilde{\psi}_\beta)$$

$$= \sum_{\alpha=1}^{n} a_{\nu\alpha} \bar{a}_{\mu\alpha}$$

If we express (3) symbolically as

$$\phi = A\tilde{\psi}$$

then Lemma 1 leads to

$$\tilde{\phi} = V^* D V \tilde{\psi}$$

whence, denoting $V\tilde{\phi}$ by ϕ and $V\tilde{\psi}$ by ψ, we obtain

$$\phi = D\psi \qquad (4)$$

The ϕ_ν and $\psi_\nu (1 \leq \nu \leq n)$ are sf's of K associated with the sv μ. Individually, they are linear combinations of the $\tilde{\phi}_\nu$ and $\tilde{\psi}_\nu (1 \leq \nu \leq n)$ with the *same* coefficients; jointly, they thus automatically satisfy (14.2-3) and (14.2-4) since their antecedents do. Moreover, since V is unitary, we have

$$(\phi_\nu, \phi_\mu) = \delta_{\nu\mu} = (\psi_\nu, \psi_\mu)$$

and the sets $\{\phi_\nu\}$ and $\{\psi_\nu\}$ therefore span the *same* spaces as the sets $\{\tilde{\phi}_\nu\}$ and $\{\tilde{\psi}_\nu\}$.

In view of (4), we can say even more. Since $A = V^* DV$ is unitary (and hence normal), the individual ϕ_ν and $\psi_\nu (1 \leq \nu \leq n)$ satisfy

$$\phi_\nu = a_\nu \psi_\nu \qquad (2')$$

where, by Lemma 2, $|a_\nu| = 1$. It follows, then, from (14.2-4) that

$$\phi_\nu = \frac{\mu_\nu}{a_\nu} K\phi_\nu \qquad (5)$$

Each ϕ_ν, $1 \leq \nu \leq n$, is thus also a cf of the kernel K associated with the cv $\lambda_\nu = \overline{a}_\nu \mu_\nu$.

The analysis presented so far is independent of the particular sv of K under consideration. As a consequence, repeating the above construction for each sv, we see that from any given original sets $\{\tilde{\phi}_\nu\}$ and $\{\tilde{\psi}_\nu\}$ of sf's, we can form new sets $\{\phi_\nu\}$ and $\{\psi_\nu\}$ which pairwise satisfy (2) and thus, by (5), are also orthonormal sets of cf's of K corresponding to the cv's λ_ν. (We assume, of course, the natural ordering for the μ_ν and hence the λ_ν.) It turns out, moreover, that the $\{\phi_\nu\}$ with their $\{\lambda_\nu\}$ comprise a full characteristic system for K. To be sure, under the above relationships the expansion (14.2-5) becomes

$$K(x, y) = \sum_{\nu} \frac{\phi_\nu(x) \overline{\phi}_\nu(y)}{\lambda_\nu} \tag{6}$$

By Theorem 14.3-1, therefore, we have for any cf ϕ of K with cv λ

$$\phi = \lambda \sum_{\nu} \frac{(\phi, \phi_\nu)}{\lambda_\nu} \phi_\nu$$

from which it follows that

$$(\phi, \phi_\nu) = \frac{\lambda}{\lambda_\nu} (\phi, \phi_\nu)$$

If ϕ is nontrivial, (ϕ, ϕ_ν) must be different from zero for some ν and thus for that (those) ν, $\lambda = \lambda_\nu$. We conclude that ϕ is a finite linear combination of those ϕ_ν associated with the cv λ, or, stated otherwise, the set $\{\phi_\nu, \lambda_\nu\}$ constitutes a full characteristic system for K.

There is a converse to the above results. If we begin alternatively with an arbitrary full characteristic system $\{\phi_\nu, \lambda_\nu\}$ for K, then for all ν

$$\phi_\nu = \lambda_\nu K \phi_\nu$$

and, by Theorem 14.4-1,

$$\phi_\nu = \overline{\lambda}_\nu K^* \phi_\nu$$

Taken together, these two relations imply that ϕ_ν is also a cf of KK^* with cv $|\lambda_\nu|^2$. Using Theorem 14.2, therefore, we find that the collections $\{\phi_\nu\}$ and $\{\psi_\nu\}$ where

$$\psi_\nu = |\lambda_\nu| K^* \phi_\nu$$

$$= \frac{\lambda_\nu}{|\lambda_\nu|} \overline{\lambda}_\nu K^* \phi_\nu$$

$$= \frac{\lambda_\nu}{|\lambda_\nu|} \phi_\nu$$

are full sets of sf's of K associated with the sv's $|\lambda_\nu|$.†

---

† Since the given characteristic set $\{\phi_\nu\}$ is linearly equivalent to the special set formed earlier, it should be clear that no linearly independent sf's have been overlooked in this construction.

We can summarize these several results as follows.

THEOREM 14.4-2

Let K be a normal $\mathfrak{L}^2$ kernel. Given any full sets of sf's of K, new full sets $\{\phi_\nu\}$ and $\{\psi_\nu\}$ can be constructed, the elements of which satisfy

$$\psi_\nu = \frac{\lambda_\nu}{|\lambda_\nu|}\,\phi_\nu$$

for certain nonvanishing complex constants λ_ν. The sv's of K are equal to the moduli $|\lambda_\nu|$. Moreover, the collection $\{\phi_\nu, \lambda_\nu\}$ comprises a full set of cf's and cv's of K.

Conversely, given a full characteristic system $\{\phi_\nu, \lambda_\nu\}$ of K, the collections $\{\phi_\nu\}$ and $\{(\lambda_\nu/|\lambda_\nu|)\phi_\nu\}$ are full sets of sf's of K corresponding to the sv's $|\lambda_\nu|$. In either case, the mean-convergent characteristic expansion

$$K(x,y) = \sum_\nu \frac{\phi_\nu(x)\,\overline{\phi_\nu}(y)}{\lambda_\nu} \tag{6'}$$

is valid, and the generalized Hilbert formula (14.3-6) assumes the familiar form

$$(Kf, g) = \sum_\nu \frac{(f, \phi_\nu)(\phi_\nu, g)}{\lambda_\nu} \tag{7}$$

14.5   THE INEQUALITIES OF SCHUR AND CARLEMAN

In an earlier chapter we mentioned the inequality

$$\sum_\nu \left|\frac{1}{\lambda_\nu}\right|^2 \le \|K\|^2 \tag{4.2-2'}$$

where the λ_ν are the cv's of K each repeated according to its (algebraic) multiplicity as a zero of the modified Fredholm determinant $\tilde{D}(\lambda)$. This result, which was originally established for continuous kernels by Schur (1909) and subsequently extended to $\mathfrak{L}^2$ kernels by Carleman (1921), clearly becomes an *equality* if the kernel K is *normal*. Goldfain (1946) first proved a converse of sorts to this observation [also see Smithies (1962, pp. 161–163)]. He verified that if equality holds above, where in the summation each cv is repeated the number of times equal to its *geometric* multiplicity (or rank),† then K must be normal. We want now to show that the approach of Goldfain easily generalizes, providing an alternate proof of the full Schur-Carleman inequality and establishing that equality holds therein *only* if the kernel is normal. Since it is instructive for the general case, we begin by considering Schur's original argument as applied to linear transformations on finite-dimensional unitary spaces.

Recall from Lemma 1 of Sec. 14.4 that for every $n \times n$ matrix A, there exists a unitary matrix V such that

$$A = V^* T V \tag{14.4-1'}$$

where T is, say, right *triangular*. It follows, then, that

$$A V^* = V^* T \tag{1}$$

which becomes, when written in terms of components,

$$\sum_k a_{ik} \bar{v}_{jk} = \lambda_j \bar{v}_{ji} + \sum_{k<j} \bar{v}_{ki} t_{kj} \tag{2}$$

Here we have made use of the easily established fact that the entries along the main diagonal of T are merely the eigenvalues λ_j of A. From (1) we have

$$AA^* = (AV^*)(AV^*)^* = (V^*T)(V^*T)^* = V^*TT^*V$$

and hence

$$\text{Tr}\,(AA^*) = \text{Tr}\,(V^*TT^*V)$$
$$= \text{Tr}\,(TT^*)$$

In terms of components this last relation has the form

$$\sum_{i,j} |a_{ij}|^2 = \sum_j |\lambda_j|^2 + \sum_j \sum_{i<j} |t_{ij}|^2$$

and we conclude that

$$\sum_j |\lambda_j|^2 \le \sum_{i,j} |a_{ij}|^2 \tag{3}$$

---

† In other words, once for each linearly independent characteristic function.

with equality iff T is *diagonal*, that is, iff A is *normal*.† The inequality (3) is, of course, the Schur result for finite-dimensional spaces.

For the general case of a square-integrable kernel K, we can argue in a fashion completely analogous to the above approach.‡ We know from Secs. 6.3 and 6.4 that for each cv λ_0 of K, the (perhaps generalized) characteristic functions satisfy

$$K\phi_{\nu 1} = \frac{1}{\lambda_0}\,\phi_{\nu 1}$$

$$\begin{aligned} &\nu = 1, 2, \ldots, p \\ &\mu = 2, 3, \ldots, m_\nu \end{aligned} \qquad (4)$$

$$K\phi_{\nu\mu} = \frac{1}{\lambda_0}\,\phi_{\nu\mu} + \phi_{\nu,\mu-1}$$

Here p is the *geometric* multiplicity of λ_0 while $\sum_{\nu=1}^{p} m_\nu$ is the *algebraic* multiplicity. The $\phi_{\nu\mu}$ do not in general form an orthonormal set, but they *are* linearly independent. Moreover, characteristic functions associated with different cv's are also linearly independent.§ It follows, then, that, utilizing the classical procedure of Gram (1883) and Schmidt (1907) [see Courant and Hilbert (1953, pp. 50–51), for example], we can construct an orthonormalized set $\{\psi_\nu\}$ of $\mathfrak{L}^2$ functions that is linearly equivalent to the set of all characteristic functions of K. We envision that this construction process is carried out in systematic fashion so that, for each cv of K, $\phi_{\nu\mu}$ is *orthonormalized* before $\phi_{\nu,\mu+1}$. As a consequence, if we now generically designate the original cf's by ϕ_ν and their associated cv's by λ_ν, then for all $\nu \geq 1$

$$\psi_\nu = \sum_{\mu=1}^{\nu} a_{\nu\mu}\phi_\mu$$

and, by (4),
$$K\psi_\nu = \sum_{\mu=1}^{\nu} a_{\nu\mu} K\phi_\mu$$

$$= \frac{a_{\nu\nu}}{\lambda_\nu}\,\phi_\nu + \sum_{\mu=1}^{\nu-1} b_{\nu\mu}\phi_\mu$$

---

† If T is diagonal, then $TT^* = T^*T$, whence, using (14.4-1), A is normal.

‡ Both Schur (1909) and Carleman (1921) established the general result by using the fact that continuous ($\mathfrak{L}^2$) kernels can be approximated in norm with arbitrary accuracy by kernels of finite rank. See Hille and Tamarkin (1931, pp. 23–25) for another proof along similar lines.

§ If M is the linear manifold spanned by the cf's belonging to a given cv of K, and N is its complement in $\mathfrak{L}^2$, then M and N are invariant under K. We commonly say that K is *decomposed* (or *reduced*) by the pair M,N. See Kato (1966, pp. 22ff) or Riesz and Sz.-Nagy (1955, pp. 179–190) for further discussion.

$$= \frac{1}{\lambda_\nu}\, \psi_\nu + \sum_{\mu=1}^{\nu-1} c_{\nu\mu}\, \psi_\mu \tag{5}$$

Equation (5) is the analogue in the general case of the earlier relations (1) and (2), the ψ_ν here playing the same role as the columns of V^* did previously. Proceeding in similar fashion, therefore, we find that for each $\nu \geq 1$

$$\| K\psi_\nu \|^2 = \left| \frac{1}{\lambda_\nu} \right|^2 + \sum_{\mu=1}^{\nu-1} |c_{\nu\mu}|^2$$

In view of (14.1-3), it follows immediately from this relation that †

$$\sum_\nu \left| \frac{1}{\lambda_\nu} \right|^2 \leq \| K \|^2 \tag{6}$$

and this is the desired Schur-Carleman result.

If equality holds in (6), then the $c_{\nu\mu}$ must perforce vanish for all ν and μ. In such a case, therefore, we have from (5) and the construction of the ψ_ν that their totality, with the associated cv's λ_ν, actually constitutes a full characteristic system for K. As a consequence of these results, we find that the series

$$\sum_{\mu,\nu} (K\psi_\nu, \psi_\mu)\psi_\mu(x)\,\overline{\psi}_\nu(y) = \sum_\nu \frac{\psi_\nu(x)\,\overline{\psi}_\nu(y)}{\lambda_\nu}$$

not only gives the best mean-square approximation to K of the form (13.1-2) but actually is mean convergent to K. It follows at once, then, that K must be normal.

The Schur-Carleman result (6), as we observed in Chap. 4, leads to the important inequality

$$|\tilde{D}(\lambda)| \leq e^{\frac{1}{2}|\lambda|^2 \|K\|^2} \tag{4.3-1'}$$

for the modified Fredholm determinant. To conclude this chapter we want to establish a comparable inequality for the modified first Fredholm

---

† For the real and imaginary parts of the cv's we have the companion inequalities

$$\sum_\nu \left| \mathrm{Re}\left(\frac{1}{\lambda_\nu}\right) \right|^2 \leq \|K_R\|^2 \qquad \sum_\nu \left| \mathrm{Im}\left(\frac{1}{\lambda_\nu}\right) \right|^2 \leq \|K_I\|^2$$

in terms of the cartesian components $K_R \equiv \frac{1}{2}(K + K^*)$ and $K_I \equiv -\frac{1}{2}i(K - K^*)$ of K (see Prob. 9).

minor, namely, †

$$\|\tilde{D}(x,y;\lambda)\| \le \|K\|\, e^{\frac{1}{2}(1+|\lambda|^2\|K\|^2)} \tag{7}$$

This result, while obviously imprecise for $\lambda = 0$, actually is sharp for all other values of the eigenparameter λ (see Prob. 12).

We first note that we need only establish (7) for *degenerate* kernels. The inequality for general kernels in $\mathfrak{L}^2$ will then follow directly since all the terms involved are continuous (see Prob. 10). For degenerate kernels, moreover, the problem is essentially matrix-theoretic, and Smithies (1962, pp. 92–93) has suggested the fundamental building blocks. We begin by writing the *degenerate* kernel K, as can always be done, in the canonical form

$$K(x,y) = \sum_{\mu,\nu=1}^{n} k_{\mu\nu}\, \phi_\mu(x)\, \overline{\phi}_\nu(y)$$

in terms of a set of orthonormalized functions $\{\phi_\nu\}$. Then, using (3.3-5), (4.1-8), and the definitions of Sec. 2.4, we find

$$\tilde{D}(x,y;\lambda) = e^{\lambda\,\mathrm{Tr}\,K}\, K\, C_\lambda(x,y) = e^{\lambda\,\mathrm{Tr}\,K}\, C_\lambda K(x,y) \tag{8}$$

where
$$C_\lambda(x,y) \equiv \sum_{\mu,\nu=1}^{n} C^{\nu\mu}\, \phi_\mu(x)\, \overline{\phi}_\nu(y) \tag{9}$$

with $C^{\mu\nu}$ the cofactor of $C_{\mu\nu}$ in the matrix

$$(C_{\mu\nu}) \equiv (\delta_{\mu\nu} - \lambda k_{\mu\nu})$$

(see Prob. 11). Note, too, that

$$\mathrm{Tr}\,K = \sum_{\nu=1}^{n} k_{\nu\nu}$$

$$\|K\|^2 = \sum_{\mu,\nu=1}^{n} |k_{\mu\nu}|^2$$

---

† This inequality, sometimes referred to as the *Carleman inequality*, is useful in the study of the completeness properties of the cf's of Hilbert-Schmidt operators. Typographical errors notwithstanding, Dunford and Schwartz (1963, sec. 11.6) have established the analogous version of the result using the usual "sup" norm of bounded linear operators. Although their result follows rather trivially in this case from our analysis, the proof of (7) involves a few more lines owing to the complexity of the more natural $\mathfrak{L}^2$ norm.

We now let ϕ, ψ be arbitrary $\mathfrak{L}^2$ functions and write $\alpha_\nu = (\phi, \phi_\nu)$ and $\beta_\nu = (\psi, \phi_\nu)$. It follows, then, from (9) that

$$(C_\lambda \phi, \psi) = \sum_{\mu,\nu=1}^{n} C^{\nu\mu} \alpha_\nu \overline{\beta}_\mu$$

$$= - \begin{vmatrix} 0 & \overline{\beta}_1 & \cdots & \overline{\beta}_n \\ \alpha_1 & C_{11} & \cdots & C_{1n} \\ \cdots & \cdots & \cdots & \cdots \\ \alpha_n & C_{n1} & \cdots & C_{nn} \end{vmatrix} \qquad (10)$$

this last result being a consequence of a well-known property of bordered determinants [see Aitken (1958, p. 75), for example]. The determinant in (10), moreover, can be estimated using Hadamard's inequality,† and this leads to

$$|(C_\lambda \phi, \psi)|^2 \le \sum_{\nu=1}^{n} |\beta_\nu|^2 \prod_{\mu=1}^{n} \left[|\alpha_\mu|^2 + \sum_{\nu=1}^{n} |C_{\mu\nu}|^2 \right]$$

$$\le \|\psi\|^2 \prod_{\mu=1}^{n} \left[|\alpha_\mu|^2 + 1 - 2 \operatorname{Re}(\lambda k_{\mu\mu}) + |\lambda|^2 \sum_{\nu=1}^{n} |k_{\mu\nu}|^2 \right]$$

$$\le \|\psi\|^2 \exp\left[\sum_{\mu=1}^{n} |\alpha_\mu|^2 \right.$$

$$\left. - 2 \operatorname{Re}\left(\lambda \sum_{\mu=1}^{n} k_{\mu\mu} \right) + |\lambda|^2 \sum_{\mu,\nu=1}^{n} |k_{\mu\nu}|^2 \right]$$

$$\le \|\psi\|^2 \exp[\|\phi\|^2 - 2 \operatorname{Re}(\lambda \operatorname{Tr} K) + |\lambda|^2 \|K\|^2]$$

---

† Hadamard (1893) first demonstrated that for the $n \times n$ matrix $(\varepsilon_{\mu\nu})$, the determinantal inequalities

$$|\det(\varepsilon_{\mu\nu})|^2 \le \begin{cases} \displaystyle\prod_{\mu=1}^{n} \sum_{\nu=1}^{n} |\varepsilon_{\mu\nu}|^2 \\[2em] \displaystyle\prod_{\nu=1}^{n} \sum_{\mu=1}^{n} |\varepsilon_{\mu\nu}|^2 \end{cases}$$

are valid. Riesz and Sz.-Nagy (1955, pp. 176ff), Smithies (1962, pp. 68ff), and Beckenbach and Bellman (1965, pp. 64 and 89), among numerous others, give modern accounts of this result.

For the particular choice $\psi = C_\lambda \phi$, we have that †

$$\|C_\lambda \phi\|^2 \leq e^{\|\phi\|^2 - 2\operatorname{Re}(\lambda \operatorname{Tr} K) + |\lambda|^2 \|K\|^2} \tag{11}$$

In this last relation (11) we now specialize ϕ by setting $\phi(x) = cK(x,y)$ for any fixed value of y. The constant c is to be selected so that the resultant inequality is as sharp as possible. We find then that $|c| = 1/\|K\|_x$ and

$$\int |C_\lambda K(x,y)|^2 \, dx \leq \|K\|_x^2 \, e^{1 - 2\operatorname{Re}(\lambda \operatorname{Tr} K) + |\lambda|^2 \|K\|^2}$$

or, using (8),

$$\int |\tilde{D}(x,y;\lambda)|^2 \, dx \leq \|K\|_x^2 \, e^{1 + |\lambda|^2 \|K\|^2}$$

where
$$\|K\|_x \equiv \left[\int |K(x,y)|^2 \, dx \right]^{\frac{1}{2}}$$

Integrating over y and taking the square root, we finally have the desired result (7).

PROBLEMS

1. a. Verify that a *partially isometric* kernel is square integrable if and only if it is degenerate.
    b. Satisfy yourself that partially isometric transformations take orthonormal sets into orthonormal sets.
2. a. Show that if K is a normal $\mathcal{L}^2$ kernel, then

$$\|K\phi\| = \|K^*\phi\| \tag{1}$$

for every $\mathcal{L}^2$ function ϕ.
    b. Conversely, show that if (1) is satisfied for every $\mathcal{L}^2$ function ϕ, then the Hermitian kernel $H \equiv KK^* - K^*K$ is such that

$$(H\phi, \phi) = 0$$

for all ϕ, from which it follows that $\|H\| = 0$ and K is normal.

---

† From this inequality and the relationship (8) it is clear that for the sup norm we have

$$\|\tilde{D}(x,y;\lambda)\|_\infty \leq \|K\|_\infty \, e^{\frac{1}{2}(1 + |\lambda|^2 \|K\|^2)}$$

$$\leq \|K\| \, e^{\frac{1}{2}(1 + |\lambda|^2 \|K\|^2)}$$

3.   **a.** Satisfy yourself that given any $\mathfrak{L}^2$ kernel K, the new kernel

$$L = \alpha K + \beta K^*$$

is normal for arbitrary complex constants α and β of the same magnitude.
   b. Use this fact to construct an example of a normal kernel that is neither Hermitian nor unitary.

4.   We know that every kernel K can be uniquely decomposed into its Hermitian *cartesian components*

$$K_R = \tfrac{1}{2}(K + K^*)$$

$$K_I = \frac{1}{2i}(K - K^*)$$

   a. Show that K is normal *iff* $K_R K_I = K_I K_R$.
   b. For the case of normal K, use Theorem 14.4-1 in order to determine the relationship between the cv's and cf's of the cartesian components K_R and K_I and those of the original kernel K.

5.   **a.** In connection with Lemma 1 of Sec. 14.4, verify that if *normal A* is given by

$$A = V^* T V$$

   where V is unitary, then T is *diagonal* with the eigenvalues of A as entries.
   b. Use this result to trivially establish Lemma 2.

6.   If K is a normal $\mathfrak{L}^2$ kernel, we have from (14.3-8) that

$$\sum_\nu \frac{\phi_\nu(x)\overline{\phi}_\nu(y)}{\mu_\nu^2} = \sum_\nu \frac{\psi_\nu(x)\overline{\psi}_\nu(y)}{\mu_\nu^2}$$

   Show directly that this implies that each ϕ_ν is unitarily related to those ψ_ν associated with the same singular value and conversely.

7.   Verify that Theorem 13.5-1 is equally valid for normal kernels.

8.   **a.** Use the Hilbert formula (14.4-7) to establish for normal kernels K an analogue to (8.4-1), namely,

$$\frac{1}{|\lambda_{n+1}|} = \max_{\phi \,\epsilon\, C_n} |(K\phi, \phi)|$$

   [For a related result see Bernau (1967).]
   b. Satisfy yourself that similar extensions of the Weyl-Courant and Poincaré characterizations (9.3-2) and (9.4-1), respectively, do not exist in general. [Turner (1969) has recently shown that there do exist valid extensions if the kernel does not differ too greatly from Hermitian form.]

9. Using (4.3-3) and (14.5-6), demonstrate that the real and imaginary parts of the cv's of a given $\mathcal{L}^2$ kernel K satisfy

$$\sum_\nu \left| \mathrm{Re}\left(\frac{1}{\lambda_\nu}\right) \right|^2 \leq \|K_R\|^2 \qquad \sum_\nu \left| \mathrm{Im}\left(\frac{1}{\lambda_\nu}\right) \right|^2 \leq \|K_I\|^2$$

in terms of K's cartesian components K_R and K_I.

10. Verify that the various terms appearing in the inequality (14.5-7) are continuous. In other words show that, for example, $\|K - K_n\| \to 0$ implies

$$\|\tilde{D}_{K_n}(x,y;\lambda)\| \to \|\tilde{D}_K(x,y;\lambda)\|$$

For this purpose, the several relations of Sec. 4.1 may be found to be useful.

11. **a.** Show that the function C_λ defined by (14.5-9) can be alternatively expressed in terms of the unmodified first Fredholm minor $D(x,y;\lambda)$ as

$$C_\lambda(x,y) = \lambda D(x,y;\lambda) + \det(C_{\mu\nu}) \sum_{\nu=1}^n \phi_\nu(x)\overline{\phi}_\nu(y)$$

b. Satisfy yourself then that the commutativity of C_λ and K is, in essence, guaranteed by the Fredholm identities (3.2-5).

12. Given $\lambda \neq 0$, construct a degenerate kernel of rank 1 for which the equality sign holds in (14.5-7).

13. Let K be an arbitrary $\mathcal{L}^2$ kernel with sv's μ_ν and sf's ϕ_ν and ψ_ν, respectively, and assume that the two related nonnegative definite kernels $[K]$ and $[K^*]$ are defined as in (14.3-9). Use the Hilbert formula (14.3-6) to establish the inequality

$$|(K\phi,\psi)|^2 \leq ([K^*]\phi,\phi)([K]\psi,\psi)$$

for all square-integrable ϕ,ψ. (This result represents the extension to the general case of the K Schwarz inequality proved earlier in Sec. 8.3 for definite kernels.)

chapter fifteen

Nuclear Kernels, Composite Kernels, and the Classes C_p

15.1 NUCLEAR KERNELS AND THE TRACE CLASS

In modern applied analysis, it occurs that various subclasses of general $\mathfrak{L}^2$ (Hilbert-Schmidt) kernels are of particular interest, and, in line with this observation, we have already devoted a substantial portion of this book to the fundamental such subclass consisting of Hermitian $\mathfrak{L}^2$ kernels. We want now to consider certain other subclasses which, in general, can be characterized by the fact that the quantities $(K\phi_\nu, \phi_\nu)$, $\|K\phi_\nu\|$, or, equivalently, $1/\mu_\nu$ go to zero more rapidly than is indicated by the relations (14.1-2), (14.1-3), or (14.2-6).

One of the more important such collections and one that has attracted considerable attention in recent years is the set of *nuclear* or *trace-class* kernels, which give rise to the so-called *trace-class* operators [Gohberg and Krein (1957 and 1969) and Lidskii (1959a)]. On the one hand, a thorough understanding of the properties of these kernels appears to be basic to the study of nuclear spaces initiated by Grothendieck (1955) and pursued by various other investigators, notably of the Russian school [see Gel'fand and Vilenkin (1964) for a survey account of some of these efforts]. On the other hand, the trace class of kernels is representative of certain other broad subclasses of the set of all $\mathfrak{L}^2$

kernels, about which one can often make definitive statements concerning, for instance, the existence of cf's, the completeness of the set of all (generalized) cf's, and the convergence of various series representations and other approximations for the kernels therein. It is essentially for this latter reason that we concern ourselves now with the nature of nuclear kernels, and the various results that we present will, by and large, be in keeping with this rationale.†

Following Grothendieck (among others), we call a kernel *nuclear* (or of *trace class*) if its *singular values* are such that

$$\sum \frac{1}{\mu_\nu} < \infty \tag{1}$$

As one might expect, there are a number of equivalent characterizations that could equally well have been used as the definition of a nuclear kernel. For instance, we have the following theorem, apparently first observed by Stinespring (1958).

THEOREM 15.1-1

An $\mathfrak{L}^2$ kernel K is nuclear iff

$$\sum_\nu \| K \phi_\nu \| < \infty$$

for *some* complete‡ orthonormalized set $\{\phi_\nu\}$ of $\mathfrak{L}^2$ functions.

Proof   Let the s f's associated with the sv's μ_ν of K be given by $\{\phi_\nu\}$ and $\{\psi_\nu\}$. If K is nuclear, then by (14.2-4)

$$\sum_\nu \| K \psi_\nu \| = \sum_\nu \frac{1}{\mu_\nu} < \infty$$

Conversely, if $\sum_\nu \| K \tilde{\phi}_\nu \| < \infty$ for some complete orthonormalized set $\{\tilde{\phi}_\nu\}$, then using various relations from Chap. 14, particularly

---

† After this chapter was essentially in final form, the Gohberg and Krein (1969) translation appeared. Readers will note some overlap with the treatment given this material in that excellent presentation.

‡ The $\{\phi_\nu\}$ actually need only be complete, of course, in the complement of the null space of K. Here and henceforth we shall assume that the reader readily recognizes such minor refinements.

(14.1-3), (14.3-9), (14.3-10), and Schwarz's inequality, we find

$$\infty > \sum_\nu \| K \tilde{\phi}_\nu \| = \sum_\nu \| U [K^*] \tilde{\vec{\phi}}_\nu \| = \sum_\nu \| [K^*] \tilde{\phi}_\nu \|$$

$$\geq \sum_\nu ([K^*] \tilde{\phi}_\nu, \tilde{\phi}_\nu) = \sum_\nu \| [K^*]^{\frac{1}{2}} \tilde{\phi}_\nu \|^2$$

$$= \sum_\nu \| [K^*]^{\frac{1}{2}} \psi_\nu \|^2 = \sum_\nu \frac{1}{\mu_\nu}$$

It is important to observe † that for nuclear kernels $\sum_\nu \| K \phi_\nu \|$ need not be convergent for all sets $\{\phi_\nu\}$. Indeed, as Prob. 1 shows, $\sum_\nu \| K \phi_\nu \|$ may diverge for even the simplest of kernels and the most natural of choices $\{\phi_\nu\}$. It is also worth noting, however, that from the sequence of inequalities appearing in the above proof,

$$\sum_\nu \frac{1}{\mu_\nu} \leq \sum_\nu \| K \phi_\nu \| \tag{2}$$

for *all* choices of the complete orthonormalized set $\{\phi_\nu\}$. In fact,

$$\sum_\nu \frac{1}{\mu_\nu} = \inf \sum_\nu \| K \phi_\nu \| \tag{3}$$

where the infimum is to be taken over all such $\{\phi_\nu\}$. For nuclear K, this finite minimum value is actually achieved *iff* the set $\{\phi_\nu\}$ is an orthonormal basis for $\mathfrak{L}^2$ consisting of cf's of $[K^*]$.

Another result in the same vein as the previous theorem is the following.

THEOREM 15.1-2

An $\mathfrak{L}^2$ kernel K is nuclear iff

$$\sum_\nu |(K \phi_\nu, \phi_\nu)| < \infty$$

for *all* complete orthonormalized sets $\{\phi_\nu\}$ of $\mathfrak{L}^2$ functions.

---

† For this and innumerable other substantive observations, suggestions, and indeed fundamental results concerning the material appearing in this and the immediately following chapter, we are indebted to D. W. Swann, who, in personal conversation and private communication, graciously shared with us his recent work in these areas.

Proof   Again using various relations from Chap. 14, we have for arbitrary orthonormal bases $\{\phi_\nu\}$ and $\{\psi_\nu\}$ of $\mathfrak{L}^2$

$$\sum_\nu |(K\phi_\nu,\psi_\nu)| = \sum_\nu |([K]^{\frac{1}{2}} U [K^*]^{\frac{1}{2}} \phi_\nu, \psi_\nu)|$$

$$\leq \sum_\nu \|[K^*]^{\frac{1}{2}} \phi_\nu\| \cdot \|[K]^{\frac{1}{2}} \psi_\nu\|$$

$$\leq \left(\sum_\nu \|[K^*]^{\frac{1}{2}} \phi_\nu\|^2\right)^{\frac{1}{2}} \left(\sum_\nu \|[K]^{\frac{1}{2}} \psi_\nu\|^2\right)^{\frac{1}{2}}$$

$$= \left(\sum_\nu \frac{1}{\mu_\nu}\right)^{\frac{1}{2}} \left(\sum_\nu \frac{1}{\mu_\nu}\right)^{\frac{1}{2}}$$

$$= \sum_\nu \frac{1}{\mu_\nu}$$

Thus, if K is nuclear, the various constructs are meaningful and, in particular, we find that

$$\sum_\nu |(K\phi_\nu, \phi_\nu)| < \infty \tag{4}$$

To prove the converse, we first note that if (4) holds for every complete orthonormalized set $\{\phi_\nu\}$, the same must be true for the *cartesian components* of K

$$K_R \equiv \tfrac{1}{2}(K + K^*)$$
$$K_I \equiv \frac{1}{2i}(K - K^*) \tag{5}$$

It follows, moreover, that if α_ν, β_ν and a_ν, b_ν are the cv's and cf's of these Hermitian kernels, respectively, then

$$\sum_\nu |(K_R a_\nu, a_\nu)| = \sum_\nu \frac{1}{|\alpha_\nu|}$$

$$\sum_\nu |(K_I b_\nu, b_\nu)| = \sum_\nu \frac{1}{|\beta_\nu|}$$

and K_R and K_I must therefore be nuclear. As a consequence, for the special choice of $\{\phi_\nu\}$ and $\{\psi_\nu\}$ as the sf's of K associated with the sv's μ_ν, we calculate, using the result already established in the first part of the theorem,

$$\sum_\nu \frac{1}{\mu_\nu} = \sum_\nu (K\psi_\nu, \phi_\nu) = \sum_\nu [(K_R\psi_\nu, \phi_\nu) + i(K_I\psi_\nu, \phi_\nu)]$$

$$\leq \sum_\nu |(K_R\psi_\nu, \phi_\nu)| + \sum_\nu |(K_I\psi_\nu, \phi_\nu)|$$

$$\leq \sum_\nu \frac{1}{|\alpha_\nu|} + \sum_\nu \frac{1}{|\beta_\nu|}$$

$$< \infty$$

Thus K is nuclear, and the proof is complete.

As an obvious corollary of this theorem we have the following.

COROLLARY

An $\mathfrak{L}^2$ kernel K is nuclear iff its (Hermitian) cartesian components K_R and K_I are both nuclear.

Yet another characterization, closely related to the above result, is contained in the following theorem.

THEOREM 15.1-3

An $\mathfrak{L}^2$ kernel K is nuclear iff

$$\sum_\nu |(K\psi_\nu, \phi_\nu)| < \infty$$

for all complete orthonormalized sets $\{\phi_\nu\}$ and $\{\psi_\nu\}$ of $\mathfrak{L}^2$ functions.

Proof   In the first part of the proof of the immediately preceding theorem we actually showed that if K is nuclear, then

$$\sum_\nu |(K\psi_\nu, \phi_\nu)| < \infty \tag{6}$$

Conversely, if (6) is valid for arbitrary orthonormal bases of $\mathfrak{L}^2$, the particular choice of $\{\phi_\nu\}$ and $\{\psi_\nu\}$ as the sets of sf's of K suffices to establish the nuclearity of K.

To conclude this section we observe that from the proof of Theorem 15.1-2

$$\sum_\nu |(K\psi_\nu, \phi_\nu)| \le \sum_\nu \frac{1}{\mu_\nu}$$

for all choices of the orthonormal bases $\{\phi_\nu\}$ and $\{\psi_\nu\}$. As a companion to the earlier relation (3), we therefore have that

$$\sum_\nu \frac{1}{\mu_\nu} = \sup \sum_\nu |(K\psi_\nu, \phi_\nu)| \qquad (7)$$

where the supremum is to be taken over all appropriate sets $\{\phi_\nu\}$ and $\{\psi_\nu\}$. In the case of a nuclear kernel K, this maximum value is finite and is actually achieved for (and only for) compatible (in the sense of Chap. 14) sets of singular functions of K.

15.2 COMPOSITE KERNELS

Theorems 15.1-2 and 15.1-3, while interesting theoretically, are obviously not of the same practical value as Theorem 15.1-1. There exists yet another characterization, however, that has perhaps even more practical significance than this last result.

THEOREM 15.2

An $\mathfrak{L}^2$ kernel K is nuclear iff it is *composite*, i.e., representable in the form

$$K = K_1 K_2$$

where K_1 and K_2 are themselves $\mathfrak{L}^2$ kernels.

Proof   Assume $K = K_1 K_2$ with K_1 and K_2 in $\mathfrak{L}^2$. Then from (14.3-9), (14.3-10)

$$\sum_\nu \frac{1}{\mu_\nu} = \mathrm{Tr}\,[K^*]$$

$$= \mathrm{Tr}\,(U^* K_1 K_2)$$

$$\le \|K_1\| \cdot \|K_2\| < \infty$$

On the other hand, if $\sum_\nu (1/\mu_\nu)$ is finite,

$$K_1 \equiv [K]^{\frac{1}{2}} \quad \text{and} \quad K_2 \equiv [K]^{\frac{1}{2}} U$$

are both square-summable kernels, and $K = K_1 K_2$.

This particular characterization is actually a special case of certain general results of Chang (1947 and 1949), about which we will have more to say subsequently. Moreover, it is precisely this characterization that Schatten (1960), Weidmann (1965), and others have used as the definition of the trace class of kernels. With this result in hand, we are able to give a *constructive* and thereby perhaps a more useful proof of our earlier Theorem 15.1-1. Indeed, following Stinespring (1958), if

$$\sum_\nu \| K\phi_\nu \| < \infty$$

for some orthonormal basis $\{\phi_\nu\}$ of $\mathfrak{L}^2$, the two kernels

$$K_1(x,y) \equiv \sum_{\mu,\nu} \alpha_{\mu\nu} \phi_\mu(x) \overline{\phi}_\nu(y)$$

$$K_2(x,y) \equiv \sum_\nu \beta_\nu \phi_\nu(x) \overline{\phi}_\nu(y)$$

(1)

where

$$\alpha_{\mu\nu} = \begin{cases} 0 & \text{if } \| K\phi_\nu \| = 0 \\ \dfrac{(K\phi_\nu, \phi_\mu)}{\| K\phi_\nu \|^{\frac{1}{2}}} & \text{otherwise} \end{cases}$$

$$\beta_\nu = \| K\phi_\nu \|^{\frac{1}{2}}$$

(2)

are easily shown to be square summable (Prob. 3). Since by direct computation,

$$K_1 K_2(x,y) = \sum_{\mu,\nu} (K\phi_\nu, \phi_\mu) \phi_\mu(x) \overline{\phi}_\nu(y)$$

it then follows that K is composite and hence nuclear, as was to be proved.

Up to this point in our discussion of nuclear (composite) kernels we have made no mention of the nature of the cv's (if they exist) associated with such kernels. From a historical as well as practical point of view, however, results in this area have considerable significance. For example, Lalesco first suggested in 1914–1915 that for composite kernels the moduli of their cv's satisfy

$$\sum_{\nu} \frac{1}{|\lambda_{\nu}|} < \infty \tag{3}$$

Almost 45 years later the implications of a condition such as (3) were still appearing as the basis of studies, say, of the completeness properties of the cf's of so-called dissipative operators [see Krein (1959), for example].†

We want to consider in the next section, therefore, what can be generally derived about the cv's of composite kernels. Before doing so, however, we note that Gohberg and Krein (1957) originally proposed saying that a kernel has *finite spectral trace* if its cv's satisfy (3).‡ This permitted them, at least for such kernels that were also Hermitian (normal), to show the equivalence of the *spectral trace*

$$\sigma_S(K) \equiv \sum_{\nu} \frac{1}{\lambda_{\nu}} \tag{4}$$

and the *matrical trace*

$$\sigma_M(K) \equiv \sum_{\nu} (K \phi_{\nu}, \phi_{\nu}) \tag{5}$$

where in the latter expression $\{\phi_{\nu}\}$ is *any* orthonormalized basis for $\mathfrak{L}^2$. Unfortunately, this equivalence does not extend to the general case (see Prob. 4 at the end of this chapter for a representative example). It does

---

† A kernel K is called *dissipative* if its imaginary cartesian component K_I is nonnegative definite. A special case of a typical result of Krein is that the (generalized) cf's of a dissipative $\mathfrak{L}^2$ kernel are complete (in the complement of the null space) if the *real* cartesian component is *nuclear*. The reader may also be interested in consulting Livsic (1954) and Lidskii (1959b), as well as the survey articles of Naimark (1956) and Brodskii and Livsic (1958) and the recent text of Gohberg and Krein (1969).

‡ This is also the Dunford and Schwartz (1963, p. 1086) definition of the trace class.

carry over, however, to composite (nuclear) kernels $K = K_1 K_2$, and since

$$\text{Tr}\,(K_1 K_2) = \text{Tr}\left[\sum_{\mu,\nu} (K_1 \phi_\nu, \phi_\mu)\phi_\mu(x) \sum_{\mu'\nu'} (K_2 \phi_{\nu'}, \phi_{\mu'})\overline{\phi}_{\nu'}(y)(\phi_\nu, \phi_{\mu'})\right]$$

$$= \sum_{\mu,\nu} (K_1 \phi_\nu, \phi_\mu)(\phi_\mu, K_2^* \phi_\nu)$$

$$= \sum_\nu (K\phi_\nu, \phi_\nu)$$

or
$$\text{Tr}(K_1 K_2) = \sigma_M(K) \tag{6}$$

for these kernels, this has important implications, as we shall see shortly.†

15.3   TRACE IDENTITIES FOR COMPOSITE KERNELS

For composite kernels $K = K_1 K_2$ we have just established the validity of (15.2-6). If, in addition, these kernels are Hermitian, then choosing the $\{\phi_\nu\}$ to be cf's of K we find

$$\sigma_M(K) = \sigma_S(K)$$

and thus
$$\text{Tr}\,(K_1 K_2) = \sigma_S(K) \tag{1}$$

also. The so-called *Lalesco result* asserts that this relation (1) holds even if the kernel is not Hermitian; i.e., (1) is valid for *all* composite kernels.

Modern proofs of this result [see Weidmann (1965),‡ for example] are based upon the fact that for composite kernels $K = K_1 K_2$, the unmodified Fredholm determinant of $[K]$ exists and majorizes, term by term, the unmodified Fredholm determinant of $K_1 K_2$. Then since the former possesses certain properties, the desired inferences can be drawn

---

† In subsequent work Krein (1959) restricted the class of finite spectral trace (absolutely convergent trace) kernels (operators) to those whose (Hermitian) cartesian components (15.1-5) satisfied (3). In view of Theorem 15.1-2, however, this is precisely the class of nuclear kernels as we have already defined it. For related results on the equivalence of σ_S and σ_M for the sum of two Hermitian kernels (operators), see Halberg and Kramer (1960) and Libin (1968).

‡ Also see Gheorghiu (1928), Hille and Tamarkin (1931), Chang (1949), Lidskii (1959a), and Gohberg and Krein (1969, pp. 101ff).

concerning the latter. It is helpful, therefore, to observe that, in general, from Chap. 4,

$$D_{[K]}(\lambda) = e^{-\lambda \, \mathrm{Tr} \, [K]} \prod_{\nu} \left(1 - \frac{\lambda}{\mu_\nu} \right) e^{\lambda/\mu_\nu}$$

Since $\mathrm{Tr} \, [K] = \displaystyle\sum_{\nu} \frac{1}{\mu_\nu} < \infty$

however, the above expression actually simplifies to

$$D_{[K]}(\lambda) = \prod_{\nu} \left(1 - \frac{\lambda}{\mu_\nu} \right) \tag{2}$$

The form of the result (2) shows that $D_{[K]}(\lambda)$ is an entire function of order $\rho = \rho_1 \leq 1$ and genus zero. If we define its *type* τ by

$$\tau \equiv \limsup_{r \to \infty} r^{-\rho} \log M(r) \tag{3}$$

where $M(r)$ is the maximum modulus achieved on the circle $|\lambda| = r$, then by a special case of a theorem of Boas (1954, pp. 27ff)† we also see that whenever $\rho = 1$, $\tau = 0$.

Let us now turn to the main result.

THEOREM 15.3

The Fredholm determinant of an arbitrary composite (nuclear) $\mathfrak{L}^2$ kernel K exists and has the convergent product expansion

$$D_K(\lambda) = \prod_{\nu} \left(1 - \frac{\lambda}{\lambda_\nu} \right) \tag{4}$$

Proof   We want to make use of a classical identity of Carleman (1921, p. 213) [see also Chang (1949, pp. 362ff)], namely, if $K = K_1 K_2$,

---

† Simply stated: An entire function $f(z)$ of order 1 is of zero (minimal) type *iff* its Hadamard factorization in terms of its zeros z_ν has the form

$$cz^m \prod_{\nu} \left(1 - \frac{z}{z_\nu} \right)$$

where c is constant.

$$K\begin{pmatrix} x_1, & \ldots, & x_\nu \\ y_1, & \ldots, & y_\nu \end{pmatrix} = \frac{1}{\nu!} \int_a^b \cdots \int_a^b K_1 \begin{pmatrix} x_1, & \ldots, & x_\nu \\ z_1, & \ldots, & z_\nu \end{pmatrix}$$

$$K_2 \begin{pmatrix} z_1, & \ldots, & z_\nu \\ y_1, & \ldots, & y_\nu \end{pmatrix} \cdot dz_1 \cdots dz_\nu$$

In order to do so we recall that the nuclear kernel K has the *canonical* decomposition (see proof of Theorem 15.2) $K = K_1 K_2$ where

$$K_1 \equiv [K]^{\frac{1}{2}}$$
$$K_2 \equiv [K]^{\frac{1}{2}} U$$

As a consequence we have

$$K_1 K_1^* = [K] = K_2 K_2^*$$

It follows, then, employing the classical formulas (see Prob. 2 of Chap. 4), that the coefficients in the power-series expansion of the Fredholm determinant of K satisfy

$$|d_\nu| = \frac{1}{\nu!} \left| \int_a^b \cdots \int_a^b K \begin{pmatrix} x_1, & \ldots, & x_\nu \\ x_1, & \ldots, & x_\nu \end{pmatrix} dx_1 \ldots dx_\nu \right|$$

$$\leq \frac{1}{(\nu!)^2} \int_a^b \cdots \int_a^b \left| K_1 \begin{pmatrix} x_1, & \ldots, & x_\nu \\ z_1, & \ldots, & z_\nu \end{pmatrix} K_2 \begin{pmatrix} z_1, & \ldots, & z_\nu \\ x_1, & \ldots, & x_\nu \end{pmatrix} \right|$$

$$dz_1 \ldots dz_\nu \, dx_1 \ldots dx_\nu$$

$$\leq \frac{1}{(\nu!)^2} \left[\int_a^b \cdots \int_a^b \left| K_1 \begin{pmatrix} x_1, & \ldots, & x_\nu \\ z_1, & \ldots, & z_\nu \end{pmatrix} \right|^2 \right.$$

$$\left. dz_1 \ldots dz_\nu \, dx_1 \ldots dx_\nu \right]^{\frac{1}{2}}$$

$$\left[\int_a^b \cdots \int_a^b \left| K_2 \begin{pmatrix} z_1, & \ldots, & z_\nu \\ x_1, & \ldots, & x_\nu \end{pmatrix} \right|^2 dx_1 \ldots dx_\nu \, dz_1 \ldots dz_\nu \right]^{\frac{1}{2}}$$

$$= \frac{1}{\nu!} \left[\int_a^b \cdots \int_a^b K_1 K_1^* \begin{pmatrix} x_1, & \ldots, & x_\nu \\ x_1, & \ldots, & x_\nu \end{pmatrix} dx_1 \ldots dx_\nu \right]^{\frac{1}{2}}$$

$$\left[\int_a^b \cdots \int_a^b K_2 K_2^* \begin{pmatrix} z_1, & \ldots, & z_\nu \\ z_1, & \ldots, & z_\nu \end{pmatrix} dz_1 \ldots dz_\nu \right]^{\frac{1}{2}}$$

or

$$|d_\nu| \leq \frac{1}{\nu!} \int_a^b \cdots \int_a^b [K] \begin{pmatrix} x_1, & \ldots, & x_\nu \\ x_1, & \ldots, & x_\nu \end{pmatrix} dx_1 \ldots dx_\nu \qquad (5)$$

In arriving at this last expression, which, save for a factor $(-1)^\nu$, should be recognized as the νth coefficient in the power-series expansion of $D_{[K]}(\lambda)$, we have twice made use of Carleman's identity as well as the Schwarz inequality.

The relation (5) clearly shows that the unmodified Fredholm determinant $D_K(\lambda)$ is majorized term by term by $D_{[K]}(\lambda)$. As a consequence of this, the order of $D_K(\lambda)$ as an entire function can be no greater than the order of $D_{[K]}(\lambda)$, and if they are the same, then the type of the former must also be no greater than the type of the latter. If the order of $D_K(\lambda)$ is strictly less than 1, then the desired expansion (4) follows immediately. On the other hand, if the order of $D_K(\lambda)$ equals 1, then so does the order of $D_{[K]}(\lambda)$. In this case, since we have already seen that $D_{[K]}(\lambda)$ is of zero (minimal) type, the same must also be true of D_K. The special case of Boas' result already mentioned then implies the validity of (4).   Q.E.D.

As an obvious by-product of this theorem we have the *Lalesco result*.

COROLLARY

For composite $\mathcal{Q}^2$ kernels $K = K_1 K_2$ and arbitrary orthonormalized bases $\{\phi_\nu\}$†

$$\text{Tr}\,(K^m) = \sum_\nu \frac{1}{\lambda_\nu{}^m} = \sum_\nu (K^m \phi_\nu, \phi_\nu) \qquad m \geq 1 \qquad (6)$$

Proof   From (4.3-3) and (15.2-6) it is clear that these equivalences are valid for all $m \geq 2$. In addition, for $m = 1$ we have $\text{Tr}\,(K) = \sigma_M(K)$ also. Since, by definition (4.1-9), (4.1-10),

$$\text{Tr}\,(K) \equiv k_1 = -d_1 \equiv -\frac{dD(\lambda)}{d\lambda}\bigg|_{\lambda=0}$$

the remaining equality follows trivially in this case from (15.3-4).

One last point should be made before we leave this section. In the

---

† Relations such as $\text{Tr}\,(K) = \sum_\nu (1/\lambda_\nu)$, which are valid whenever the order ρ of $D_K(\lambda)$ is less than 1 [provided $\text{Tr}\,(K)$ is properly interpreted] often can be of value in establishing the existence of cv's of rather complicated kernels [see Cochran (1965) and Kwan (1965), for example]. Also compare Theorem 13.4-1.

above relation (6) and the subsequent expressions for the case $m = 1$, it should be clearly understood that Tr (K) is merely a shorthand way of writing

$$\text{Tr } (K_1 K_2) \equiv \int_a^b \int_a^b K_1(x,z) K_2(z,x) \, dz \, dx$$

where K_1 and K_2 are any two $\mathfrak{L}^2$ kernels whose composition equals K (in the almost-everywhere sense). It could be true, of course, that the given $\mathfrak{L}^2$ kernel K is one for which the conventional trace

$$\int_a^b K(x,x) \, dx$$

is not even defined.†

15.4   THE RESULTS OF CHANG AND WEYL

In the preceding section we have seen that composite (nuclear) kernels have finite spectral traces. In other words, we have established the fact that

$$\sum_\nu \frac{1}{\mu_\nu} < \infty \qquad \text{implies} \qquad \sum_\nu \frac{1}{|\lambda_\nu|} < \infty \ddagger$$

This implication, important as it is, is actually only a special case of a more general result, first demonstrated by Chang (1949) using a function-theoretic approach, namely,

$$\sum_\nu \frac{1}{\mu_\nu^{\, p}} < \infty \qquad \text{implies} \qquad \sum_\nu \frac{1}{|\lambda_\nu|^p} < \infty \qquad (1)$$

for general $\mathfrak{L}^2$ kernels and $0 < p < 2$. Weyl subsequently showed (1949) that for *arbitrary n* and positive p the polynomial

$$\prod_{\nu=1}^n \left(1 - \frac{\lambda}{\lambda_\nu^{\, p}}\right)$$

---

† N. B.: The line $y = x$ is a set of *planar* measure zero.
‡ As Prob. 4 clearly demonstrates, the converse of this implication is not generally valid.

is majorized term by term by the polynomial

$$\prod_{\nu=1}^{n} \left(1 - \frac{\lambda}{\mu_\nu{}^p}\right)$$

(the special case $p = 1$ was treated in Theorem 15.3), and, in fact,

$$\sum_{\nu=1}^{n} \frac{1}{|\lambda_\nu|^p} \leqq \sum_{\nu=1}^{n} \frac{1}{\mu_\nu{}^p} \qquad (2)$$

[See also Fan (1949, 1950, and 1951), Silberstein (1953), and Chang (1954).]

A year after the appearance of Weyl's paper, Horn (1950) published an interesting extension of certain of Weyl's inequalities. As a special case his results included the following.

HORN'S LEMMA   Let K, L be two $\mathfrak{L}^2$ kernels with sv's α_ν and β_ν, respectively. For arbitrary n and positive p the sv's γ_ν of the composition KL satisfy

$$\sum_{\nu=1}^{n} \frac{1}{\gamma_\nu{}^p} \leqq \sum_{\nu=1}^{n} \frac{1}{(\alpha_\nu\beta_\nu)^p} \qquad (3)$$

This result leads to an easy demonstration of a significant generalization of an earlier characteristic-value error estimate (see Prob. 6). It also permits a simplified proof of the extension of Theorem 15.2, originally due to Chang (1947 and 1949), which we mentioned earlier [see also Visser and Zaanen (1952)], namely, the following.

THEOREM 15.4

An $\mathfrak{L}^2$ kernel K is *(m — 1)-fold composite*, i.e., representable in the form

$$K = K_1 K_2 \cdots K_m$$

where K_1, K_2, . . . , K_m are themselves $\mathfrak{L}^2$ kernels, iff

$$\sum_{\nu} \left(\frac{1}{\mu_\nu}\right)^{2/m} < \infty$$

Proof   Let $K = K_1 K_2 \cdots K_m$ with K_1, K_2, . . . , K_m in $\mathfrak{L}^2$. For the case $m = 1$, the desired result is trivially true. Proceeding inductively, we assume it to be valid for all compositions with less than m factors. Thence, designating the sv's of K_1 by α_ν and those of $K_2 K_3 \cdots K_m$ by β_ν, we find

$$\sum_\nu \left(\frac{1}{\mu_\nu} \right)^{2/m} \leq \sum_\nu \left(\frac{1}{\alpha_\nu \beta_\nu} \right)^{2/m}$$

$$\leq \left[\sum_\nu \left(\frac{1}{\alpha_\nu} \right)^2 \right]^{1/m} \left[\sum_\nu \left(\frac{1}{\beta_\nu} \right)^{2/(m-1)} \right]^{(m-1)/m}$$

$$< \infty$$

Here we have made use of the fundamental result (3) and Hölder's inequality [see Beckenbach and Bellman (1965, p. 19), for example].

To establish the converse result, we merely note from Sec. 14.3 that if $\sum_\nu (1/\mu_\nu)^{2/m}$ is finite, then the equations

$$K_i \equiv [K]^{1/m} \qquad 1 \leq i \leq m - 1$$
$$K_m \equiv [K]^{1/m} U$$

define m square-summable kernels whose composition is equivalent to K.   Q.E.D.

15.5   KERNELS OF THE CLASS C_p

It is appropriate that at least some mention be made of the other classes (spaces) that Dunford and Schwartz (1963, pp. 1088–1144) designate as C_p and McCarthy (1967) calls c_p. For our purposes,† these are the collections of kernels K whose singular values μ_ν satisfy

$$\sum_\nu \frac{1}{\mu_\nu{}^p} < \infty \qquad 0 < p < \infty \qquad (1)$$

---

† The operator spaces appear to have been first introduced essentially by Schatten and Von Neumann (1948) and Schatten (1950), although some of the related results date as far back as 1937 (Von Neumann). See Dunford and Schwartz (1963, pp. 1163–1164) for remarks regarding modern Russian efforts in this area. Also see the recent text of Gohberg and Krein (1969).

Clearly, then, C_2 is the class of all $\mathfrak{L}^2$ (Hilbert-Schmidt) kernels while C_1 is the so-called *trace class* which has occupied our attention heretofore in this chapter.

In dealing with trace-class kernels, it was essentially a trivial observation to note (as we did in Prob. 2) that

$$K \in C_1 \qquad \text{iff} \qquad [K]^{\frac{1}{2}} \in C_2$$

If for arbitrary real positive r we define $[K]^r$ in the obvious manner as

$$[K]^r \equiv \sum_{\nu} \frac{\phi_\nu(x)\,\overline{\phi}_\nu(y)}{\mu_\nu{}^r}$$

[cf (14.3-9)], we can make an analogous elementary characterization statement in this more general situation, namely,

$$K \in C_p \qquad \text{iff} \qquad [K]^{p/2} \in C_2 \tag{2}$$

Other characterizations, however, may be of more practical value. Fortunately, a number of the results that we have established in earlier sections have their analogues here, at least for $1 \le p \le 2$, in which case the kernels in C_p behave very much like nuclear kernels. For instance, we find that a kernel K is in C_p iff

$$\sum_{\nu} \|K\phi_\nu\|^p < \infty$$

for *some* complete orthonormalized set $\{\phi_\nu\}$ of $\mathfrak{L}^2$ functions. Similarly, a kernel K is in C_p iff

$$\sum_{\nu} |(K\psi_\nu, \phi_\nu)|^p < \infty$$

for *all* orthonormalized bases $\{\phi_\nu\}, \{\psi_\nu\}$ of $\mathfrak{L}^2$. As before, we have the identities†

$$\sum_{\nu} \frac{1}{\mu_\nu{}^p} = \inf \sum_{\nu} \|K\phi_\nu\|^p \tag{3}$$

$$\sum_{\nu} \frac{1}{\mu_\nu{}^p} = \sup \sum_{\nu} |(K\psi_\nu, \phi_\nu)|^p \tag{4}$$

---

† Actually (4) is valid for all $p \ge 1$ (see Prob. 8); (3) holds for $0 < p \le 2$, while for $p > 2$ sup must replace inf.

where the inf and sup are to be taken over all appropriate choices of $\{\phi_\nu\}$ and $\{\psi_\nu\}$.

For finite $p > 2$ we no longer are dealing with kernels that are necessarily square summable. On the other hand, we have not left the space of kernels that give rise to completely continuous transformations.† There are kernels, however, that are in this more general class and yet not in C_p for any finite p. A simple example of such a kernel is

$$K(x,y) = \sum_\nu \frac{\phi_\nu(x)\,\overline{\phi_\nu}(y)}{\log(\nu+1)} \qquad (\phi_\mu, \phi_\nu) = \delta_{\mu\nu}$$

Much of the interest in the classes C_p stems from the fact that (at least for $p \geq 1$) they actually form normed linear spaces with the norm‡

$$\|K\|_p \equiv \left[\sum_\nu \frac{1}{\mu_\nu^p}\right]^{1/p} \tag{5}$$

These spaces, moreover, are metrizable and complete. In fact, C_p may be viewed as the Banach space formed by completion, with respect to the appropriate norm, of the class of all degenerate kernels. By Jensen's inequality [Beckenbach and Bellman (1965, p. 18), for example] we have

$$\left(\sum_\nu \frac{1}{\mu_\nu^q}\right)^{1/q} \leq \left(\sum_\nu \frac{1}{\mu_\nu^p}\right)^{1/p} \qquad p \leq q$$

---

† Such kernels may be characterized as those having a *polar representation* (14.3-10) in which $\mu_\nu \to \infty$ [see Schatten (1960, pp. 18–19)].

‡ If we let p increase without limit, we obtain the familiar *operator norm*

$$\|K\|_\infty \equiv \frac{1}{\mu_1}$$

which is commonly associated with the transformation (or operator) generated by the kernel K (also see the footnote on page 108). By virtue of (3) and (4) we have the additional well-known relations

$$\|K\|_\infty = \sup_{\|\phi\|=1} \|K\phi\| = \sup_{\substack{\|\phi\|=1 \\ \|\psi\|=1}} |(K\psi, \phi)|$$

so that C_p is contained in C_q whenever $p \leq q$. The Hermitian kernel

$$K(x,y) = \sum_{\nu} \frac{\phi_\nu(x)\overline{\phi}_\nu(y)}{\nu^{1/p}} \qquad (\phi_\mu, \phi_\nu) = \delta_{\mu\nu}$$

suffices as an example to show that the inclusion is proper if p is strictly less than q. Finally, we note without proof that the earlier Theorem 15.2 has the following extension.

THEOREM 15.5-1

A kernel K belongs to the class C_p iff it is representable as the composition

$$K = K_1 K_2$$

where K_1 and K_2 belong to the class C_{2p}.

Indeed, the following deeper and more interesting result is actually valid.

THEOREM 15.5-2

If the kernel K_1 belongs to C_p and K_2 to C_q, then the composition

$$K = K_1 K_2$$

belongs to the class C_r where

$$\frac{1}{r} = \frac{1}{p} + \frac{1}{q}$$

and $$\|K\|_r \leq \|K_1\|_p \|K_2\|_q$$

PROBLEMS

1. Consider the degenerate kernel

$$K = xy \qquad 0 \leq x,y \leq \pi$$

    Verify that the series $\sum_{\nu} \|K\phi_\nu\|$ diverges for the complete orthonormalized set $\{\phi_\nu\} \equiv \{\sqrt{2/\pi} \sin \nu x\}$.

2. Show that, for an $\mathfrak{L}^2$ kernel K, the following statements are equivalent:
    a. K is nuclear.

b. K^* is nuclear.

c. $[K]$ is nuclear.

d. $[K]^{\frac{1}{2}}$ is square summable.

3. Demonstrate that the kernels K_1 and K_2 given by (15.2-1) and (15.2-2) are both square integrable.

4. Consider the Volterra kernel

$$
K(x,y) = \begin{cases} 1 & y \le x \\ 0 & x < y \end{cases} \qquad 0 \le x,y \le \pi
$$

a. We know that the spectral trace (15.2-4) of this kernel vanishes. Show that the matrical trace (15.2-5) of this kernel is generally *different* from zero. For example, verify that

$$
\sum_\nu (K\phi_\nu, \phi_\nu) = \frac{4}{\pi} \sum_\nu \frac{1}{(2\nu - 1)^2} = \frac{\pi}{2}
$$

for the particular complete orthonormalized set

$$
\{\phi_\nu\} \equiv \left\{ \sqrt{\frac{2}{\pi}} \, \sin \nu x \right\}
$$

b. Verify that

$$
KK^* = \min (x,y) \qquad 0 \le x,y \le \pi
$$

so that the min kernel is composite and hence nuclear.

c. The preceding result could equivalently have been deduced from the behavior of the cv's themselves, which were essentially obtained earlier in Chap. 7, Prob. 4, as

$$
\lambda_\nu = \tfrac{1}{4}(2\nu - 1)^2
$$

with the associated cf's

$$
\phi_\nu = \sqrt{\frac{2}{\pi}} \, \cos (2\nu - 1) \frac{x}{2}
$$

Satisfy yourself that the Volterra kernel under consideration is therefore *not* nuclear.

5. Give an example of a nuclear $\mathfrak{L}^2$ kernel whose conventional trace

$$
\int_a^b K(x,x) \, dx
$$

is *not* defined.

6. Let K and L be two normal $\mathfrak{L}^2$ kernels with characteristic values λ_ν and μ_ν, respectively.
    a. Show that the singular values of the composite kernels KL, KL^*, K^*L, and K^*L^* are identical.
    b. Verify that Horn's lemma of Sec. 15.4 implies that the trace of each of these composite kernels is bounded from above, in magnitude, by

    $$\sum_\nu \left| \frac{1}{\lambda_\nu \mu_\nu} \right|$$

    c. Use (4.2-2) and the result of part b to establish that

    $$\sum_\nu \left(\left| \frac{1}{\lambda_\nu} \right| - \left| \frac{1}{\mu_\nu} \right| \right)^2 \leq \| K - L \|^2$$

    [This generalization of the cv error estimate (9.3-4) can be further improved; for example, see Cochran and Hinds (1972) where the Hoffman-Wielandt result (1953) for normal matrices is extended to this more general setting.]

7. [Gohberg and Krein (1969, p. 104).] Show that for a nuclear kernel K with cv's λ_ν and sv's μ_ν, respectively,

    $$\left| \sum_\nu \frac{1}{\lambda_\nu} \right| \leq \sum_\nu \frac{1}{\mu_\nu}$$

    with equality *iff* $e^{i\alpha}K$ is nonnegative definite for some real constant α.

8. Using 2×2 matrices, construct a counterexample to (15.5-4) for the case $0 < p < 1$.

9. Show by explicit example that for $0 < p < 1$ the expression (15.5-5) need not satisfy the triangle inequality and, therefore, is inappropriate as a norm in this case.

10. Discuss the nature of the kernel

    $$K(x,y) \equiv \sum_{\nu=0}^\infty \frac{\sin(2\nu+1)x \cos(2\nu+1)y}{(2\nu+1)}$$

    $$= \begin{cases} \frac{1}{4}\pi & 0 < y < \min(\pi-x,x) \\ 0 & \min(\pi-x,x) < y < \max(\pi-x,x) \\ -\frac{1}{4}\pi & \max(\pi-x,x) < y < \pi \end{cases}$$

    In particular, calculate the norm of K, determine its cv's and sv's and decide to which classes C_p it belongs.

chapter sixteen

The Effect of Smoothness Conditions on General Kernels

16.1 HÖLDER CONTINUITY

We turn now to a series of results that complement many of those discussed in the preceding chapter. Although our hypotheses will, on the whole, be of a different formal nature than those considered previously, many of our conclusions will appear pleasantly familiar. To avoid the needless consideration of various essentially pathological cases, however, we shall assume throughout this chapter that the kernels $K(x,y)$ with which we are dealing are all measurable along their main diagonals $x = y$.

The oldest known result of interest here appears to be one concerning Hölder continuous kernels which, in its original form, is due essentially to Fredholm (1903) [see also Lalesco (1907a† and 1912, pp. 86–89) and Cochran (1965)]. In analogy with the notion of relative uniformity as it pertains to convergence of series of functions (see Sec.

---

† Here and in 1907b Lalesco also briefly discusses conditions under which a kernel possesses only a *finite* number of cv's. Swann (1971a) has recently considered these matters more fully.

3.1), we shall say that a function $K(x,y)$ of two variables is *relatively uniformly Hölder continuous* (with exponent α) with respect to, say, the second variable if it satisfies the following inequality:[†]

$$| K(x,y) - K(x,z) | < A(x) | y - z |^{\alpha} \tag{1}$$

where $A(x)$ is nonnegative and summable. We note that if $K \in \mathfrak{L}^2$,

$$\int_a^b | K(x,y) | \, dx \tag{2}$$

must be finite for *some* y in the fundamental interval. For such functions it then follows from (1) that $K(x,y)$ is summable, as a function of x, for *all* y. Moreover, (1) also ensures that

$$\int_a^b | K(x,x) | \, dx < \infty \tag{3}$$

in these cases, too (see Prob. 1).

We are now ready to prove the following theorem.

THEOREM 16.1[‡]

Let the $\mathfrak{L}^2$ kernel K (with well-defined trace) be relatively uniformly Hölder continuous in either variable, with exponent $\alpha > \frac{1}{2}$. Then the Fredholm determinant of K exists and has the convergent product expansion

$$D_K(\lambda) = \prod_{\nu} \left(1 - \frac{\lambda}{\lambda_\nu} \right) \tag{4}$$

Proof   We want to determine an upper bound for the order ρ of $D_K(\lambda)$ as given by the classical formulas (cf Chap. 4, Prob. 2), viz.,

$$D_K(\lambda) = 1 + \sum_{\nu=1}^{\infty} d_\nu \lambda^\nu \tag{5}$$

where $$d_\nu = \frac{(-1)^\nu}{\nu!} \int_a^b \cdots \int_a^b K\binom{x_1, \ldots, x_\nu}{x_1, \ldots, x_\nu} dx_1 \cdots dx_\nu \tag{6}$$

[†] A similar definition holds if the function K is relatively uniformly Hölder continuous with respect to the first variable.

[‡] See Weidmann (1966) for a closely related result in which (1) is replaced by integrated Hölder conditions suggested by earlier work of Hammerstein (1923 and 1928).

with

$$K\begin{pmatrix} x_1, \ldots, x_\nu \\ x_1, \ldots, x_\nu \end{pmatrix} \equiv D \equiv \begin{vmatrix} K(x_1,x_1) & K(x_1,x_2) & \ldots & K(x_1,x_\nu) \\ K(x_2,x_1) & K(x_2,x_2) & \ldots & K(x_2,x_\nu) \\ \cdots \cdots \cdots \cdots \cdots \cdots \cdots \cdots \cdots \\ K(x_\nu,x_1) & K(x_\nu,x_2) & \ldots & K(x_\nu,x_\nu) \end{vmatrix}$$

(7)

Since we could just as easily be working with the adjoint kernel $K^* = \overline{K(y,x)}$ as with K itself, we can assume, without loss of generality, that the Hölder continuity stated in the hypothesis is with respect to the second variable and that K satisfies (1) and (3).

In order to obtain a bound for the coefficients d_ν of (6), we first transform the determinant in (7) by subtracting the second column from the first, the third from the second, and so on. Having done this, we next perform a second series of operations on the transformed determinant, dividing the first column by $(x_1 - x_2)^\alpha$, the second by $(x_2 - x_3)^\alpha$, and so on here, too, up to and including the $(\nu - 1)$st column. In this manner we arrive at the representation

$$D = \begin{vmatrix} \varepsilon_{11} \ldots \varepsilon_{1\nu} \\ \cdots \cdots \cdots \\ \varepsilon_{\nu 1} \ldots \varepsilon_{\nu\nu} \end{vmatrix} \prod_{j=1}^{\nu-1} (x_j - x_{j+1})^\alpha$$

(8)

where

$$\varepsilon_{ij} \equiv \frac{K(x_i,x_j) - K(x_i,x_{j+1})}{(x_j - x_{j+1})^\alpha} \qquad i = 1, 2, \ldots, \nu$$
$$\varepsilon_{i\nu} \equiv K(x_i,x_\nu) \qquad\qquad j = 1, 2, \ldots, \nu - 1$$

The determinant in (8) can be estimated using Hadamard's inequality in the form

$$|\det(\varepsilon_{ij})|^2 \le \prod_{i=1}^\nu \sum_{j=1}^\nu |\varepsilon_{ij}|^2$$

We then find, using (1) and employing an obvious inequality, that

$$|D| < B \prod_{j=1}^{\nu-1} |x_j - x_{j+1}|^\alpha$$

(9)

where $$B \equiv \prod_{i=1}^\nu \left[(\nu - 1)^{\frac{1}{2}} A(x_i) + |K(x_i,x_\nu)| \right]$$

(10)

Inasmuch as the original determinant (7) was symmetric in the x_i, we may assume in deriving a further upper bound that

$$b \geq x_1 \geq x_2 \geq \cdots \geq x_\nu \geq a$$

The second factor on the right-hand side of (9) is then maximized by spacing the x_i uniformly between a and b, and as a consequence we obtain

$$|D| < C^\nu (\nu - 1)^{-\alpha(\nu-1)} B \qquad \nu > 1 \qquad (11)$$

Here and henceforth C will be used to designate a generic constant, independent of ν, but not necessarily the same each time it appears.

In order to complete our estimation of the coefficients of the power series (5) we note that

$$\int_a^b \cdots \int_a^b B \, dx_1 \cdots dx_\nu = \int_a^b \left\{ \int_a^b [(\nu - 1)^{\frac{1}{2}} A(x) + |K(x,y)|] \, dx \right\}^{\nu-1}$$

$$[(\nu - 1)^{\frac{1}{2}} A(y) + |K(y,y)|] \, dy$$

$$\leq C^\nu \nu^{\nu/2}$$

by virtue of (3) and the earlier observation concerning the summability of $K(x,y)$ for fixed y. Combining this result with that of (11) and making use of Stirling's expansion for the gamma (factorial) function [Courant and Hilbert (1953, pp. 522–524)], we finally find

$$|d_\nu| < C^\nu \nu^{-\nu(\alpha+\frac{1}{2})} \qquad (12)$$

The relation (12) implies that the order of the entire function $D_K(\lambda)$ satisfies

$$\rho \leq \frac{1}{\alpha + \frac{1}{2}}$$

For $\alpha > \frac{1}{2}$, ρ is then less than 1 and the desired product expansion (4) follows immediately.   Q.E.D.

The above result is sharp in the following sense: there exist $\mathfrak{L}^2$ kernels K which are relatively uniformly Hölder continuous with exponent α precisely equal to $\frac{1}{2}$ and for which

$$\sum_\nu \frac{1}{|\lambda_\nu|} = \infty \qquad (13)$$

that is, $\rho = 1$. To be sure, Bernstein (1914a and b, 1934) [see also Bary (1964, vol. 2, pp. 153–171), for example] proved that if $\alpha \leq \frac{1}{2}$, then it was possible to construct continuous functions $k(x)$ of a single real variable, satisfying Hölder conditions of order α, and whose classical Fourier series were not absolutely convergent. If we use such a function to generate a *periodic difference kernel* according to

$$K(x,y) = k(x-y) \qquad 0 \leq x, y \leq 2\pi$$

then the Hölder continuity of k combined with the equivalence

$$2\pi \sum_\nu |a_\nu| = \sum_\nu \frac{1}{|\lambda_\nu|} \tag{14}$$

where the a_ν are the Fourier coefficients of k, leads to the announced result (13).

It is worth observing, before we move on to the next section, that the relation (14) is generally valid for periodic difference kernels (see Chap. 13, Prob. 1).† This means that these kernels provide an important connecting link between integral equations, on the one hand, and the theory of Fourier series, on the other.‡ As such they are well suited, therefore, for sharpening results and/or constructing counterexamples, a fact which has been exploited previously by numerous researchers.

16.2 BOUNDED VARIATION

For functions of two (or more) variables the concept of bounded variation can take many forms. Indeed, Clarkson and Adams (1933 and 1934) compare and discuss the merits of seven different definitions proposed by various authors (at various times and for various purposes). From our point of view, however, it will be sufficient to restrict attention to the behavior of the functions under consideration as it pertains (essentially) to any one of its variables alone. In keeping with earlier terminology we shall say that a function $K(x,y)$ of two variables is *relatively uniformly of bounded variation* with respect to, say, the second variable

---

† Periodic difference kernels are also *normal*, and thus not only is $\lambda_\nu = 1/(2\pi a_\nu)$ but also the singular values coincide with the moduli of these characteristic values.

‡ The existence of this link led Swann in 1966 to initiate a careful investigation of certain remarkable analogies that exist between these two disciplines (see Sec. 16.5).

if for all $\nu \geq 2$ and arbitrary choice of partition $b \geq y_1 \geq y_2 \geq \cdots \geq y_\nu \geq a$,†

$$\sum_{j=1}^{\nu-1} |K(x,y_j) - K(x,y_{j+1})| < B(x) \tag{1}$$

where $B(x)$ is nonnegative and summable.

In the preceding section we saw that $\mathfrak{L}^2$ kernels that were relatively uniformly Hölder continuous in either variable, with exponent $\alpha \geq \frac{1}{2}$, gave rise to Fredholm determinants of order $\rho \leq 1$. The analogous result here is the following theorem [also see Swann (1971b)].

THEOREM 16.2

Let the $\mathfrak{L}^2$ kernel K (with well-defined trace) be relatively uniformly of bounded variation in either variable. Then the Fredholm determinant of K is an entire function of order $\rho \leq 1$.

Proof   The technique of proof is similar to that used in establishing Theorem 16.1. For the same reason as stated there, we can assume, without loss of generality, that the bounded variation stated in the hypothesis is with respect to the second variable and that K satisfies (1). We again transform the determinant in (16.1-7) by subtracting the second column from the first, the third from the second, and so on, thus obtaining the representation

$$D = \det(\varepsilon_{ij})$$

where now

$$\varepsilon_{ij} \equiv K(x_i,x_j) - K(x_i,x_{j+1})$$
$$\varepsilon_{i\nu} \equiv K(x_i,x_\nu)$$

$$\begin{aligned} i &= 1, 2, \ldots, \nu \\ j &= 1, 2, \ldots, \nu-1 \end{aligned}$$

Employing Hadamard's inequality, we then find

$$|D| \leq \prod_{i=1}^{\nu} \left[\sum_{j=1}^{\nu} |\varepsilon_{ij}|^2 \right]^{\frac{1}{2}}$$

$$\leq \prod_{i=1}^{\nu} \sum_{j=1}^{\nu} |\varepsilon_{ij}|$$

$$\leq \prod_{i=1}^{\nu} [B(x_i) + |K(x_i,x_\nu)|] \tag{2}$$

---

† As before, a similar definition holds here, too, if the function K is relatively uniformly of bounded variation with respect to the first variable. [Compare the definition of Tonelli (1926).]

Here, of course, we have made use of (1), taking advantage, whenever necessary, of the symmetry of the original determinant (16.1-7).

Continuing our estimation of the coefficients d_ν as given by (16.1-6), we note that

$$\int_a^b \cdots \int_a^b |D| \, dx_1 \cdots dx_\nu \le \int_a^b \left\{ \int_a^b \left[B(x) + |K(x,y)| \right] dx \right\}^{\nu-1}$$

$$\left[B(y) + |K(y,y)| \right] dy$$

$$\le C^\nu$$

where C is a generic constant independent of ν. This last relation is a consequence of the summability of B, the square summability of K, and an observation analogous to that immediately preceding (16.1-3) (see Prob. 2). Again making use of Stirling's expansion for the gamma function, we finally obtain

$$|d_\nu| < C^\nu \nu^{-\nu}$$

and thus the order ρ of $D_K(\lambda)$ satisfies

$$\rho \le 1 \qquad\qquad\qquad \text{Q.E.D.}$$

This result is also sharp in the sense that there exist $\mathfrak{L}^2$ kernels that are relatively uniformly of bounded variation and for which $\rho = 1$; that is, (16.1-13) is valid. As might be expected, we need only observe the behavior of periodic difference kernels generated by functions of a single real variable which are, on the one hand, of bounded variation but, on the other, have nonabsolutely convergent classical Fourier series. Bary (1964, vol. 2, pp. 162–165) shows how such functions can be easily constructed.

16.3   KERNELS WITH COMBINED PROPERTIES

Theorems 16.1 and 16.2 actually represent separate and distinct results. Although it is easily established that if $K(x,y)$ is relatively uniformly Hölder continuous in either variable with exponent $\alpha > 1$, then it is a function solely of the *other* variable and that if $\alpha = 1$, then it is of bounded variation in the *same* variable, it is equally easily established that there are functions satisfying (16.2-1) that do not satisfy (16.1-1) for any positive α; similarly there exist functions that satisfy (16.1-1) with given $\alpha < 1$ which have unbounded variation.

If we combine the two properties of Hölder continuity and bounded variation we obtain the following.

THEOREM 16.3-1

Let the $\mathfrak{L}^2$ kernel K (with well-defined trace) be relatively uniformly Hölder continuous in either variable with exponent $\alpha > 0$ and relatively uniformly of bounded variation in the *same* variable. Then the Fredholm determinant of K is an entire function of order

$$\rho \leq \frac{1}{\frac{1}{2}\alpha + 1} < 1 \tag{1}$$

and therefore the convergent product expansion

$$D_K(\lambda) = \prod_\nu \left(1 - \frac{\lambda}{\lambda_\nu}\right) \tag{2}$$

is valid.

Proof   Following the outline of the two previous proofs, we assume, without loss of generality, that K satisfies (16.1-1) and (16.2-1) and begin, as usual, by transforming the determinant in (16.1-7) by subtracting the second column from the first, and so on. We also perform a second series of operations on the transformed determinant, this time, however, dividing the first column by $(x_1 - x_2)^{\alpha/2}$, the second by $(x_2 - x_3)^{\alpha/2}$, and so on, up to and including the $(\nu - 1)$st column. We thus have

$$D = \det(\varepsilon_{ij}) \prod_{j=1}^{\nu-1} (x_j - x_{j+1})^{\alpha/2} \tag{3}$$

with

$$\varepsilon_{ij} \equiv \frac{K(x_i, x_j) - K(x_i, x_{j+1})}{(x_j - x_{j+1})^{\alpha/2}} \qquad \begin{array}{l} i = 1, 2, \ldots, \nu \\ j = 1, 2, \ldots, \nu - 1 \end{array}$$

$$\varepsilon_{i\nu} \equiv K(x_i, x_\nu)$$

At this stage we again make use of the symmetry of the original determinant (16.1-7) and assume

$$b \geq x_1 \geq x_2 \geq \cdots \geq x_\nu \geq a \tag{4}$$

It then follows from Hadamard's inequality and (16.1-1), (16.2-1) that

$$|\det(\varepsilon_{ij})| \le \prod_{i=1}^{\nu} \left[\sum_{j=1}^{\nu} |\varepsilon_{ij}|^2\right]^{\frac{1}{2}}$$

$$\le \prod_{i=1}^{\nu} \left[A(x_i)\sum_{j=1}^{\nu-1} |K(x_i,x_j) - K(x_i,x_{j+1})| + |K(x_i,x_\nu)|^2\right]^{\frac{1}{2}}$$

$$\le \prod_{i=1}^{\nu} \left[A^{\frac{1}{2}}(x_i)B^{\frac{1}{2}}(x_i) + |K(x_i,x_\nu)|\right]$$

If we designate this last product by E, we find, as before,

$$|D| \le E\prod_{j=1}^{\nu-1} |x_j - x_{j+1}|^{\alpha/2}$$

$$< C^\nu \nu^{-\frac{1}{2}\alpha\nu} E \qquad\qquad (5)$$

Performing the required integrations of E leads to

$$\int_a^b \cdots \int_a^b E\, dx_1 \cdots dx_\nu \le \int_a^b \left\{ \int_a^b [A^{\frac{1}{2}}(x)B^{\frac{1}{2}}(x) + |K(x,y)|]dx\right\}^{\nu-1}$$
$$[A^{\frac{1}{2}}(y)B^{\frac{1}{2}}(y) + |K(y,y)|]\, dy$$

$$\le C^\nu$$

and thus from (16.1-6),

$$|d_\nu| < C^\nu \nu^{-\nu(\frac{1}{2}\alpha+1)}$$

This last relation implies that $D_K(\lambda)$ is an entire function of order

$$\rho \le \frac{1}{\frac{1}{2}\alpha + 1}$$

Since this is less than 1 for all positive α, the desired product expansion (2) follows directly.   Q.E.D.

A somewhat more surprising result, which indicates the interplay between the variables that occurs in characteristic-value calculations, is the following.

THEOREM 16.3-2

For bounded $\mathfrak{L}^2$ kernels the conclusions of Theorem 16.3-1 are equally valid even if the Hölder continuity and the bounded variation are with respect to *opposite* variables.

Proof   Our proof follows the now familiar pattern. We assume that the Hölder continuity is with respect to the first variable while the bounded variation concerns the second. Thus, K satisfies

$$|K(x_1,y) - K(x_2,y)| < A|x_1 - x_2|^\alpha \tag{6}$$

and for all $\nu \geq 2$ and arbitrary, but ordered, choice of partition,

$$\sum_{j=1}^{\nu-1} |K(x,y_j) - K(x,y_{j+1})| < B \tag{7}$$

where A and B are positive constants. We again transform the determinant in (16.1-7) by subtracting the second column from the first, and so on. This time, for our second series of operations on the transformed determinant, we subtract the second row from the first, the third from the second, and so on, again. The representation we obtain is

$$D = \det(\varepsilon_{ij})$$

where for $i,j = 1, 2, \ldots, \nu - 1$,

$$\varepsilon_{ij} \equiv K(x_i,x_j) - K(x_i,x_{j+1}) - K(x_{i+1},x_j) + K(x_{i+1},x_{j+1})$$

$$\varepsilon_{i\nu} \equiv K(x_i,x_\nu) - K(x_{i+1},x_\nu)$$

$$\varepsilon_{\nu j} \equiv K(x_\nu,x_j) - K(x_\nu,x_{j+1})$$

$$\varepsilon_{\nu\nu} \equiv K(x_\nu,x_\nu)$$

Note especially that the two relations

$$|\varepsilon_{ij}| \leq |K(x_i,x_j) - K(x_i,x_{j+1})| + |K(x_{i+1},x_j) - K(x_{i+1},x_{j+1})|$$

and

$$|\varepsilon_{ij}| \leq |K(x_i,x_j) - K(x_{i+1},x_j)| + |K(x_i,x_{j+1}) - K(x_{i+1},x_{j+1})|$$

are valid for $1 \leq i,j \leq \nu - 1$.

We now employ Hadamard's inequality and take cognizance of

(6), (7), and the symmetry of the original determinant (16.1-7). Thus

$$|D| \le \left[\sum_{j=1}^{\nu-1} |\varepsilon_{\nu j}|^2 + |\varepsilon_{\nu\nu}|^2 \right]^{\frac{1}{2}} \prod_{i=1}^{\nu-1} \left[\sum_{j=1}^{\nu-1} |\varepsilon_{ij}|^2 + |\varepsilon_{i\nu}|^2 \right]^{\frac{1}{2}}$$

$$\le [B + |K(x_\nu, x_\nu)|] \prod_{i=1}^{\nu-1} \left[2A |x_i - x_{i+1}|^\alpha \sum_{j=1}^{\nu-1} |\varepsilon_{ij}| + A^2 |x_i - x_{i+1}|^{2\alpha} \right]^{\frac{1}{2}}$$

$$\le C \prod_{i=1}^{\nu-1} [2A^{\frac{1}{2}} B^{\frac{1}{2}} |x_i - x_{i+1}|^{\frac{1}{2}\alpha} + A |x_i - x_{i+1}|^\alpha]$$

$$\le C^\nu \prod_{i=1}^{\nu-1} |x_i - x_{i+1}|^{\frac{1}{2}\alpha}$$

$$\le C^\nu \nu^{-\frac{1}{2}\alpha\nu}$$

It then follows that

$$|d_\nu| < C^\nu \nu^{-\nu(\frac{1}{2}\alpha+1)}$$

which, in essence, concludes our proof.

16.4   DIFFERENTIABLE KERNELS

In the preceding section we saw that Hölder continuity (with exponent α) plus bounded variation of an $\mathfrak{L}^2$ kernel K lead to the inequality

$$\rho \le \frac{1}{\frac{1}{2}\alpha + 1} \tag{1}$$

for the order ρ of the Fredholm determinant associated with K. This, in turn, implies that the cv's λ_ν of K must be such that

$$\lim_{\nu \to \infty} \frac{\nu^{\frac{1}{2}\alpha+1-\varepsilon}}{|\lambda_\nu|} = 0 \tag{2}$$

for arbitrary positive ε. In particular, we have for $\alpha > 0$

$$\lim_{\nu \to \infty} \frac{\nu}{|\lambda_\nu|} = 0$$

a result that necessitates $\alpha > \frac{1}{2}$ in the absence of the bounded variation.

At the other extreme, we find for $\alpha = 1$

$$\lim_{\nu \to \infty} \frac{\nu^{\frac{3}{2}-\varepsilon}}{|\lambda_\nu|} = 0 \tag{3}$$

Here the bounded variation contributes nothing new since it automatically occurs in this case.

Weyl (1911 and 1912a) first showed that for *continuously differentiable* Hermitian kernels, (3) is valid with $\varepsilon = 0$. Mazurkiewicz (1915) subsequently extended Weyl's conclusion to the non-Hermitian case. Neither of these results, however, is sharp. In fact, as a special case of a theorem of Hille and Tamarkin (1931) we find that if the general $\mathfrak{L}^2$ kernel $K(x, y)$ is such that

1. $K, \dfrac{\partial K}{\partial x}, \ldots, \dfrac{\partial^{s-1} K}{\partial x^{s-1}}$ are continuous, say, in x, for almost all y

2. $\dfrac{\partial^s K(x, y)}{\partial x^s} = \displaystyle\int_a^x g(t, y)\, dt + A(y)$

    with $\displaystyle\int_a^b \left[\int_a^b |g(x, y)|^p\, dx \right]^{1/(p-1)} dy < \infty \qquad 1 < p \le 2$

then
$$\lim_{\nu \to \infty} \frac{\nu^{s+2-1/p}}{|\lambda_\nu|} = 0 \tag{4}$$

We note from (4), therefore, that simply a *square-summable* first derivative of K suffices for the validity of (3) with $\varepsilon = 0$.

Although we choose not to dwell here on the proof of the Hille-Tamarkin results, we shall establish the following theorem, which is closely related, albeit slightly weaker.

THEOREM 16.4

If the $\mathfrak{L}^2$ kernel K (with well-defined trace) satisfies conditions 1 and 2 above for some nonnegative integer s, then the Fredholm determinant of K is an entire function of order

$$\rho \le \frac{1}{s + 2 - 1/p} \tag{5}$$

Proof   In our usual fashion, we operate upon the determinant (16.1-7), assuming, in view of its symmetry, the natural ordering (16.3-4) for the points x_i. We begin by subtracting each row from the one above, proceeding in a consecutive manner from the first row. We then divide the newly formed ith row $(1 \le i \le \nu - 1)$ by $(x_i - x_{i+1})$, using the mean-value theorem to reexpress the resulting quotient (for almost all x_j) as

$$\frac{\partial K(x_i', x_j)}{\partial x} \qquad \text{with } x_i \ge x_i' \ge x_{i+1}$$

The entire process is then repeated, at this stage, however, only operating on the first $(\nu - 1)$ rows and dividing the new ith row $(1 \le i \le \nu - 2)$ by $(x_i' - x_{i+1}')$, and so on, for a total of s times (we assume $\nu > s$). One final row-from-row subtraction is then performed, followed by a division of the ith row $(1 \le i \le \nu - s - 1)$ by

$$(x_i^{(s)} - x_{i+1}^{(s)})^{(1-1/p)}$$

with the result that

$$D = \prod_{r=0}^{s} \prod_{i=1}^{\nu-r-1} (x_i^{(r)} - x_{i+1}^{(r)})^{\alpha_r} \det (\varepsilon_{ij}) \tag{6}$$

where $\quad \alpha_r = 1 \qquad\qquad r \ne s$

$$\alpha_s = \left(1 - \frac{1}{p}\right)$$

$$K^{(r+1)}(x_i^{(r+1)}, x_j) \equiv \frac{K^{(r)}(x_i^{(r)}, x_j) - K^{(r)}(x_{i+1}^{(r)}, x_j)}{x_i^{(r)} - x_{i+1}^{(r)}} \qquad \text{a.e. in } x_j \quad r \ne s\dagger$$

and for all $1 \le j \le \nu$ $\qquad\qquad (x_i^{(r)} \ge x_i^{(r+1)} \ge x_{i+1}^{(r)})$

$$\varepsilon_{ij} \equiv \frac{K^{(s)}(x_i^{(s)}, x_j) - K^{(s)}(x_{i+1}^{(s)}, x_j)}{(x_i^{(s)} - x_{i+1}^{(s)})^{(1-1/p)}} \qquad 1 \le i \le \nu - s - 1$$

$$\varepsilon_{ij} \equiv K^{(\nu-i)}(x_i^{(\nu-i)}, x_j) \qquad\qquad \nu - s \le i \le \nu$$

In estimating $|D|$ from (6), the product factors are easily dispensed with in the conventional fashion, yielding as an upper bound

$$C^\nu \nu^{-\nu(s+1-1/p)} \tag{7}$$

$\dagger$ Here, for convenience, we introduce the shorthand notation

$$K^{(r)}(x, y) \equiv \frac{\partial^r K(x, y)}{\partial x^r}$$

with C designating a generic constant independent of ν. Then, using Hadamard's, Hölder's, and Jensen's inequalities [see Beckenbach and Bellman (1965, pp. 18–19 and 64), for example], we find

$$|\det(\varepsilon_{ij})|^2 \leq \prod_{j=1}^{\nu} \sum_{i=1}^{\nu} |\varepsilon_{ij}|^2$$

$$\leq \prod_j \left[\sum_{i=1}^{\nu-s-1} |x_i^{(s)} - x_{i+1}^{(s)}|^{2(1/p-1)} \left| \int_{x_{i+1}^{(s)}}^{x_i^{(s)}} g(t,x_j)\,dt \right|^2 \right.$$
$$\left. + \sum_{i=\nu-s}^{\nu} |K^{(\nu-i)}(x_i^{(\nu-i)}, x_j)|^2 \right]$$

$$\leq \prod_j \left\{ \sum_{i=1}^{\nu-s-1} \left[\int_{x_{i+1}^{(s)}}^{x_i^{(s)}} |g(t,x_j)|^p\,dt \right]^{2/p} + \sum_{i=0}^{s} |K^{(i)}(x_{\nu-i}^{(i)}, x_j)|^2 \right\}$$

$$\leq \prod_j \left\{ \left[\int_a^b |g(t,x_j)|^p\,dt \right]^{2/p} + \sum_{i=0}^{s} |K^{(i)}(x_{\nu-i}^{(i)}, x_j)|^2 \right\}$$

Noting from conditions 1 and 2 that for each $r \leq s$

$$|K^{(r)}(x,y)| \leq C \left[\int_a^b |g(t,y)|^p\,dt \right]^{1/p} + A_r(y)$$

where A_r is a nonnegative summable function of y (since K is also in $\mathfrak{L}^2$ and hence must be summable in y), this last result gives rise to

$$|\det.(\varepsilon_{ij})| \leq C^\nu \prod_{j=1}^{\nu} \left\{ \left[\int_a^b |g(t,x_j)|^p\,dt \right]^{1/p} + B(x_j) \right\} \qquad (8)$$

with positive, integrable $B(x_j)$. Combining (7) and (8), performing the indicated integrations, employing Stirling's approximation, and using Hölder's inequality once more finally yields

$$|d_\nu| \leq C^\nu \nu^{-\nu(s+2-1/p)} \left\{ \int_a^b \left[\int_a^b |g(x,y)|^p\,dx \right]^{1/p} dy + B \right\}^\nu$$

$$\leq C^\nu \nu^{-\nu(s+2-1/p)} \left\{ C_1 \left[\int_a^b \left[\int_a^b |g(x,y)|^p\,dx \right]^{1/(p-1)} dy \right]^{1-1/p} + B \right\}^\nu$$

$$\leq C^\nu \nu^{-\nu(s+2-1/p)}$$

from which the desired relation (5) immediately follows.

The result that we have just demonstrated is sharp in the sense that there exist $\mathfrak{L}^2$ kernels satisfying conditions 1 and 2 above for which equality holds in (5). The periodic difference kernel $K(x - y)$ generated by the function

$$k(x) = \sum_{\nu=2}^{\infty} \frac{e^{i\nu x}}{\nu^{s+2-1/p} \log \nu} \qquad 0 \le x \le 2\pi$$

is, for $1 < p \le 2$ and integer s, a simple example of such a kernel.†

The Hille and Tamarkin theorem mentioned earlier is a generalization of (4) (which also, therefore, improves upon our earlier Theorems 16.1 and 16.3-1) to the case wherein condition 2 is replaced by a *fractional integral* of order $0 < \alpha < 1$ similar to (1.4-1). Under other less restrictive conditions, Hille and Tamarkin also have derived related results, including growth estimates for the cv's of kernels whose sth derivative satisfies an integrated Hölder (Lipschitz) condition à la Szász (1922 and 1928) or Hardy and Littlewood (1928a) [also see Silberstein (1953)].

Chang (1952)‡ has likewise considered the ramifications of conditions 1 and 2, at least in the case $p = 2$. Adapting an earlier argument of Krein (1937), he has shown that relation (4) is equally valid for the *singular* values of $\mathfrak{L}^2$ kernels satisfying these assumptions. As a consequence, we see from Theorem 15.4 that such kernels must at least be $(2s + 1)$-*fold composite*. Roughly speaking, therefore, we have the "rule of thumb" that the smoother a kernel is, the greater is the number of factors into which it can be decomposed.

16.5 ANALOGIES WITH FOURIER SERIES

Throughout the several sections of this chapter, whenever we have needed examples to clarify the sharpness of a particular result, we have turned to the theory of Fourier series, selecting therefrom functions having the necessary properties and then using them to generate periodic difference kernels exhibiting the desired behavior. The rationale for this procedure has as its basis the remarkable analogies that exist between growth estimates for the classical Fourier coefficients of functions (of one variable) satisfying various smoothness conditions, on the one hand, and similar estimates for the characteristic values of (two-variable) kernels having appropriately corresponding properties, on the other.§ As

---

† The convergence exponent of the zeros of $D_K(\lambda)$ for this kernel is precisely

$$1/(s + 2 - 1/p)$$

‡ See also Smithies (1937).

§ In a series of unpublished manuscripts covering the period 1966–1968, Swann laid much of the groundwork for the systematic analysis of the analogies that we have presented in this chapter.

we have already noted (at least implicitly) the hypothesis and conclusion (though, naturally, not the proof) of Theorem 16.1 mirror a Fourier-series result of Bernstein (1914a and b) [see also Bary (1964, vol. II, p. 154)]. The order estimate established in the proof of Theorem 16.1 is, in fact, exactly the same as the convergence exponent determined later by Szász (1922) for the analogous Fourier series situation. Theorems 16.3-1, 16.3-2 have distinct parallels in similar theorems of Zygmund (1928) [Bary (1964, p. 161)] and Waraszkiewicz (1929). The case $s = 0$ of Theorem 16.4 is closely related to a result of Tonelli (1925) [see also Zygmund (1928) and Bary (1964, p. 161)], while the higher-order cases, and especially the generalizations of Hille and Tamarkin already mentioned, are analogous to a theorem of Szász (1928) [see also Hardy and Littlewood (1928b), Theorem 8].

The strengthening of (essentially) Theorem 16.4 by Chang (1952), however, throws interesting new light on the above relationships. His result suggests that perhaps there is an even more powerful analogy to be exploited, namely, between growth estimates of Fourier coefficients of functions and similar estimates for *singular* values of kernels. Indeed, if we recall the theorem of M. Riesz [see Hardy and Littlewood (1928c) or Bary (1964, pp. 184ff)] that "the Fourier series of an $\mathfrak{L}^2$ function f converges absolutely *iff* f can be represented as the *convolution* of two other $\mathfrak{L}^2$ functions" and use the nature of periodic difference kernels to provide the key as to what best constitutes the analogous result in the theory of integral equations, we logically arrive at Theorem 15.2, namely, "the series of reciprocal *singular* values of an $\mathfrak{L}^2$ kernel K converges *iff* K can be represented as the *composition* of two other $\mathfrak{L}^2$ kernels."[†] On the basis of these two analogies, which, as far as integral equation theory is concerned, focus on the singular values of given kernels rather than their characteristic values, we conjecture that all of the integral equation results previously enunciated in this chapter remain valid when the sv's are considered in place of the cv's.

In closing we remark that a suitable modification of the proof of Theorem 16.1 enables one to establish that K is nuclear if $\alpha > \frac{3}{4}$. It also follows from recent work of Weidmann (1966) that $K(x,y)$ is nuclear if it is (essentially) Hölder continuous in *both* variables with exponent $\alpha > \frac{1}{2}$. It is our feeling, of course, that these results can be sharpened in the manner suggested above.

[†] It is interesting to note that whereas Riesz's theorem is essentially of a more theoretical nature, the analogous Theorem 15.2 has considerable practical value.

PROBLEMS

1. Verify that if a function $K(x,y)$ of two variables is measurable for $x = y$, satisfies (16.1-1), and, for *some* y, is a summable function of x, then $K(x,x)$ and, for *all* y, $K(x,y)$ are summable functions of x.

2. Show that conclusions similar to those in Prob. 1 are also valid if the function $K(x,y)$ satisfies (16.2-1) instead of (16.1-1).

3. Consider the $\mathfrak{L}^2$ kernel given by

$$K(x,y) = 2 \sum_{\nu=1}^{\infty} \frac{\sin \nu \pi x \cos \nu \pi y}{\nu^2 \pi^2} \qquad 0 \le x, y \le 1$$

   a. Show that this series representation is indeed a singular function expansion and discuss its convergence.

   b. Verify the closed-form expressions

$$[K] = \begin{cases} x(1-y) & x \le y \\ y(1-x) & y \le x \end{cases}$$

$$U = \frac{\sin \pi x}{\cos \pi y - \cos \pi x}$$

   and hence

$$K = \frac{x}{\pi} \int_x^1 \log |\cos \pi y - \cos \pi z| \, dz - \frac{(1-x)}{\pi} \int_0^x \log |\cos \pi y - \cos \pi z| \, dz$$

4. Given an $\mathfrak{L}^2$ kernel K with cv's λ_ν and sv's μ_ν, respectively, one of the Weyl (1949) results mentioned in Sec. 15.4 is that for arbitrary n

$$\prod_{\nu=1}^{n} \left| \frac{1}{\lambda_\nu} \right| \le \prod_{\nu=1}^{n} \left(\frac{1}{\mu_\nu} \right)$$

   a. Use this result and Stirling's formula for the factorial function to show that for positive p

$$\frac{1}{\lambda_\nu} = O(\nu^{-p}) \qquad \text{if} \qquad \frac{1}{\mu_\nu} = O(\nu^{-p})$$

   b. What does this imply about the order of the Fredholm determinant of the kernel of Prob. 3?

5. a. Exhibit two $\mathfrak{L}^2$ kernels whose composition yields the kernel of Prob. 3.
    b. Verify that although its trace vanishes this kernel does possess characteristic values.

6. Satisfy yourself that the kernel of Prob. 3 is relatively uniformly Hölder continuous in both variables with exponent $\alpha > \frac{1}{2}$. Can $\alpha = 1$?

7. a. Show that the kernel of Prob. 3 satisfies (16.2-1).
    b. What is the best upper bound for the order ρ of the Fredholm determinant of this kernel provided by the several theorems of Secs. 16.1, 16.2, and 16.3?

8. a. Discuss how Theorem 16.4 is weaker than the Hille-Tamarkin result (16.4-4).
   b. Determine which values of s and p are appropriate in (16.4-4) for the kernel of Prob. 3.

chapter seventeen

Other Types of Special Kernels

17.1 GENERAL DECOMPOSITION OF AN $\mathfrak{L}^2$ KERNEL

In this chapter we shall investigate several more types of $\mathfrak{L}^2$ kernels which have properties that often prove beneficial when they are encountered in the analysis of various applied problems. Before pursuing this discussion, however, it will be helpful to consider anew some of the implications of the singular decomposition treatment given earlier in Chap. 6.

We know that an $\mathfrak{L}^2$ kernel K possesses at most denumerably many characteristic values, and corresponding to each of these cv's is a (singular) portion of the kernel having the form (6.4-2). Strictly formally, then, without detailing the convergence properties of the possibly infinite series, we have

$$K = \sum_{\nu} K_{\nu} + L \tag{1}$$

In this representation, each K_{ν} is assumed to be the degenerate kernel associated with a distinct cv λ_{ν} while L is taken to have no finite cv's at all.

The K_ν's together are mutually orthogonal and each individually is orthogonal to L.

The expansion (1) generates a natural decomposition of the function space $\mathfrak{L}^2$ into the vector sum of linearly independent subspaces $M_\nu(\nu \geq 1)$ and N. Each of these subspaces is transformed into itself by action of the kernel K; in fact, for each of the finite-dimensional subspaces M_ν we have, symbolically,

$$KM_\nu = K_\nu M_\nu \subset M_\nu$$
$$K_\mu M_\nu = 0 \qquad\qquad \mu \neq \nu$$
$$LM_\nu = 0$$

while
$$KN = LN \subset N$$
$$K_\nu N = 0$$

We also observe that, in view of the linear independence of the $M_\nu(\nu \geq 1)$ and N, the (generalized) characteristic functions of K_ν^* lie in the *orthogonal complement* of the space $N \oplus \sum_{\mu \neq \nu} M_\mu$. These cf's, however, need not be entirely in M_ν.† It follows, then, that each of the degenerate kernels K_ν can be further split uniquely into two parts, namely,

$$K_\nu = K_\nu' + K_\nu'' \tag{2}$$
where
$$K_\nu'^* M_\nu \subset M_\nu$$

K_ν'' has no finite cv's and the two kernels K_ν' and K_ν'' taken together are what may be called *quasi-orthogonal*, i.e.,

$$K_\nu'' K_\nu' = 0 \qquad \text{but perhaps} \qquad K_\nu' K_\nu'' \neq 0$$

(see Probs. 1 and 2).

17.2 ANTI-HERMITIAN KERNELS

For Hermitian kernels, the above expansions (17.1-1) and (17.1-2) simplify considerably, as we have essentially noted in earlier chapters. The kernel L does not occur, while each of the K_ν is itself Hermitian and

---

† In other words, the expansion for the adjoint kernel K^* analogous to (1) need not (and, in general, does not) induce the same decomposition of $\mathfrak{L}^2$ as does the expansion of the kernel K itself.

hence cannot be further decomposed. The situation is very similar to this in the case of anti-Hermitian kernels.

We call an $\mathfrak{L}^2$ kernel *anti-Hermitian* if

$$K^*(x,y) = -K(x,y)$$

As a consequence, if we define a new kernel $\tilde{K}$ by

$$\tilde{K} \equiv iK$$

then
$$\tilde{K}^* = -iK^* = iK = \tilde{K}$$

and thus $\tilde{K}$ is Hermitian. The various properties of anti-Hermitian kernels (which, by the way, are also normal) follow directly from this observation, and we list them below for the sake of completeness.

Property 1   The characteristic values of anti-Hermitian kernels are purely imaginary.

Property 2   An anti-Hermitian kernel has at least one cv. If the kernel is real valued, then there must be at least two. A kernel without any cv's, therefore, cannot be anti-Hermitian.

Property 3   Characteristic functions of anti-Hermitian kernels belonging to distinct cv's are orthogonal to one another. The cf's of *real* anti-Hermitian kernels must be complex valued. On the other hand, the cf's of *purely imaginary* anti-Hermitian kernels may be selected to be real.

Property 4   The cv's of an anti-Hermitian kernel and its *transpose* are identical, while the cf's may be chosen as complex conjugates. If ϕ is a cf of the anti-Hermitian kernel K associated with the cv λ, then ϕ is also a cf of K^* belonging to $-\lambda$.

Property 5   The *even*-order iterates of anti-Hermitian kernels are Hermitian, while the *odd*-order iterates remain anti-Hermitian. For purely imaginary values of the eigenparameter λ, therefore, the resolvent of an anti-Hermitian kernel is anti-Hermitian. The poles of the resolvent kernel are all *simple*, and thus there are no generalized cf's.

Property 6   The cv's of anti-Hermitian kernels form a nonvoid denumerable sequence with no finite limit point. The associated cf's may be chosen to be orthonormal. Closed anti-Hermitian kernels, moreover, have a countable infinity of cv's.

Property 7   All the cv and cf estimation techniques discussed earlier for Hermitian kernels, including the procedures of Schmidt, Rayleigh and Ritz, Courant, Kellogg, Weinstein and Aronszajn, and Weinstein and Kato, have their analogues for anti-Hermitian kernels.

Property 8   For anti-Hermitian kernels the expansion (17.1-1) assumes the mean-convergent form

$$K(x,y) = \sum_\nu \frac{\phi_\nu(x)\overline{\phi}_\nu(y)}{i\mu_\nu} \tag{1}$$

where the μ_ν are real. If K is real valued, then to each μ_ν there corresponds the two eigenpairs $(i\mu_\nu, \phi_\nu)$ and $(-i\mu_\nu, \overline{\phi}_\nu)$ so that (1) becomes

$$K(x,y) = 2 \sum_\nu \frac{\text{Im}\,[\phi_\nu(x)\overline{\phi}_\nu(y)]}{\mu_\nu}$$

in this case. Bessel's inequality and the Hilbert expansion theorems also have their natural analogues for anti-Hermitian kernels.

17.3   SYMMETRIZABLE KERNELS

It often occurs that although a given $\mathcal{L}^2$ kernel K itself is not necessarily Hermitian, the composition HK formed with another $\mathcal{L}^2$ kernel H does have this property. If there exists a Hermitian symmetrizing kernel H, then K is often said to be (left) symmetrizable. There is a corresponding definition of right-symmetrizable kernels, but these need not be discussed in detail here in view of their completely analogous behavior.

We have already encountered symmetrizable kernels in Sec. 6.4. Other examples include the generalizations of the classical *polar kernels* of Hilbert (1906, pp. 462ff; 1912, pp. 195–204)† wherein

$$K(x,y) = A(x)G(x,y)$$

---

† Other references of interest are Boggio (1907), Goursat (1908), Marty (1910, pp. 515–518), Fubini (1910), Garbe (1915), and Zaanen (1946, V; 1953, chap. 17).

with real A and Hermitian G, the further extensions suggested by Goursat (1908 and 1909) [see also Zaanen (1946, III; 1953, chap. 15)] in which

$$K(x,y) = A(x)G(x,y)B(y)$$

with real A and B and Hermitian G, and the class of composite kernels of the form

$$K(x,y) = K_1 K_2(x,y)$$

where K_1 and K_2 are both Hermitian. This last group of kernels contains the special case in which K_2 is semidefinite; such kernels were first studied in detail by Anna Pell (1911a and b) and named in her honor [also see Zaanen (1946, VI; 1953, chap. 17) and Chang (1949)].

Unfortunately, as the example

$$K(x,y) = 1 - 2x \qquad 0 \le x,y \le 1$$

for which $H(x,y) = 1$ is a symmetrizing kernel, shows, symmetrizable kernels in general need not possess cv's. These pathological cases can be eliminated from consideration, however, by the requirement that

$$\|HK\| \ne 0$$

A more subtle difficulty occurs, nevertheless, with kernels typified by

$$K(x,y) = 1 - 2y \qquad 0 \le x,y \le 1$$

where
$$HK = 1 - 2x - 2y + 4xy$$

for $H(x,y) \equiv 1 - 4xy$. In this case H symmetrizes K in a nontrivial fashion and yet K still has no cv's. Marty (1910, pp. 1031ff) first demonstrated that this last situation does not occur if H is semidefinite, i.e., either nonnegative or nonpositive definite. Following Marty, therefore, we agree to call a given $\mathfrak{L}^2$ kernel (left) *symmetrizable* only if there exists a *semidefinite Hermitian* kernel H with the property that HK is both *Hermitian* and *nonnull*.†

Marty used a modified Schmidt approach (see Sec. 7.4) to establish the existence of cv's for symmetrizable kernels. The proof we give here

---

† Clearly H may be selected to be nonnegative definite, and we shall always assume such to be the case.

is similar to the more modern demonstration of Reid (1951). [See also Zaanen (1943).]

THEOREM 17.3

An $\mathfrak{L}^2$ symmetrizable kernel K has at least one characteristic value.

Proof We first assume that K is symmetrized by the nonnegative definite $\mathfrak{L}^2$ kernel H. Moreover, since $\|HK\| \neq 0$, we can further assume, without loss of generality, that there is an $\mathfrak{L}^2$ function ϕ such that

$$(HK\phi, \phi) > 0$$

Making use of the inequality†

$$|(HK\phi, \phi)| \leq \|K\|(H\phi, \phi) \tag{1}$$

we then know that among the functions ϕ for which $(H\phi, \phi) = 1$, $(HK\phi, \phi)$ has a finite positive least upper bound M. As a consequence, there exists a sequence of $\mathfrak{L}^2$ functions $\{\phi_\nu\}$ such that

$$(H\phi_\nu, \phi_\nu) = 1 \quad \text{and} \quad M_\nu \equiv (HK\phi_\nu, \phi_\nu) \to M \quad \text{as } \nu \to \infty \tag{2}$$

If $(1/M)$ is a cv of K, we are done. Otherwise, the resolvent R_K is well defined and square integrable for $\lambda = 1/M$, and we can form the meaningful sequence $\{\psi_\nu\}$ where

$$\psi_\nu \equiv \phi_\nu - \frac{R_K \phi_\nu}{\|R_K\|}$$

It follows, then, that

$$M(H\psi_\nu, \psi_\nu) - (HK\psi_\nu, \psi_\nu) = M - M_\nu - \frac{2MM_\nu}{\|R_K\|}$$

$$+ \frac{M^2}{\|R_K\|^2} \left[(HR_K\phi_\nu, \phi_\nu) - M_\nu \right]$$

$$\leq \left(M - M_\nu - \frac{MM_\nu}{\|R_K\|} \right)\left(1 + \frac{M}{\|R_K\|} \right) \tag{3}$$

---

† This is an extension of the K Schwarz inequality (8.3-3) to the symmetrizable case and was first introduced by Reid (1951). We lead the reader to its verification in Prob. 4 at the end of this chapter.

Here we have made use of the Fredholm identity (3.2-5), the definitions (2), and the inequality (1). As $\nu \to \infty$, however, the right-hand side of (3) becomes negative, which means that the left-hand side of (3) must also be negative, contradicting the definition of M. Hence, $(1/M)$ must be a bona fide cv of K.†

The following corollary is an immediate consequence of the above result.

COROLLARY

An $\mathfrak{L}^2$ kernel without any cv's cannot be symmetrizable.

We also remark that in the course of the proof of Theorem 17.3 we made use of a valuable property of symmetrizable kernels, namely, the following.

Property 1   The iterates of a kernel K symmetrizable by a semidefinite Hermitian kernel H are also symmetrizable with the same H. For real values of the eigenparameter λ, therefore, the resolvent R_K is likewise symmetrizable by H. (See Prob. 4.)

A further property, which may be easily established by the reader, is the following.

Property 2   The cf's of a kernel K, symmetrizable by H, which are associated with *nonconjugate* cv's, are *H-orthogonal* to one another. In other words, if $\phi = \lambda K \phi$, $\psi = \mu K \psi$, and $\lambda \neq \overline{\mu}$, then

$$(H\phi, \psi) = 0 = (HK\phi, \psi)$$

In particular, cf's belonging to nonreal cv's are *H*-orthogonal to *themselves*.

The cv that we obtained in the course of the proof of Theorem 17.3 was *real valued*. As Property 2 suggests, however, this need not be the case for all the cv's of a general symmetrizable kernel. The example

$$K(x,y) = \frac{\phi_1(x)\overline{\phi}_1(y)}{\lambda_1} + \frac{\phi_2(x)\overline{\phi}_2(y)}{\lambda_2}$$

---

† It should be noted that for a cf ϕ belonging to the cv $(1/M)$, we need not have $(H\phi, \phi) = 1$, even though the extremizing sequence $\{\phi_\nu\}$ employed in the above proof had this property. (See Prob. 5.)

with λ_1 real, λ_2 arbitrary, and $(\phi_\mu, \phi_\nu) = \delta_{\mu\nu}$, amply demonstrates this point. In order to proceed further, therefore, we need yet another hypothesis.

Definition [Reid (1951), Zaanen (1953, chap. 12).] A given $\mathfrak{L}^2$ kernel K is termed *fully symmetrizable*† (by H) if $\|H\phi\| \neq 0$ for every cf ϕ of K.

In particular, we observe that kernels for which $H\phi = 0$ implies $K\phi = 0$, which is the case, for instance, whenever H is positive (or negative) definite, are fully symmetrizable. Such K have been called *Marty kernels* by Zaanen (1946, IV) since it was Marty who first suggested many of their properties.

We are now in a position to establish the following.

Property 3 The cv's of *fully* symmetrizable kernels are *real*.

If this were not so, then, by Property 2, cf's ϕ belonging to nonreal cv's would be H-orthogonal to themselves. The inequality (1), with K replaced by H, would imply therefore that $\|H\phi\| = 0$, contrary to the definition of full symmetrizability.

Property 2 now can be reworded as follows.

Property 2′ The cf's of a kernel K, fully symmetrizable by H, that are associated with *distinct* cv's are H-orthogonal to one another.

We also have the following.

Property 4 If a kernel K is fully symmetrizable by (nonnegative definite) H, then the cf's $\phi_\nu (1 \leq \nu \leq n)$ associated with an arbitrary (but fixed) cv λ of K may be chosen to be H-orthonormal. The cv's of K^* are identical with the cv's of K. Indeed, a full set of cf's $\psi_\nu (1 \leq \nu \leq n)$ belonging to λ can be formed from the ϕ_ν by means of the relation $\psi_\nu \equiv H\phi_\nu$.

The first part of this property is a consequence of the definition of full symmetrizability and the fact that the ϕ_ν are therefore H-independent; that is, the $H\phi_\nu$ are linearly independent. The H-orthonormalization may be accomplished by a Gram-Schmidt procedure, modified in the obvious fashion (see Prob. 6). The last part above follows from the nonnull char-

---

† Mercer (1920) calls such kernels *completely symmetrizable*.

acter of the n linearly independent functions $\psi_\nu \equiv H\phi_\nu$ $(1 \leq \nu \leq n)$ for which we have

$$\psi_\nu = H\phi_\nu = \lambda HK\phi_\nu = \lambda K^* H\phi_\nu = \lambda K^* \psi_\nu \qquad (4)$$

Property 5 The poles of the resolvent of a fully symmetrizable kernel are *simple*, and thus there are no generalized cf's. If H can be chosen to be positive definite, then the expansion (17.1-1) for K assumes the form

$$K(x,y) = \sum_{\nu=1}^{n} \frac{\phi_\nu(x)\overline{\psi}_\nu(y)}{\lambda_\nu} \qquad (5)$$

whenever K has only finitely many cv's.

The simplicity of the poles follows from the fact that the cf's of K generate the cf's of K^* through the relation $\psi_\nu = H\phi_\nu$. We then merely have to take cognizance of the biorthonormality relation (6.4-4) that must hold between these functions and reason as we did earlier in arriving at (6.4-8). As a consequence, that portion of the kernel which gives rise to the cv's (when there are only finitely many) must have the form (5). Clearly, a kernel orthogonal to K and possessing no finite cv's may be added to (5) without invalidating the full symmetrizability of K, except, of course, in the case when H is positive definite.

If we did not have the full symmetrizability of K, these constructions could not be generally carried out. Lalesco (1918) gives a complicated example of a symmetrizable kernel whose resolvent has a pole of second order. The kernel

$$K(x,y) = \frac{1}{\lambda_0} [\phi_1(x)\overline{\phi}_1(y) + \phi_2(x)\overline{\phi}_2(y)] + \phi_1(x)\overline{\phi}_2(y) \qquad (6)$$

with real λ_0 and orthonormal ϕ_1 and ϕ_2 provides a much simpler example (see Prob. 7).

We close this section with the observation that if we consider the Hilbert space generated by the inner product $[f,g] \equiv (Hf,g)$ as the underlying space (when dealing with a given kernel K fully symmetrizable by the nonnegative definite Hermitian kernel H), then the classical cv and cf estimation techniques developed earlier have valid analogues in this symmetrizable situation. In particular, the cv's of fully symmetrizable kernels may be characterized in a *min-max* fashion à la Weyl-Courant (9.3-2) as well as with a *max-min* relation similar to the Poincaré result

(9.4-1). Moreover, Bessel's inequality and the Hilbert formula (13.3-4) take the obvious forms

$$[f,f] \geq \sum_{\nu} |[f,\phi_\nu]|^2$$

$$[Kf,g] = \sum_{\nu} \frac{[f,\phi_\nu][\phi_\nu,g]}{\lambda_\nu} = \sum_{\nu} [Kf,\phi_\nu][\phi_\nu,g]$$

in terms of the H-orthonormalized cf's ϕ_ν of K. On the other hand, although the Hilbert expansion (13.2-3) for Kf

$$\sum_{\nu} \frac{[f,\phi_\nu]}{\lambda_\nu} \phi_\nu$$

and its higher-order iterative analogues converge in the new Hilbert space, they need not do so in $\mathfrak{L}^2$. Zaanen (1953, pp. 412ff) carefully discusses an example of Toeplitz [Hellinger and Toeplitz (1927, pp. 1571–1573)] that exhibits this behavior. This means that the expansion formulas for the kernels themselves similar to (13.3-5), for instance, will have to assume a weaker form in the general fully symmetrizable case.†
For further details on this and related questions, the reader is referred to the relevant literature.‡

17.4 COMPLEX-SYMMETRIC KERNELS AND VARIATIONAL PRINCIPLES

In an increasing number of applied fields § investigators are encountering integral equations with complex-valued kernels K, which have the property that

$$K(y,x) = K(x,y) \tag{1}$$

Such kernels, of which

$$K(x,y) = e^{ik(x-y)^2}$$

---

† In certain situations, as, for example, with a Pell kernel, these difficulties by and large do not occur [Zaanen (1953, chaps. 16 and 17)]. See also Mercer (1920).

‡ Reid (1931 and 1951), Wilkins (1944), Zaanen (1947), and Zimmerberg (1948) have treated various types of fully symmetrizable transformations that occur in certain vector integral equations. Zaanen (1950) has also extensively studied the more general case of *normalizable* transformations.

§ See Slepian and Pollak (1961), Fox and Li (1961 and 1963), Jones (1956), and Dolph (1956), for example.

[see (A-1)], with k a given complex constant, may be considered representative, belong to the class of so-called *complex-symmetric* kernels. For this class it is well known that one can generally derive a variational expression, the *extremization* of which is in a sense equivalent to solving a given Fredholm equation.† This has encouraged researchers to use modified Rayleigh-Ritz procedures‡ in order to obtain estimates of the cv's and cf's of complex-symmetric kernels. Related techniques have also been employed for the approximate solution of *non*homogeneous equations of both the first and second kind with such functions as kernels, and often the results yielded by this approach have been quite good from an engineering point of view.

The overall situation regarding complex-symmetric kernels, however, is not as clearly defined as the (fortuitous?) results mentioned above might tend to indicate. In saying this, we do not mean to deprecate undeservedly the work that has appeared heretofore; rather, we mean to suggest that one ought not to generalize too hastily from isolated individual examples. The analysis of integral equations with general complex-symmetric kernels remains, at present, an art form in which each separate equation appears to necessitate treatment based almost solely on its own individual features and peculiarities.

This is not really a particularly surprising turn of events. Recall, for instance, that given any complex matrix $A \equiv (A_{\mu\nu})$, $1 \leq \mu, \nu \leq n$, one can find a *complex-symmetric* matrix to which A is *similar* [see Gantmacher (1959, pp. 10–14), for example]. In other words, as far as finite-dimensional vector spaces are concerned, a complex-symmetric representation can be found for the most general of transformations.

In function spaces, the analogous result is not entirely valid. The $\mathfrak{L}^2$ kernel

$$K(x,y) = \sum_\nu \frac{\phi_{\nu+1}(x)\phi_\nu(y)}{\nu^2} \tag{2}$$

where the real functions $\phi_\nu (\nu \geq 1)$ form a complete orthonormalized set, for example, cannot be put in complex-symmetric form. § If it could, then there would be an operator P such that, symbolically, $P^{-1}KP$ was

---

† Actually, only in the Hermitian case has the general equivalence of *stationary values* and characteristic values been established.

‡ Some authors call this approach the *method of moments*; others term it *Galerkin's method*. The reader may find the short account of Kantorovich and Krylov (1958, pp. 150–154) or the discussion of Harrington (1968, chap. 1) to be of interest.

§ D. E. Bzowy first brought this observation to our attention.

complex symmetric. From this it would follow that[†]

$$(PP^T)^{-1}K(PP^T) = K^T \tag{3}$$

and hence
$$K^T\phi_1 = 0 = (PP^T)^{-1}K(PP^T)\phi_1$$
$$= K(PP^T)\phi_1$$

which implies
$$0 = PP^T\phi_1$$
$$= \phi_1$$

which is absurd. By modifying the kernel (2) in the obvious manner, it is clear how other kernels, which possess any number of cv's and yet which are not similar to complex-symmetric kernels, can be constructed.

To be sure, however, the class of complex-symmetric kernels does exhibit the fullest "spectrum" of behavior. As the kernel

$$K(x,y) = (1 + ix\sqrt{3})(1 + iy\sqrt{3}) \qquad -1 \le x,y \le 1$$

demonstrates, there exist complex-symmetric kernels without any cv's (see Prob. 8). At the other extreme, since purely imaginary anti-Hermitian kernels satisfy (1), we can form complex-symmetric kernels having an arbitrary (countable) number of cv's. We can even construct complex-symmetric kernels that are unitarily related to other complex-symmetric kernels of different rank (see Prob. 9).

The generality that thus characterizes the class of complex-symmetric kernels can (and will, on occasion) lead to rather unexpected results. To illustrate this point let us recall the *variational expression*

$$I \equiv \frac{\langle K\phi,\psi \rangle}{\langle \phi,\psi \rangle} = \frac{1}{\lambda} \tag{4}$$

which is naturally associated with the characteristic-value problem for a given kernel K.[‡] Since K is taken to be complex symmetric, we assume that the expressions $\langle\,,\rangle$ in (4) designate real, rather than complex, inner products. Following a standard procedure [see Cairo and Kahan (1965,

---

† Here the superscript T denotes *transpose*.

‡ Expression (4) is only one of several variational expressions which could equally well have been used.

chap. 4) or Moiseiwitsch (1966, chap. 4), for instance], then, if we let

$$\phi \equiv \phi_0 + \varepsilon \phi_1$$

$$\psi \equiv \psi_0 + \delta \psi_1$$

where ε and δ are complex parameters and

$$\phi_0 = \lambda_0 K \phi_0$$

$$\psi_0 = \lambda_0 K^T \psi_0$$

it follows that

$$I = \frac{1}{\lambda_0} + \varepsilon \delta \frac{\langle \phi_1, \psi_1 \rangle}{\langle \phi_0, \psi_0 \rangle} \left[\frac{\langle K \phi_1, \psi_1 \rangle}{\langle \phi_1, \psi_1 \rangle} - \frac{1}{\lambda_0} + o(1) \right]$$

In the special case $\psi = \phi$, we have

$$I = \frac{1}{\lambda_0} + \varepsilon^2 \frac{\langle \phi_1, \phi_1 \rangle}{\langle \phi_0, \phi_0 \rangle} \left[\frac{\langle K \phi_1, \phi_1 \rangle}{\langle \phi_1, \phi_1 \rangle} - \frac{1}{\lambda_0} + o(1) \right] \tag{5}$$

and this expression shows, indeed, that if the difference between ϕ and ϕ_0 is of order ε, then the difference between I and $1/\lambda_0$ is of order ε^2 (provided that $\langle \phi_0, \phi_0 \rangle$ does not vanish). Since all the terms are in general complex, however, we have no a priori knowledge concerning the "direction" of this error. Moreover, note what can happen if one proceeds in a formal fashion to analyze examples such as the following.[†]
  Consider the complex-symmetric kernel

$$K(x,y) = 1 + \tfrac{1}{4}i \sqrt{3} \, (x + y) \qquad -1 \le x, y \le 1 \tag{6}$$

and the two functions

$$\psi_1(x) = \tfrac{2}{5}\sqrt{5} \, (2 + \sqrt{3}) - 1 + 3x^2$$

$$\psi_2(x) = i \frac{\sqrt{70}}{28} \left[\frac{2\sqrt{5}}{5} (1 - \sqrt{3}) - 1 - \sqrt{3} + 3x^2(1 + \sqrt{3}) \right] - 3x + 5x^3 \tag{7}$$

We propose to determine that linear combination of ψ_1 and ψ_2 which

---

[†] Patterned after an illustration of Morgan (1964); see also Kaplan and Morgan (1964).

yields the "best" value of the variational expression I. Toward that end, we calculate

$$
\begin{aligned}
\langle K\psi_1, \psi_1 \rangle &= \tfrac{16}{5}(7 + 4\sqrt{3}) \\
\langle K\psi_1, \psi_2 \rangle &= -i\,\frac{4\sqrt{70}}{35}\,(1 + \sqrt{3}) \\
\langle K\psi_2, \psi_2 \rangle &= -\tfrac{4}{7}(2 - \sqrt{3}) \\
\langle \psi_1, \psi_1 \rangle &= \tfrac{32}{5}(2 + \sqrt{3}) \\
\langle \psi_1, \psi_2 \rangle &= 0 \\
\langle \psi_2, \psi_2 \rangle &= 0
\end{aligned}
\tag{8}
$$

The Rayleigh-Ritz procedure (see Sec. 9.1) then suggests that among the stationary values of $I(\psi_1 + \alpha\psi_2)$ we should find the most appropriate choice of the parameter α. Substituting (8) into the expression for I and equating $dI/d\alpha$ to zero, we obtain the single root

$$
\alpha_0 = -i\,\frac{\sqrt{70}}{5}\,(5 + 3\sqrt{3})
\tag{9}
$$

for which the corresponding value of the variational expression is

$$
I(\psi_1 + \alpha_0\psi_2) = 0!!
\tag{10}
$$

Here, then, is a simple example wherein straightforward application of techniques that have proven beneficial with other types of kernels yield totally unacceptable results in the complex-symmetric case. Moreover, the numerical procedure even fails to take note of the fact that ψ_1 has been carefully selected so that

$$
I(\psi_1) = 1 + \tfrac{1}{2}\sqrt{3}
$$

a value that coincides with the reciprocal of the smallest characteristic value of K (even though ψ_1 is not a cf associated with that cv). Although these specific difficulties may not occur in the analysis of other problems having complex-symmetric form, on the basis of the evidence presented in this section, a rather cautious approach does seem to be called for in such situations.

17.5 POSITIVE KERNELS

Far more can be said when the integral equation with which we are confronted has a *positive (nonnegative)* kernel,[†] that is, an $\mathcal{L}^2$ kernel K for which

$$K(x,y) > 0 \qquad (\geq 0)$$

almost everywhere in the fundamental domain. We conclude this chapter with a brief account of several of the more interesting results associated with such kernels and an indication of the trends that modern research is taking in closely related fields.

Actually positive kernels have been around for decades. The kernels appearing in early applications, however, particularly those occurring in gas dynamics and classical radiation theory, for example, had additional structure as well. Hilbert (1912) attributes to his student Hecke essentially the following uniqueness result.

THEOREM 17.5-1

If the characteristic function ϕ associated with some one of the characteristic values λ of a *positive symmetric* kernel K can itself be chosen to be *positive*, then that cv λ is simple; i.e., it has unit multiplicity.

Proof   Let

$$\phi(x) = \lambda \int K(x,y)\phi(y)\,dy$$

where both $K(x,y) = K(y,x) > 0$ and $\phi(x) > 0$ almost everywhere. Moreover, assume there exists another cf ψ, which we may take to be real, also belonging to this cv λ of K. It follows, then, that

$$\lambda \int \frac{K(x,y)}{\phi(x)}\,\phi(y)\psi(x)\,dy = \psi(x) = \lambda \int \frac{K(x,y)}{\phi(x)}\,\phi(x)\psi(y)\,dy$$

or

$$\int \frac{K(x,y)}{\phi(x)}\left[\phi(y)\psi(x) - \phi(x)\psi(y)\right]dy = 0$$

[†] There should be no confusion with the class of *positive-(nonnegative-) definite* kernels considered earlier. In particular, positive kernels need not even be Hermitian.

Performing obvious manipulations on this relation, we obtain

$$\int \int \frac{K(x,y)}{\phi(x)\phi(y)} \left[\phi(y)\psi(x) - \phi(x)\psi(y) \right] \phi(y)\psi(x) \, dy \, dx = 0$$

The symmetry of K then leads to

$$\int \int \frac{K(x,y)}{\phi(x)\phi(y)} \left[\phi(y)\psi(x) - \phi(x)\psi(y) \right]^2 dx \, dy = 0$$

from which, in view of the positivity of K and ϕ, we infer that

$$\phi(y)\psi(x) - \phi(x)\psi(y) = 0$$

or, equivalently,

$$\frac{\psi(x)}{\phi(x)} = \frac{\psi(y)}{\phi(y)} = \text{const} \qquad\qquad \text{Q.E.D.}$$

As an immediate corollary we have the following.

COROLLARY

The above theorem is equally valid for positive kernels K having the polar form

$$K(x,y) = A(x)G(x,y)B(y)$$

of Sec. 17.3 where G is symmetric and A and B are each of one sign.

Proof   We assume, without loss of generality, that both $A > 0$ and $B > 0$ almost everywhere in the interval of interest. The substitution

$$\psi(x) = \sqrt{\frac{B(x)}{A(x)}} \; \phi(x)$$

first suggested by Goursat (1908), then transforms the integral equation

$$\phi(x) = \lambda \int A(x)G(x,y)B(y)\phi(y) \, dy$$

with polar kernel into the integral equation

$$\psi(x) = \lambda \int \sqrt{A(x)B(x)} \; G(x,y) \sqrt{A(y)B(y)} \; \psi(y) \, dy$$

with symmetric kernel. The original theorem now applies directly to the kernel of this new equation.

The characteristic value referred to in the statement of the above theorem actually turns out to be the *smallest* among all the cv's.† In order to help demonstrate this fact and establish related conclusions, we recall the following easily verified result due essentially (in the real case) to Abel (1826b) and Pringsheim (1897) [see Hobson (1957, vol. II, pp. 175 and 182), Whittaker and Watson (1958, pp. 57–58), and especially Titchmarsh (1939, pp. 214–215)].

LEMMA   Let $|\lambda| \leq r$ be the circle of convergence of the analytic function

$$f(\lambda) \equiv \sum_{\nu=0}^{\infty} a_\nu \lambda^\nu$$

If the coefficients a_ν are all nonnegative, then the function f has a singularity at the point $\lambda = r$.

Using this lemma we can provide an elementary proof of the following theorem.

THEOREM 17.5-2

If the *nonnegative* kernel K has characteristic values, then these cv's can be ordered so that the *least* among them is *positive*. To this least cv there corresponds a *nonnegative* characteristic function.

Proof $K(x,y) \geq 0$ implies that all of the coefficients in the Neumann series

$$R_K(x,y;\lambda) \equiv \sum_{\nu=1}^{\infty} \lambda^{\nu-1} K^\nu(x,y) \qquad (3.1\text{-}4')$$

for the resolvent kernel are nonnegative almost everywhere. If λ_1 is the least characteristic value of K, then, by the preceding Abel-Pringsheim result, $|\lambda_1|$ must likewise be a singularity of R_K and hence also a cv of K. Under the assumption that the order of this

---

† See Cryer (1967) where two methods, based upon modifications of approximations used earlier by Wielandt (1956), are given for numerically estimating this smallest cv.

(pole) singularity is n, we may now expand R_K in a Laurent series as in Sec. 6.1, viz.,

$$R_K(x,y;\lambda) = \sum_{\nu=1}^{n} \frac{r_\nu(x,y)}{(\lambda - |\lambda_1|)^\nu} + r(x,y;\lambda) \qquad (6.1\text{-}1')$$

It follows that for at least one fixed value y_0 of y, $r_n(x,y_0)$ is a nontrivial $\mathfrak{L}^2$ function of the single variable x and thus, properly normalized, is a bona fide characteristic function of K associated with the cv $|\lambda_1|$. (For at least one fixed value x_0 of x, moreover, $r_n(x_0,y)$ is proportional to a cf of K^* belonging also to the cv $|\lambda_1|$.)

In order to show that an almost-everywhere nonnegative cf of K associated with $|\lambda_1|$ can be selected, we merely note from (3.1-4') that, owing to the nonnegative character of K, R_K is itself nonnegative for all $0 \le \lambda < |\lambda_1|$ and approaches positive infinity as real λ tends to $|\lambda_1|$ from below. For almost all x,y in the domain of interest, therefore, $(-1)^n r_n(x,y) \ge 0$ and thus $|r_n(x,y_0)|$, properly normalized, constitutes the desired cf. Q.E.D.

COROLLARY

If the kernel K of the above theorem is *positive*, then the characteristic function associated with the least cv is also *positive*.

Proof   Since

$$r_n(x,y_0) = |\lambda_1| \int K(x,z) r_n(z,y_0)\, dz$$

$r_n(x,y_0)$ cannot vanish on a set of positive measure in the fundamental interval without leading to a contradiction.

The results immediately above and the theorem that follows are analogous, in the integral equations setting, to classical results of Perron (1907) and Frobenius (1908 and 1912) concerning matrices with either positive or nonnegative entries. Jentzsch (1912) was the first to establish similar conclusions for integral equations, but he assumed that the kernels of interest were continuous functions in the fundamental domain [see also Hopf (1928)]. Subsequently Rutman (1938 and 1940) gave these investigations a more modern flavor by using topological methods and extending consideration to completely continuous operators which leave invariant certain cones in Banach spaces. Further generalizations and a number of interesting applications can be found in Krein and Rutman (1948), Birkhoff (1957), Karlin (1959 and 1964), and Gantmacher and Krein (1961).

We are now ready to establish our final (and strongest) results. The conclusions are essentially those of the Jentzsch theorem (1912), but the hypotheses are much weaker. The method of proof follows that of Krein and Rutman (1948).

THEOREM 17.5-3

A *positive* kernel K has a *unique positive* characteristic function. The corresponding characteristic value is *positive* and is strictly *smaller* (in magnitude) than all other cv's of K. Moreover, this cv appears as a *simple* pole of the resolvent of K, and thus there are no generalized cf's associated therewith.

Proof   We first note that the second-order trace

$$k_2 \equiv \int \int K(x,y)K(y,x) \, dy \, dx$$

of K must be positive, and therefore, by (4.3-3), K has characteristic values. It follows then from Theorem 17.5-2 and its corollary that these cv's can be so ordered that the least one λ_1 is positive and has a corresponding positive characteristic function ϕ_1.

Let us now assume that there is another (perhaps complex) cv λ_2 with $|\lambda_2| = \lambda_1$. The Fredholm alternative (Sec. 3.6) ensures that the homogeneous adjoint equation

$$\psi(x) = \lambda_2 \int K(y,x)\psi(y) \, dy \tag{1}$$

has a nontrivial solution, from which it follows that

$$|\psi(x)| \le \lambda_1 \int K(y,x)|\psi(y)| \, dy \tag{2}$$

and hence, multiplying by the positive function ϕ_1 and integrating,

$$\int \phi_1(x)|\psi(x)| \, dx \le \lambda_1 \int \int K(y,x)\phi_1(x)|\psi(y)| \, dy \, dx$$

$$= \int \phi_1(y)|\psi(y)| \, dy$$

In view of this last relation, however, the equality sign in (2) must hold almost everywhere, and $|\psi(x)|$ is perforce proportional to a cf of K^* associated with the cv λ_1. Moreover, as a consequence of the corollary of the previous theorem applied to the positive kernel K^*, $|\psi(x)| > 0$ almost everywhere.

For each value of x for which equality holds in (2) and $\psi(x)$ does not vanish, we now choose real $\alpha(x)$ such that

$$\int K(y,x)\psi(y)\,dy = e^{i\alpha(x)}\left|\int K(y,x)\psi(y)\,dy\right|$$

Owing to the nature of $\psi(x)$, it follows that

$$\int K(y,x)\psi(y)\,dy = e^{i\alpha(x)}\frac{|\psi(x)|}{\lambda_1} = e^{i\alpha(x)}\int K(y,x)\,|\psi(y)|\,dy$$

and, therefore,

$$\int K(y,x)\,[\psi(y) - e^{i\alpha(x)}\,|\psi(y)|\,]\,dy = 0$$

The positivity of K then implies that

$$\psi(y) = e^{i\alpha(x)}\,|\psi(y)|$$
$$= e^{i\alpha}\,|\psi(y)| \tag{3}$$

with constant α for almost all y in the fundamental interval. After proper normalization, therefore, the function $\psi(x)$ is positive almost everywhere, and thus, from (1),

$$0 < \lambda_2 = \lambda_1$$

To conclude our proof, we note that there can only be the one cf ϕ_1 associated with the cv λ_1. If there were additional cf's, then there would also have to exist additional solutions Ψ of the adjoint equation

$$\Psi(x) = \lambda_1\int K(y,x)\Psi(y)\,dy \tag{4}$$

that were orthogonal to ϕ_1 (see Sec. 6.4). Equation (3) above, however, would hold for these Ψ and would preclude such behavior. Likewise there are no generalized cf's belonging to λ_1 since our earlier analysis does imply the existence of a positive solution ψ_1 of (4) with the property, therefore, that $(\phi_1,\psi_1) \neq 0$ (see the footnote on page 82). Finally, we observe that ϕ_1 must be the only positive cf of K. Other such functions Φ would have to be associated with positive cv's greater than λ_1, and for these functions it would follow that $(\Phi,\psi_1) \neq 0$ which is an impossibility (see the proof of Property 4 in Sec. 7.1).   Q.E.D.

It should be remarked that the assumed positivity of the kernel K in the above theorem can be replaced by the following slightly weaker hypothesis: For each $\varepsilon > 0$, there exists an integer $N(\varepsilon)$ such that the iterated kernel $K^N(x,y)$ vanishes on a set of points whose measure is at most ε. This permits the application of these results to nonnegative kernels, say, which are continuous in the domain of definition and positive on the main diagonal $x = y$.†

In recent years the interest in positive kernels and special classes thereof has greatly expanded, the impetus coming from numerous applications of such functions in mathematics, statistics, and mechanics. The broad range of endeavor includes, for instance, the work of Chandrasekhar and Bellman and his colleagues on problems in radiative transfer [see Chandrasekhar (1950) or Bellman, Kalaba, and Prestrud (1963), for example] and the investigations of Schoenberg, Gantmacher, Krein, Karlin, and others concerning *totally* positive matrices and kernels. Karlin's recent book (1968) provides an excellent overview and a comprehensive bibliography of this latter subject.

PROBLEMS

1. Discuss the decompositions considered in Sec. 17.1 as they pertain to the kernel

$$K(x,y) = xy + x \qquad -1 \leq x,y \leq 1$$

2. Let $K = K_1 + K_2$ where K_1 and K_2 are *quasi-orthogonal* kernels; that is, $K_2 K_1 = 0$.
   a. By using the expression

$$\tilde{D}(\lambda) = \exp\left(-\sum_{\nu=2}^{\infty} \frac{\lambda^\nu}{\nu} k_\nu\right)$$

   demonstrate that, just as in the fully orthogonal case, the modified Fredholm determinant of K is equal to the product of the modified Fredholm determinants of K_1 and K_2.
   b. If $K_2^2 = 0$, show that the resolvent of K satisfies the relation

$$R_K = R_{K_1} + \lambda R_{K_1} K_2 + K_2$$

---

† Sarymsakov (1949) has investigated the case where $K(x,y)$ is continuous and nonnegative and is such that for every pair (x,y) there exist positive integers m and n for which $K^m(x,y) > 0$ and $K^n(y,x) > 0$.

3. Verify that real-valued anti-Hermitian kernels have at least two distinct characteristic values.

4. [Reid (1951).]
   a. Recall the K Schwarz inequality (8.3-3) for a *nonnegative definite* kernel H, viz.,

   $$|(H\psi,\phi)| \leq (H\psi,\psi)^{\frac{1}{2}}(H\phi,\phi)^{\frac{1}{2}}$$
   $$\leq \tfrac{1}{2}[(H\psi,\psi) + (H\phi,\phi)] \tag{1}$$

   Show that if K is an $\mathfrak{L}^2$ kernel such that HK is Hermitian, then HK^ν is also Hermitian for all integer $\nu \geq 1$, and hence, using (1),

   $$|(HK^\nu\phi,\phi)| \leq \tfrac{1}{2}\|K\|^\nu(H\phi,\phi) + \tfrac{1}{2}\|K\|^{-\nu}(HK^{2\nu}\phi,\phi)$$

   b. Apply induction to the result in part a to establish that

   $$|(HK\phi,\phi)| \leq \|K\|(H\phi,\phi) \sum_{\nu=1}^{\infty} 2^{-\nu} + \lim_{\nu\to\infty} \|K\|^{1-2^\nu}(HK^{2^\nu}\phi,\phi)2^{-\nu}$$
   $$\leq \|K\|(H\phi,\phi)$$

5. Using two orthonormal functions a_1 and a_2, construct a degenerate kernel K of second order that is symmetrizable by a semidefinite kernel H but whose cf's are orthogonal to H.

6. Assume that a given kernel K is *fully* symmetrizable by a nonnegative definite kernel H. Let $\{\phi_\nu\}(1 \leq \nu \leq n)$ be a complete set of cf's associated with an arbitrary (but fixed) cv λ of K.
   a. Verify that the linear independence of the ϕ_ν implies that the functions $\psi_\nu = H\phi_\nu$ are also linearly independent, i.e., the ϕ_ν are H-*independent*.
   b. [Zaanen (1953, p. 219).] Using a procedure analogous to that of Gram and Schmidt, show that the ϕ_ν $(1 \leq \nu \leq n)$, or, in fact, any countable set of H-independent functions, may be H-orthonormalized.

7. a. Verify that

   $$H(x,y) = \phi_2(x)\overline{\phi}_2(y)$$

   is a symmetrizing kernel for the example (17.3-6), viz.,

   $$K(x,y) = \frac{1}{\lambda_0} [\phi_1(x)\overline{\phi}_1(y) + \phi_2(x)\overline{\phi}_2(y)] + \phi_1(x)\overline{\phi}_2(y)$$

   where λ_0 is real and $(\phi_\mu,\phi_\nu) = \delta_{\mu\nu}$.
   b. Deduce that the resolvent of K is given by

   $$R_K(x,y;\lambda) = \frac{1}{\lambda_0 - \lambda} [\phi_1(x)\overline{\phi}_1(y) + \phi_2(x)\overline{\phi}_2(y)] + \left(\frac{\lambda_0}{\lambda_0 - \lambda}\right)^2 \phi_1(x)\overline{\phi}_2(y)$$

8. Show that the complex-symmetric kernel

$$K(x,y) = (1 + ix\sqrt{3})(1 + iy\sqrt{3}) \qquad -1 \leq x,y \leq 1$$

does not have any characteristic values.

9. It is a well-known result of Schur (1945) that, given a *complex-symmetric* matrix S, one can always find a *unitary* matrix U such that $U^T S U$ is diagonal. Although the entries in the diagonal matrix $D \equiv U^T S U$ do not, in general, correspond to the eigenvalues of the original matrix S, it is true, however, that D and S have the same rank. Show that this last result does not carry over to complex function spaces. In other words, construct a simple complex-symmetric kernel, say, *without* any cv's, which is unitarily related as above to a "diagonal" kernel *with* cv's. (Recall from Sec. 14.3 that unitary operators have the form

$$\sum_\nu \phi_\nu(x)\overline{\psi}_\nu(y)$$

where both $\{\phi_\nu\}$ and $\{\psi_\nu\}$ are complex orthonormal sets of functions.)

10. For the example (17.4-6), verify the calculations (17.4-8), (17.4-9), and (17.4-10).

11. In Theorem 17.5-2 we see that for kernels that are merely nonnegative, we must assume that they possess characteristic values or add a condition to the hypotheses which ensures that behavior in order that the conclusions remain valid. Give an example why this should be the case.

12. [Jentzsch (1912).] Let λ and μ be the least (in magnitude) cv's of two $\mathfrak{L}^2$ kernels K and L, respectively. Use the Neumann series representation for the resolvent (3.1-4) to prove that

$$0 < |K| \leq L \qquad \text{implies} \qquad |\mu| \leq |\lambda|$$

[*Hint*: Show that the assumption $|\mu| > |\lambda|$ leads to a contradiction.]

chapter eighteen

Difference Kernels and the Method of Wiener and Hopf

No modern book on integral equations is complete without at least a modicum of consideration of techniques applicable to equations having those specialized kernels of the form

$$K(x,y) \equiv k(x - y)$$

Equations with such *difference kernels* occur rather naturally in the study of various physical phenomena, and their analysis plays a central role in fields as disparate as statistical communication theory and mathematical astrophysics. We have already encountered these functions in several earlier sections of the text (see, for example, Secs. 1.3, 5.2, and 16.1). We now devote this final chapter to a more comprehensive discussion of the procedures developed to cope with integral equations having them as kernels.

18.1 THE INFINITE INTERVAL

Although we have previously restricted our attention to bounded domains, it is convenient to relax that condition in the case of difference kernels.† Thus, for instance, using the integral Fourier transform (1.3-1) in a manner completely analogous to that with which the Laplace transform was employed in Sec. 5.2, the nonhomogeneous integral equation

$$\phi(x) = f(x) + \lambda \int_{-\infty}^{\infty} k(x-y)\phi(y)\, dy \qquad (1)$$

can be solved. For this situation we obtain the classical result [see Wiener (1932 and 1933) and Krein (1958)].

THEOREM 18.1-1

Let $k(t)$ be integrable over $(-\infty, \infty)$ and have a Fourier transform given by

$$K(\omega) \equiv \frac{1}{\sqrt{2\pi}} \int_{-\infty}^{\infty} k(t) e^{i\omega t}\, dt \qquad (2)$$

For arbitrary integrable‡ f and given λ, (1) has a unique integrable solution *iff*

$$1 - \sqrt{2\pi}\,\lambda K(\omega) \neq 0 \qquad -\infty < \omega < \infty \qquad (3)$$

When this relation is satisfied, the solution may be represented as

$$\phi(x) = f(x) + \lambda \int_{-\infty}^{\infty} r(x-y; \lambda) f(y)\, dy \qquad (4)$$

where $r(t; \lambda)$ is an integrable function with the Fourier transform

$$R(\omega; \lambda) = \frac{K(\omega)}{1 - \sqrt{2\pi}\,\lambda K(\omega)} \qquad (5)$$

In the case of the homogeneous version of (1), the solutions are essentially exponential in character; their precise form is detailed in Wiener

---

† This leads, of course, to a class of *singular* integral equations to which the classical Fredholm-Carleman theory need not apply. Characteristic values no longer need to be isolated and a so-called "continuous spectrum" may well occur.

‡ Under certain conditions, a wider class of inhomogeneous terms can be effectively handled using a generalized approach; see Morse and Feshbach (1953, sec. 8.5), for instance.

(1933) or Titchmarsh (1948, sec. 11.2), for example.

The analysis of the equation

$$\phi(x) = f(x) + \lambda \int_0^\infty k(x - y)\phi(y)\, dy \tag{6}$$

in which the fundamental interval is only semi-infinite, is considerably more complex. Its solution typically involves a heavy reliance upon complex function theory, including, in particular, the notion of the *factorization* of functions regular analytic in certain strip domains of the complex plane. We shall discuss in Sec. 18.2 the celebrated procedure of Wiener and Hopf (1931) [see also Hopf (1934), Paley and Wiener (1934), Smithies (1939), Reissner (1941), Fock (1942), Titchmarsh (1948), and Noble (1958), for example], which is applicable to equations such as (6). In the meantime, however, it is worth mentioning that recent work of Krein (1958) [see also Rapoport (1948 and 1949)] has shown that a sufficiently general and complete theory concerning the integral equation (6) can be developed. For given λ and h, the integer [see Ahlfors (1953, pp. 92–94)]

$$N(\lambda, h) \equiv -\frac{1}{2\pi} \int_{-\infty+ih}^{\infty+ih} d_\omega \{\arg [1 - \sqrt{2\pi}\, \lambda K(\omega)]\} \tag{7}$$

which may be termed the *index* of (6), plays an important role. If we introduce the notation $L(a,b;h)$ for the space of all functions $f(t)$ such that $e^{-ht}f$ is integrable over (a,b), then, in analogy with Theorem 18.1-1, we have the following typical theoretical results, which we quote without proof.

THEOREM 18.1-2

Given λ and h, let $k(t)$ be in $L(-\infty, \infty; h)$ and have a Fourier transform $K(\omega)$ given by (2).† For arbitrary f in $L(0, \infty; h)$, (6) has a unique solution ϕ in $L(0, \infty; h)$ *iff*

$$1 - \sqrt{2\pi}\, \lambda K(u + ih) \neq 0 \qquad -\infty < u < \infty \tag{8}$$

and $N(\lambda, h) = 0$. When these conditions are satisfied, the solution may be represented by a relation analogous to (4) wherein the

---

† For the case wherein $k(t)$ has polynomial rather than exponential decay at infinity, see Pao (1967).

resolvent kernel $r(x,y;\lambda,h)(0 \le x,y < \infty)$ satisfies

$$r(x,y;\lambda,h) = r(x - y,0;\lambda,h) + r(0,y - x;\lambda,h)$$

$$+ \int_0^\infty r(x - z,0;\lambda,h)r(0,y - z;\lambda,h)\ dz$$

The functions $r(t,0;\lambda,h)$ and $r(0,t;\lambda,h)$ are in $L(0,\infty;h)$ and $L(0,\infty;-h)$, respectively, and vanish for $t < 0$.

THEOREM 18.1-3

Under the same conditions as in Theorem 18.1-2, if (8) holds, the homogeneous version of (6) has nontrivial solutions in $L(0,\infty;h)$ *iff* the index (7) is *positive*. When this situation prevails, there actually exist $N(\lambda,h)$ linearly independent solutions $\phi_\nu(x)(1 \le \nu \le N)$ of the homogeneous equation, which behave like $o(e^{hx})$ as $x \to \infty$. In this case, moreover, for arbitrary f in $L(0,\infty,h)$, (6) has an infinity of solutions in $L(0,\infty-h)$.

On the other hand, if the index (7) is *negative*, then, for given f, (6) either has no solution or possesses a unique solution in $L(0,\infty;h)$. The latter situation occurs *iff* $e^{-2ht} f$ is orthogonal over $(0,\infty)$ to every solution (of which there are $|N|$ that are linearly independent) of the homogeneous adjoint equation associated with (6).

18.2   THE APPROACH OF WIENER AND HOPF

The existence of the integral defining the Fourier transform of a given function implies certain growth restrictions on that function which may not be satisfied in a given practical problem. In such situations, however, the *generalized* Fourier transforms

$$F_+(\omega) \equiv \frac{1}{\sqrt{2\pi}} \int_0^\infty f(t)e^{i\omega t}\ dt$$

$$F_-(\omega) \equiv \frac{1}{\sqrt{2\pi}} \int_{-\infty}^0 f(t)e^{i\omega t}\ dt \tag{1}$$

where $\omega = u + iv$, may exist, F_+ for sufficiently large positive v and F_- for sufficiently large negative v. This is the case, for instance, whenever the given function grows at most exponentially. In fact, under these circumstances the following fundamental result may be readily established [see, for example, Paley and Wiener (1934, p. 3), Titchmarsh (1948, p. 340)]:

LEMMA 18.2-1   Let $f(t)$ be a measurable function of the real variable t square integrable over any finite interval and satisfying

$$f(t) = \begin{cases} O(e^{T_- t}) & t \to +\infty \\[2mm] O(e^{T_+ t}) & t \to -\infty \end{cases} \tag{2}$$

Then $F_+(\omega)$, as given by (1), not only exists for $v > T_-$ but is a regular analytic function of ω throughout this half-plane. In similar fashion, $F_-(\omega)$ is regular analytic in ω for all $v < T_+$. Moreover, by virtue of the familiar Riemann-Lebesgue lemma, both F_+ and F_- tend uniformly to zero as $|\omega| \to \infty$ in any closed domain interior to their respective half-planes of regularity.

As one might expect, there is an important converse to this result. For our purposes, however, it suffices merely to note that whenever the relations (1) are meaningful, the reciprocal expressions can be put in the form

$$\frac{1}{\sqrt{2\pi}} \int_{-\infty+ia}^{\infty+ia} F_+(\omega) e^{-i\omega t}\, d\omega = \begin{cases} f(t) & t > 0 \\[2mm] 0 & t < 0 \end{cases}$$

$$\tag{3}$$

$$\frac{1}{\sqrt{2\pi}} \int_{-\infty+ib}^{\infty+ib} F_-(\omega) e^{-i\omega t}\, d\omega = \begin{cases} 0 & t > 0 \\[2mm] f(t) & t < 0 \end{cases}$$

where $a > T_-$ and $b < T_+$. Having made these observations, we are now ready to proceed.

The Wiener-Hopf method was originally designed to handle the homogeneous version of (18.1-6), namely,

$$\phi(x) = \lambda \int_0^\infty k(x - y)\phi(y)\, dy \tag{4}$$

We assume in analyzing this expression that the right-hand side of (4) defines $\phi(x)$ for negative values of x, as well as positive, and that both the kernel k and the solution ϕ we seek have at most exponential growth characteristics of the form (2) with $T_- < T_+$. It follows, then, taking Fourier transforms, that

$$\Phi_+(\omega) + \Phi_-(\omega) = \sqrt{2\pi}\ \lambda \Phi_+(\omega) K(\omega)$$

or $\qquad \Phi_+(\omega)[1 - \sqrt{2\pi}\ \lambda K(\omega)] = -\ \Phi_-(\omega) \qquad\qquad (5)$

where Φ_+, Φ_- are the generalized transforms (1) of ϕ and $K = K_+ + K_-$ is the regular transform (18.1-2) of the kernel k (see Prob. 2).

Equation (5) is to be recognized as a relation between analytic functions. In view of the assumed growth behavior of ϕ, Φ_+ is a regular analytic function of ω at least in the half-plane $v > T_-$. Likewise Φ_- is regular analytic in ω for $v < T_+$. K itself is regular analytic at least in a strip $\alpha < v < \beta$, which we presume overlaps a portion of each of these half-planes of regularity of Φ_+ and Φ_-. The essence of the approach of Hopf and Wiener at this point is to apply a method of *factorization* (which we shall discuss later) to the bracketed term in (5). Using analytic continuation arguments and certain elementary notions from entire function theory, both Φ_+ and Φ_- can then be independently determined and hence ϕ may be obtained from (3). To get a feel for the technique, let us consider, for a moment, the specific example in which

$$k(t) = e^{-|t|} \qquad\qquad (6)$$

(see Prob. 3).†

The kernel (6) has the Fourier transform

$$\sqrt{2\pi}\ K(\omega) = \frac{2}{1 + \omega^2}$$

The bracketed term in (5), therefore, may be factored by inspection as

$$[1 - \sqrt{2\pi}\ \lambda K(\omega)] = \frac{1 + \omega^2 - 2\lambda}{1 + \omega^2}$$

$$= \frac{(1 + \omega^2 - 2\lambda)/(1 - i\omega)}{1 + i\omega} \qquad\qquad (7)$$

Substituting into (5) and rearranging, we then obtain

$$\Phi_+(\omega)\left(\frac{1 + \omega^2 - 2\lambda}{1 - i\omega}\right) = -\ \Phi_-(\omega)(1 + i\omega) \qquad\qquad (8)$$

---

† This example was already analyzed by Lalesco (1912, pp. 121–123), using a different approach, of course.

The right-hand side of this new relation has the half-plane $v <$ min $(1, T_+)$ as its domain of regularity, while the left-hand side is regular for $v > $ max $(-1, T_-)$. If we assume that these two half-planes overlap, then we can assert that each side of (8) is the analytic continuation of the other side and thus, in actuality, both sides must describe an analytic function regular throughout the entire complex plane. We call this entire function $P(\omega)$.

In order to determine $P(\omega)$ more completely we merely need recall the extended form of Liouville's theorem [see Titchmarsh (1939, p. 85), for example], namely, the following.

LEMMA 18.2-2   If the entire function $f(\omega)$ satisfies

$$f(\omega) = O(|\omega|^p)$$

as $|\omega| \to \infty$, then it must be a polynomial of at most the pth degree.†

For our special case, it follows then from (8) and the vanishing of Φ_+, Φ_- for appropriate large $|\omega|$, by Lemma 18.2-1, that $P(\omega) = $ constant. As a result we find

$$\Phi_+(\omega) = \frac{C(1 - i\omega)}{\omega^2 + 1 - 2\lambda} \qquad \Phi_-(\omega) = \frac{-C}{1 + i\omega}$$

and
$$\phi(x) = \begin{cases} C \displaystyle\int_{-\infty+ia}^{\infty+ia} \frac{1 - i\omega}{\omega^2 + 1 - 2\lambda} e^{-i\omega x} \, d\omega & x \geq 0 \\[4mm] -C \displaystyle\int_{-\infty+ib}^{\infty+ib} \frac{1}{1 + i\omega} e^{-i\omega x} \, d\omega & x < 0 \end{cases}$$

The last two integrals may be evaluated in the standard manner using residue calculus, closing off the contour with semicircles in the lower and upper half-planes, respectively. This yields

$$\phi(x) = \begin{cases} D\,(\sinh \sqrt{1 - 2\lambda}\, x + \sqrt{1 - 2\lambda} \cosh \sqrt{1 - 2\lambda}\, x) & x \geq 0 \\[3mm] D \sqrt{1 - 2\lambda}\, e^x & x < 0 \end{cases}$$

with D a new constant, from which it is clear that $T_+ = 1$ and

---

† In the case $p = 0$, this result appears actually to have been first established by Cauchy (1844). It was subsequently popularized by Liouville in his lectures [see Liouville (1880)].

$T = \text{Re} \sqrt{1 - 2\lambda}$; hence if λ is real, then it must also be positive in order to validate our analysis. †

In the general situation, the analogue of the factorization (7) is the expression

$$\frac{\psi_+(\omega)}{\psi_-(\omega)} \prod_{\nu=1}^{p} (\omega - \omega_\nu) \qquad (9)$$

Here the $\omega_\nu (1 \leq \nu \leq p)$ are the zeros of $[1 - \sqrt{2\pi} \, \lambda K(\omega)]$ in the strip $\alpha < \nu < \beta$, of which there can only be finitely many [Paley and Wiener (1934, pp. 51ff) and Titchmarsh (1948, pp. 399ff)]. The function $\psi_-(\omega)$ is selected to be regular analytic *and free from zeros* in the half-plane $\nu < \beta$, while $\psi_+(\omega)$ is chosen to be regular analytic *and free from zeros* for $\nu > \alpha$. Furthermore, and this is essential from a practical point of view, both ψ_+ and ψ_- are selected so as to have algebraic, rather than exponential, growth as $|\omega| \to \infty$ in their respective domains of regularity. That this is possible was originally demonstrated by Wiener and Hopf (1931) [see Swann (1960, chap. 2), however, for a correct determination of the asymptotic growth estimates]. As in the special case, the substitution of (9) into (5) and subsequent rearrangement‡ leads to a relation between analytic functions regular in two overlapping half-planes, which together exhaust the entire complex ω plane. An analytic continuation argument, then, followed by an application of the generalized Liouville result Lemma 18.2-2 and an inversion using (3), finally suffices for the solution of the desired equation.

18.3   GENERALIZATIONS OF THE WIENER-HOPF PROCEDURE

The original approach of Hopf and Wiener was extended to the full inhomogeneous equation (18.1-6) by Reissner.(1941) and Fock (1942). Subsequently Krein (1958) clarified and enlarged upon the work of these two authors and in the process analyzed rather thoroughly several practical examples. [For two other specific generalizations, see Heins and Wiener (1946) and Carrier (1949).] Krein also briefly discussed the theory applicable to integral equations that have Wiener-Hopf form but are of the first kind.

---

† Observe that this conclusion is in keeping with Theorem 18.1-3. The given kernel k is in $L(-\infty, \infty; h)$ for all $|h| < 1$ and has a transform K such that $[1 - \sqrt{2\pi} \, \lambda K(\omega)]$ vanishes for $\omega = \pm i \sqrt{1 - 2\lambda}$. It follows, then, that (4) has a nontrivial solution for real λ iff $\lambda > 0$ and h satisfies $\text{Re} \sqrt{1 - 2\lambda} < h < 1$, in which case $N(\lambda, h) = 1$.

‡ This rearrangement is done in such a manner that the largest possible region of overlap is maintained.

More recently, Hutson (1965) has considered Wiener-Hopf-like equations with kernels k given by

$$k(x,y) = \int_{-\infty}^{\infty} A(\omega)u_1(x,\omega)u_2(y,\omega)\, d\omega$$

where $A(\omega)$ is a suitably chosen function and u_1 and u_2 satisfy

$$\frac{d^2u}{dx^2} + [\omega^2 - q(x)]u = 0$$

with appropriate q. This formulation includes the standard Wiener-Hopf equation ($q = 0$) and leads to theoretical results akin to those given earlier in Sec. 18.1.

Also of interest are the results of Gohberg and Fel'dman (1968) concerning integro-difference equations of the form

$$\sum_{\nu=-\infty}^{\infty} a_\nu \phi(x - x_\nu) = f(x) + \lambda \int_0^{\infty} k(x - y)\phi(y)\, dy$$

wherein x_ν and a_ν are given real and complex constants, respectively. For such equations, which generalize the *discrete* Wiener-Hopf problems as well as the classical, it appears that a theory similar to Krein's in the standard situation can be developed. The same is true for integro-differential equations of the form

$$\sum_{\nu=0}^{n} a_\nu \frac{d^\nu \phi}{dx^\nu} = \lambda \int_0^{\infty} k(x - y)\phi(y)\, dy$$

as shown by Swann (1960).

Alternate approaches to the solution of the Wiener-Hopf equation (18.1-6) have been proposed by Pincus (1966), Shinbrot (1969), and Noble (1964), among others. Using the orthogonal projection of the space of functions square integrable over $(-\infty, \infty)$ onto the space of functions square integrable over $(0, \infty)$ and introducing certain new Hilbert spaces, Pincus is able to apply earlier work of his on barrier-related spectral problems to the Wiener-Hopf situation, at least in the case of a Hermitian kernel. Shinbrot also has employed various projection operators so as to be able to view the classical Wiener-Hopf equations as special cases of more general problems. In this generalized setting, which makes little distinction between ordinary kernels and kernels that are distributions, he has shown that the solution to Wiener-Hopf-like problems can be constructed even for cases in which no classical factorization is possible.†

Noble's contribution, like Shinbrot's, was motivated by a desire to avoid reliance upon the factorization procedure, particularly in those cases wherein numerical solutions are the desired end result. Employing a method similar to one suggested earlier by Ambarzumian (1943) [see also Chandrasekhar (1950) and especially Busbridge (1960, sec. 28)] for the specific equations encountered in radiative transfer theory, Noble has formally demonstrated the equivalence of Wiener-Hopf-like equations and certain (perhaps coupled) nonlinear equations with well-behaved kernels [see also Case (1957), Sobolev (1963), and DePree (1969)]. These last equations have the putative advantage of being easier to solve numerically.

18.4 DIFFERENCE KERNELS ON A FINITE INTERVAL

Already in Carleman's thesis (1916) there appeared discussion pertinent to the analysis of so-called *finite convolution equations*, i.e., integral equations with difference kernels valid over a fixed finite interval. Subsequently, Carleman (1922) showed how to solve equations of the first kind such as

$$f(x) = \int_{-a}^{a} k(x - y)\phi(y)\,dy \qquad a < \infty \tag{1}$$

in the case where k is either a logarithm or a power. This work was improved upon by Latta (1956) [also see Peters (1969)], who developed an operational technique that leads to the unique solution of (1) whenever the existence of such a solution is known a priori and the kernel k satisfies a linear differential equation (of any order) with linear coefficients. Shinbrot (1958–1959) [see also Pearson (1957–1958)] provided a further generalization of this approach by demonstrating that for kernels k having the form

$$k(t) = p(t)j(t) + q(t)$$

where $j(t)$ is a *Latta kernel* and p and q are polynomials, the solution of (1) can be reduced to the solution of an equivalent equation with the kernel j to which the method of Latta then applies.

---

†The classical factorization of the Wiener-Hopf method has been generalized so as to apply to certain abstract operators in Shinbrot (1964).

Since finite convolution equations occur rather often, for example, in the study of statistical communication theory, radiative transfer, electromagnetic diffraction, and linearized gas dynamics, it is appropriate that we briefly sketch the fundamentals of Latta's approach. We begin by introducing the operator notation

$$\Gamma\psi \equiv \int_a^b k(x-y)\psi(y)\,dy \tag{2}$$

where it is assumed that k is integrable over the interval (a,b), ψ is differentiable, and for all x in $[a,b]$

$$k(x-a)\psi(a) = k(x-b)\psi(b) \tag{3}$$

It follows, then, that for such k and ψ, $\Gamma\psi$ can be alternatively expressed as

$$\Gamma\psi = \psi(b)\int_a^b k(x-y)\,dy - \int_a^b \psi'(y)\int_a^y k(x-z)\,dz\,dy$$

and hence, differentiating with respect to x and using (3),

$$\frac{d}{dx}(\Gamma\psi) = \int_a^b k(x-y)\psi'(y)\,dy$$

This last relation, which has the symbolic form

$$\Gamma'\psi = \Gamma\psi' \tag{4}$$

plays a fundamental role in the solution of the integral equation (1). In any given practical problem, however, (4) must be complemented by an additional expression that is a consequence of the specific differential equation satisfied by the kernel in question. For instance, if $k(t) = \log|t|$, then $tk'(t) - 1 = 0$, from which it follows that

$$\Gamma'(x\psi) = \int_a^b k'(x-y)y\psi(y)\,dy$$

$$= x\int_a^b k'(x-y)\psi(y)\,dy - \int_a^b k'(x-y)(x-y)\psi(y)\,dy$$

or $$\Gamma'(x\psi) = x\Gamma'\psi - \mu_0 \tag{5}$$

where μ_0 is the zero-order moment of ψ, namely,

$$\mu_0 \equiv \int_a^b \psi(y)\,dy$$

For the validity of (5), moreover, ψ need only be integrable.

As an example illustrative of Latta's approach, let us discuss how the above relations can be used in order to solve the integral equation

$$\pi = \int_{-1}^1 \log|x - y|\,\phi(y)\,dy \tag{6}$$

[see also Stakgold (1968, pp. 191–192)]. Assuming the existence and uniqueness of a solution, we note that ϕ must be an even function of its argument. It follows, then, from (5) and (6) that

$$\Gamma(x\phi) = -\mu_0 x$$

and
$$\Gamma(x^2\phi) = -\tfrac{1}{2}\mu_0 x^2 + C$$

where C is constant. Applying (4) to the function

$$\psi(x) = \int_{-1}^x y\phi(y)\,dy$$

therefore, which satisfies (3) since ϕ is even, we obtain

$$\Gamma'\psi = \Gamma\psi' = \Gamma(x\phi) = -\mu_0 x$$

and upon integration

$$\Gamma\psi = -\tfrac{1}{2}\mu_0 x^2 + D$$
$$= \Gamma(x^2\phi) + \Gamma(E\phi)$$

As a consequence of the assumed uniqueness of the solution of (6), it follows that

$$\psi = (x^2 + E)\phi$$
$$\psi' = x\phi = 2x\phi + (x^2 + E)\phi'$$

and therefore,
$$\phi = \frac{F}{\sqrt{x^2 + E}}$$

The form of ψ dictates that the constant $E = -1$, while substitution in (6) itself determines F as $-i/\log 2$.

Other investigators have also considered (1) or the analogous equation of the second kind. For the most part, however, these analyses are concerned either with approximate solutions for general classes of kernels or exact solutions for very special kernels. Bellman and Latter (1952), for instance [see also Wing (1968)], consider the homogeneous equation of the second kind with nonnegative definite, even, monotone kernels and discuss the behavior of the smallest characteristic value as $a \rightarrow \infty$. Latter (1958–1959) obtains approximate characteristic values and functions for large a, when the kernel is Hermitian, by modification of the classical Wiener-Hopf approach [see also Kac, Murdock, and Szegö (1953), Widom (1958 and 1961), Hutson (1964–1965), and Wilf (1970)]. Rosenblatt (1963a and b) performs his asymptotic analysis under the assumptions that the kernels in question are positive definite, even, and otherwise have an appropriately smooth form. Widom (1963 and 1964) concerns himself with real, even kernels whose Fourier transforms $K(\omega)$ are nonnegative and sufficiently regular, while Shinbrot (1960) makes other assumptions on the nature and growth properties of K.[†] Recently Stakgold (1969) has shown how the classical approach of Weinstein (see Chap. 11) may be adapted in order to yield uniformly improvable estimates for each of the characteristic values of a real, even difference kernel on a finite interval. Boland (1969b), meanwhile, has suggested how convexity considerations may be employed to yield bounds for at least the lower-order cv's of a subclass of such kernels.

Prompted by computational considerations, Leonard and Mullikin (1965a and b) have analyzed the Fredholm equations of the second kind with kernels satisfying

$$k(x) = \int_\alpha^\infty \frac{m(\tau)}{\tau} e^{-|x|\tau} \, d\tau \qquad 0 \leq \alpha < \infty$$

where the function m is an appropriately normalized, nonnegative, Hölder continuous function. This work, which has direct applicability in neutron transport and radiation theory, has been complemented by recent efforts of Wing (1967b), Kagiwada et al. (1968, 1969, and 1970), and others, who have adapted *imbedding* procedures to reduce integral equations on both the finite and semi-infinite interval to equivalent, but computationally simpler, initial-value problems.

[†] The case wherein K is a rational function of its argument is a familiar one to communication theorists. This situation, which occurs in the analysis of certain stationary random processes, can be solved exactly in principle [see Helstrom (1965), for example]. The kernels considered by Rosenblatt (1963a and b) were locally of this type. See also Pal'cev (1970).

Also of interest is the discussion of Ursell (1969) for the situation wherein the kernel is highly oscillatory. A special case, namely, that in which the kernel can be written as a *finite* Fourier transform, had been considered earlier by Roark and Wing (1965) [see also Wing (1965), Roark and Shampine (1965 and 1968), and the recent generalizations of Burgmeier (1970)].

Lastly, we mention that Williams (1963) has shown that equations of the first kind with kernels that can be expressed as an integral of the Weber-Schafheitlin type can be systematically handled, while the general approach of Shinbrot (1969 and 1970) appears capable of yielding the solution to (1) in a number of interesting cases heretofore intractable.

PROBLEMS

1. **a.** The characteristic values of the so-called *bandpass* kernel [Slepian and Pollak (1961)]

$$L(x,y) = \frac{\sin 2ka(y-x)}{\pi(y-x)} \qquad -a \le x,y \le a \qquad \text{real } k$$

   are all real and greater than unity. Show then that the cv's of the closely related complex-symmetric kernel

$$K(x,y) = \sqrt{\frac{k}{\pi}} \, e^{ik(x-y)^2} \qquad -a \le x,y \le a$$

   all lie outside the unit circle.
   b. Verify that for $a = \infty$, K is no longer square integrable and every λ of unit modulus is a "cv" with an associated "cf" e^{ibx} with real b. [See (18.1-3).]

2. Verify that (18.2-5) follows by taking Fourier transforms of both sides of (18.2-4). Compare with Prob. 2 of Chap. 5.

3. The homogeneous Wiener-Hopf equation (18.2-4) with the so-called Lalesco kernel

$$k(t) = e^{-|t|}$$

   can also be solved by analyzing the corresponding boundary-value problem. Carry out the necessary steps in order to demonstrate the equivalence, in this case, of these two approaches.

4. **a.** Show that for the Carleman kernel

$$k(t) = |t|^{-\alpha}$$

   with $0 < \alpha < 1$, the analogue of (18.4-5) is

$$\Gamma'(x\psi) = x\Gamma'\psi + \alpha\Gamma\psi$$

b. Use Latta's method to verify that the solution of the integral equation

$$\alpha \pi x = \int_{-1}^{1} \frac{\phi(y)}{|x-y|^\alpha}\,dy \qquad -1 < x < 1$$

for $0 < \alpha < 1$ is

$$\phi(x) = x(1-x^2)^{\frac{1}{2}(\alpha-1)}\cos\tfrac{1}{2}\pi\alpha$$

[The asymptotic distribution of the cv's of the Carleman kernel has recently been derived by Ukai (1971).]

5. For the kernel

$$k(t) = K_0(|t|)$$

where K_0 is the modified Hankel function of zero order, the analogue of (18.4-5) is

$$\Gamma''(x\psi) - \Gamma(x\psi) = x\Gamma''\psi + \Gamma'\psi - x\Gamma\psi$$

a. Demonstrate that the approach of Latta can be used to solve the equation

$$e^{-x} = \int_0^{\infty} K_0(|x-y|)\phi(y)\,dy \tag{1}$$

b. Alternatively, try using the Wiener-Hopf technique to effect the solution of (1). [The general integral equation of the first kind on a finite interval, with this modified Hankel function kernel, has been recently solved by Belward (1969).]

6. Compare Latta's method and a more general classical Fourier transform approach to the solution of

$$e^{-\alpha x^2} = \int_{-\infty}^{\infty} e^{-(x-y)^2}\phi(y)\,dy \qquad 0 < \alpha < 1$$

Appendix A

The Correspondence Between
Differential and Integral Equations

Every linear differential equation can be transformed, in principle, along with its auxiliary conditions (e.g., boundary values), into an equivalent integral equation. Boundary-value problems associated with equations of elliptic type give rise generally to Fredholm integral equations, while initial-value problems lead to Volterra integral equations [see Tricomi (1957, pp. 127ff), Courant and Hilbert (1953, pp. 351ff), and Garabedian (1964), for example]. The converse, however, is not always true. For instance, the integral equation

$$\phi(x) = \lambda \int_{-1}^{1} e^{ik(x-y)^2} \phi(y) \, dy \tag{1}$$

which occurs in the theoretical description of an optical maser with rectangular plane reflecting surfaces [Morgan (1963) and Cochran (1965)], has no known differential equation equivalent. In this appendix we want to briefly examine the correspondence between differential and integral equations, where it exists, using several illustrative examples.

BOUNDARY-VALUE PROBLEMS

Consider the differential equation

$$\frac{d^2u}{dx^2} = f(x,u) \qquad 0 \leq x \leq L \tag{2}$$

with the boundary conditions

$$u(0) = 0 = u(L) \tag{3}$$

If
$$G(x,x') = \begin{cases} \dfrac{x(L-x')}{L} & x \leq x' \\[2mm] \dfrac{x'(L-x)}{L} & x' \leq x \end{cases}$$

is the Green's function (impulse response) associated with the differential operator in (2) and the auxiliary conditions (3), then there is an integral equation equivalent to (2) and (3) given by

$$u(x) = -\int_0^L G(x,x') f\left(x', u(x')\right) \, dx'$$

[compare (1.1-7)].

A similar result is valid for higher-order linear ordinary differential equations. In particular, if

$$L[u] + \lambda r(x)u = f(x) \qquad a \leq x \leq b$$

is a given differential equation with a real linear operator L of nth order and if the n boundary conditions are linear, independent, and homogeneous,† then

$$u(x) = -\int_a^b G(x,x') f(x') \, dx' + \lambda \int_a^b G(x,x') r(x') u(x') \, dx'$$

---

† Recall that such a boundary-value problem with nonhomogeneous conditions can always be reduced to this case by subtracting from u a polynomial of at most nth degree. The analogous reduction for initial-value problems is obtained by making the substitution

$$U(x) = u(x) - \sum_{\nu=0}^{n-1} \frac{u^{(\nu)}(0)}{\nu!} x^\nu$$

Here $G(x,x')$ is the Green's function, which satisfies, as a function of x, the homogeneous equation

$$L[G] = 0 \qquad x \neq x' \qquad a \leq x \leq b$$

with the homogeneous boundary conditions. In addition $G(x,x')$ has the appropriate behavior at the *influence point* $x = x'$; that is,

$$\frac{\partial^\nu G(x'+0,x')}{\partial x^\nu} - \frac{\partial^\nu G(x'-0,x')}{\partial x^\nu} = \begin{cases} 0 & \nu = 0,\,1,\,\ldots,\,n-2 \\ -\dfrac{1}{a_0(x')} & \nu = n-1 \end{cases}$$

where the operator $L = a_0(x)\,d^n/dx^n +$ lower-order terms, with $a_0(x) \neq 0$ for $a \leq x \leq b$. If both L and the boundary conditions are *self-adjoint* [see Ince (1956), Coddington and Levinson (1955), and Cole (1968)], then the Green's function is a symmetric function of x and x'.

A correspondence analogous to that outlined above is valid in the case of partial differential equations when the boundary-value problem is well posed.

Before leaving this section, the following interesting relationship is worth noting: The Green's function associated with the problem in (2) and (3) has the uniformly convergent Fourier series expansion

$$G(x,x') = \frac{2}{L} \sum_{\nu=1}^{\infty} \frac{\sin(\nu\pi x/L)\sin(\nu\pi x'/L)}{(\nu\pi/L)^2}$$

If this is viewed as the kernel of a Fredholm integral equation on $0 \leq x,x' \leq L$, then there is associated with G the resolvent kernel

$$R_G(x,x';\lambda) = \frac{2}{L} \sum_{\nu=1}^{\infty} \frac{\sin(\nu\pi x/L)\sin(\nu\pi x'/L)}{(\nu\pi/L)^2 - \lambda}$$

(see Theorem 13.3-4). In this form R_G may be recognized as the Green's function for the differential operator $(d^2/dx^2) + \lambda$. There are analogues of this result for more complicated linear equations of the same as well as higher order and in more than one dimension.

INITIAL-VALUE PROBLEMS

We move now to a discussion of initial-value problems and the integral equations equivalent thereto. Let us begin by considering the linear differential equation

$$u^{(n)}(x) = a_0(x)u^{(n-1)} + a_1(x)u^{(n-2)} + \cdots + a_{n-1}(x)u + f(x) \qquad (4)$$

for, say, $0 \le x \le L$, with initial values

$$u(0) = u'(0) = \cdots = u^{(n-1)}(0) = 0$$

If we define $\phi(x) \equiv u^{(n)}(x)$, then u may be expressed as the n-tuple integral of ϕ (see Chap. 1, Sec. 4), and thus

$$u^{(\nu)}(x) = \frac{1}{(n-\nu-1)!} \int_0^x (x-y)^{n-\nu-1} \phi(y) \, dy \qquad 0 \le \nu \le n-1$$

From this relation it follows immediately that the initial-value problem (4) has the equivalent representation

$$\phi(x) = f(x) + \int_0^x K(x,y)\phi(y) \, dy \qquad 0 \le x \le L \qquad (5)$$

where $\qquad K(x,y) = \sum_{\nu=0}^{n-1} \frac{a_\nu(x)}{\nu!} (x-y)^\nu \qquad (6)$

We see, therefore, that initial-value problems of the type (4) can be converted into *Volterra* integral equations with kernels $K(x,y)$ which are polynomials in y. Conversely, given any polynomial in y which has coefficients that are functions of x, it is clear that it can be put in the form (6). Continuing the argument in reverse, (5) where $K(x,y)$ is such a polynomial in y is thus completely equivalent to an initial-value problem of the form (4).

It is interesting to inquire if there is an analogous result whenever the kernel of a given Volterra integral equation is a polynomial in the first variable x. The answer, in this case, clearly shows some of the beauty inherent in the theoretical development of integral equations. To see this we let

$$K(x,y) = \sum_{\nu=0}^{n-1} \frac{b_\nu(y)}{\nu!} (x-y)^\nu \qquad (7)$$

and consider the resolvent kernel $R_K(x, y; \lambda)$ of K as a function of y for *fixed* x and λ. Define a function $u(y)$ by an n-tuple integral of the form

$$u(y) = \frac{(-1)^n}{(n-1)!} \int_y^x (z-y)^{n-1} R_K(x, z; \lambda) \, dz + \alpha \frac{(x-y)^{n-1}}{(n-1)!} (-1)^{n-1}$$

where α is yet to be determined. It follows that

$$u(x) = u'(x) = \cdots = u^{(n-2)}(x) = 0$$

$$u^{(n-1)}(x) = \alpha \tag{8}$$

and

$$u^{(n)}(y) = R_K(x, y; \lambda) \tag{9}$$

If $K(x, y)$ given by (7) is taken to be a Volterra kernel (vanishing for $y > x$), R_K is also Volterra in nature, as is discussed in detail in Chap. 5, and correspondingly the second Fredholm identity (3.2-5) takes the form

$$R_K(x, y; \lambda) = K(x, y) + \lambda \int_y^x R_K(x, z; \lambda) K(z, y) \, dz$$

In view of the above development this relation may be equivalently expressed as

$$u^{(n)}(y) = \sum_{\nu=0}^{n-1} \frac{b_\nu(y)}{\nu!} \left[(x-y)^\nu + \lambda \int_y^x R_K(x, z; \lambda)(z-y)^\nu \, dz \right]$$

$$= \sum_{\nu=0}^{n-1} \frac{b_\nu(y)}{\nu!} \left[(x-y)^\nu + \lambda \alpha (x-y)^\nu - \nu!(-1)^\nu \lambda u^{(n-\nu-1)} \right]$$

and if we choose $\alpha = -1/\lambda$, this becomes

$$u^{(n)}(y) = -\lambda \sum_{\nu=0}^{n-1} b_\nu(y)(-1)^\nu u^{(n-\nu-1)}(y) \tag{10}$$

We recognize (10) as a linear ordinary differential equation, which, coupled with the conditions (8), forms a *backwards* initial-value problem. Equation (9) then shows that the resolvent kernel associated with the (polynomial in x) kernel (7) is the nth derivative of the solution of this problem.

For initial-value problems associated with linear integro-differential equations of the form

$$u^{(n)}(x) = \sum_{\nu=0}^{n-1} a_\nu(x) u^{(n-\nu-1)} + f(x) + \sum_{\nu=1}^{m} \int_0^x K_\nu(x,y) u^{(\nu)}(y) \, dy$$

the substitution

$$\phi(x) = u^{(\gamma)}(x) \qquad \gamma = \max\,(n,m)$$

also leads to equivalent Volterra integral equations. (If $m > n$, these are of the first kind.) The situation is somewhat more complicated for generalized initial-value problems (Cauchy problems) associated with partial differential equations of hyperbolic type. Garabedian (1964, chap. 4) discusses the correspondence for such equations in two independent variables using the notion of the Riemann function.

Appendix B

Relevant Fundamental Concepts

Collected in this appendix for the use of interested readers are various fundamental concepts and results that are relevant to the study of integral equations as it is pursued in this book. These concepts and results have been grouped under the three headings: Linear Algebra, Functions of a Real Variable, and Functions of a Complex Variable. Those readers desirous of discussions more complete than those given herein are encouraged to consult, for example, the excellent accounts of these topics provided by Halmos (1958), Burkill (1951) or Natanson (1955), and Copson (1935) or Ahlfors (1953).

LINEAR ALGEBRA

For our purposes, a **complex vector** (or **linear**) **space** is a set V of elements called *vectors* with the property that linear combinations with arbitrary complex coefficients (**scalars**) of elements of V are also in V. We assume here, of course, the validity of the familiar properties of commutativity, associativity, and distributivity with respect to the operations of addition and scalar multiplication. A **linear manifold** is a nonempty subset U of a vector space V which is a vector space in its own right. V is the **direct sum** of the linear manifolds $U_1, U_2, \ldots, U_n$, if every vector v in V has a *unique* representation $v = u_1 + u_2 + \cdots + u_n$ where u_ν is in U_ν.

A finite set $\{v_\nu\}$ of vectors is **linearly independent** if for all complex numbers α_ν, $\sum_\nu \alpha_\nu v_\nu = 0$ implies $\alpha_\nu = 0$ for each ν. An *infinite* set of vectors is linearly independent if every finite subset has this property. A set of vectors that is not linearly independent is called **linearly dependent**. Any linearly independent set U of vectors in a vector space V that is contained in no larger such set constitutes a **basis** for V. Every basis of a given vector space has the same number of elements, and that number is the **dimension** of the space. If the number is finite, the space is called **finite dimensional**. In finite-dimensional vector spaces each basis U has the property that every vector in the space can be *uniquely* represented as a linear combination of elements of U.

A **linear transformation** (or **operator**) $\mathscr{A}$ on a vector space V is a correspondence that assigns to each vector v in V a unique vector $\mathscr{A}v$ in V in such a way that

$$\mathscr{A}(\alpha u + \beta v) = \alpha \mathscr{A}u + \beta \mathscr{A}v$$

for all choices of vectors u and v in V and scalars α and β. One of the simplest such operators is the **identity transformation** $\mathscr{I}$, which associates with each vector v the vector v itself. For each v in V the vector $\mathscr{A}v$ is called the **image** of v under the transformation $\mathscr{A}$; the set of all images constitutes the **range** of $\mathscr{A}$. The **null space** of $\mathscr{A}$ is the set of all v in V whose images are the zero vector. The **rank** and **nullity** of $\mathscr{A}$ are the dimensions of the range and null space, respectively. In a space of finite dimension n, the sum of these two quantities is precisely equal to n.

The **sum** $\mathscr{C}$ of two linear transformations $\mathscr{A}$ and $\mathscr{B}$ on a vector space V is determined by the assignment rule $\mathscr{C}v = \mathscr{A}v + \mathscr{B}v$ for all v in V. The **product transformation** $\mathscr{A}\mathscr{B}$ is analogously given by $(\mathscr{A}\mathscr{B})v = \mathscr{A}(\mathscr{B}v)$. A **projection** on some linear manifold of V is a linear transformation $\mathscr{A}$ with the property that $\mathscr{A}^2 v = \mathscr{A}v$ for all v in V. For such a transformation, the direct sum of its range and its null space yields the entire space.

A scalar λ is an **eigenvalue** of a linear transformation $\mathscr{A}$ if there exists a *nonzero* vector v such that $\mathscr{A}v = \lambda v$. The vector v is called an **eigenvector** of $\mathscr{A}$ associated with the eigenvalue λ. The **geometric multiplicity** of λ is the number of linearly independent such eigenvectors. A linear manifold M of a vector space V is **invariant** under $\mathscr{A}$ if for each v in M the image $\mathscr{A}v$ is also in M. If the zero vector is included, the set of all eigenvectors associated with a given eigenvalue forms an invariant linear manifold.

A linear transformation $\mathscr{B}$ on a vector space V is **nilpotent** if there exists a positive integer p such that $\mathscr{B}^p v = 0$ for all v in V. If $\mathscr{A}$ is a linear transformation on V with *distinct* eigenvalues λ_ν, there exist linear manifolds M_ν of V that are invariant under $\mathscr{A}$ and on which the transformation $\mathscr{A} - \lambda_\nu \mathscr{I}$ is nilpotent. Indeed, if V is finite dimensional, the M_ν can be chosen so that taken together their direct sum constitutes the whole space V.

Given a linear transformation $\mathscr{A}$ on a finite-dimensional vector space V, we have for an arbitrary basis v_ν, $\nu = 1, 2, \ldots, n$,

$$\mathscr{A} v_\nu = \sum_{\mu=1}^{n} a_{\mu\nu} v_\mu$$

The square array $A = (a_{\mu\nu})$ of complex numbers (scalars) that details the action of the linear transformation on V with respect to the particular basis $\{v_\nu\}$ is called a **matrix**. The scalars $[a_{\mu 1}, \ldots, a_{\mu n}]$ form a **row**, and $(a_{1\nu}, \ldots, a_{n\nu})$ a **column**, of A. Such $1 \times n$ or $n \times 1$ arrays, if standing alone, are termed **row** or **column vectors**, respectively. In keeping with the definitions of operator addition and multiplication, linear combinations $C = \alpha A + \beta B$ of matrices, where $A = (a_{\mu\nu})$, $B = (b_{\mu\nu})$, and $C = (c_{\mu\nu})$, are formed in accordance with the relation

$$c_{\mu\nu} = \alpha a_{\mu\nu} + \beta b_{\mu\nu}$$

while the matrix multiplication $C = AB$ follows the rule

$$c_{\mu\nu} = \sum_{\sigma=1}^{n} a_{\mu\sigma} b_{\sigma\nu}$$

A **positive** (**nonnegative**) matrix is one that has *positive* (*nonnegative*) elements. A **triangular** matrix A is one for which $a_{\mu\nu}$ is zero whenever $\mu < \nu$ (or $\mu > \nu$); if $a_{\mu\nu}$ is zero for all $\mu \neq \nu$, the matrix is **diagonal**. The **identity** matrix I is a *diagonal* matrix with unit diagonal elements; that is, $a_{\mu\nu} = \delta_{\mu\nu}$, the familiar Kronecker delta.

The **conjugate** $\overline{A}$, **transpose** A', and **conjugate transpose** or **adjoint** A^* of a given matrix $A = (a_{\mu\nu})$ are given by

$$\overline{A} \equiv (\overline{a_{\mu\nu}}) \qquad A' \equiv (a_{\nu\mu})$$
$$A^* \equiv (\overline{A})'$$

If $A^* = A$, the matrix is **Hermitian** (**symmetric** if also $\overline{A} = A$). If $A' = A$ but $\overline{A} \neq A$, the matrix is **complex-symmetric**. **Anti-Hermitian** (**skew-symmetric**) matrices are those satisfying $A^* = -A$ $(A' = -A)$. A **normal** matrix satisfies $AA^* = A^*A$. A **unitary** (**orthogonal**) matrix is a normal matrix that, in addition, satisfies $AA^* = I$ $(AA' = I)$. Hermitian and anti-Hermitian matrices are also normal, but complex-symmetric matrices generally are not.

The **determinant** $\det A$ associated with a square matrix A has the following inductive definition: the determinant of a 1×1 matrix is simply that single element; more generally,

$$(\det A)\delta_{\mu\nu} = \begin{cases} \sum_{\sigma=1}^{n} a_{\sigma\mu} a^{\sigma\nu} \\ \\ \sum_{\sigma=1}^{n} a_{\mu\sigma} a^{\nu\sigma} \end{cases} \quad \mu, \nu = 1, 2, \ldots, n$$

where $a^{\mu\nu}$, the **cofactor** of the element $a_{\mu\nu}$ in A, is $(-1)^{\mu+\nu}$ times the determinant obtained from $\det A$ by deleting its μth row and νth column. A square matrix A is **singular** or **nonsingular** according as its determinant does or does not vanish. Using the transpose of the matrix of cofactors, an **inverse** matrix A^{-1}, with the property that $AA^{-1} = I = A^{-1}A$, can be defined for *nonsingular* matrices by the formula

$$A^{-1} \equiv \frac{(a^{\nu\mu})}{\det A}$$

Cramer's rule follows, namely, if n linear equations $\sum_{\nu} a_{\mu\nu} x_\nu = y_\mu$ in n unknowns have a *nonsingular* coefficient matrix $A = (a_{\mu\nu})$, then there is a *unique* solution representable as

$$x_\nu = \frac{\sum_{\mu=1}^{n} a^{\mu\nu} y_\mu}{\det A} \quad \nu = 1, 2, \ldots, n$$

More generally, since the representations (as column vectors) of the vectors in the null space of a linear transformation $\mathscr{A}$ with matrix representation A lie in the null space of the adjoint matrix A^*, in the case of *singular* A the matrix equation $Ax = y$ has solutions *if and only if $A^*y = 0$.*

In view of the above, the scalar λ can appear as an eigenvalue of any matrix representation A of a given linear transformation $\mathscr{A}$ on a finite-dimensional vector space *if and only if* it is a root of the **characteristic polynomial** (in λ) $\det(A - \lambda I) = 0$. It follows, then, that

$$\det A = \prod_{\nu=1}^{n} \lambda_\nu$$

where each eigenvalue λ_ν of A is included as many times as its **algebraic multiplicity** as a zero of the characteristic polynomial. Note that the *geometric* multiplicity of λ_ν is no greater than its *algebraic* multiplicity. The **trace** of A is defined as

$$\mathrm{Tr}\, A \equiv \sum_{\nu=1}^{n} a_{\nu\nu} \equiv \sum_{\nu=1}^{n} \lambda_\nu$$

Two matrices A and B are **similar** if there exists a *nonsingular matrix* P such that

$$B = P^{-1}AP$$

This similarity transformation can be viewed as nothing more than a change of coordinates (or basis) in the underlying vector space V. Rank, nullity, and eigenvalues are all invariant under similarity transformations. Every matrix A is similar to a triangular matrix, and the entries along the main diagonal of the latter are the eigenvalues of the former. In the case of normal A, the triangular matrix becomes diagonal. This diagonal representation is sometimes referred to as the **canonical form** of the original matrix. If the eigenvalues of a normal matrix A are positive (nonnegative), A is termed **positive (nonnegative) definite**. Alternatively, these are the coefficient matrices which give rise to positive (nonnegative) **quadratic forms** of scalars $\sum_{\mu,\nu=1}^{n} a_{\mu\nu} \alpha_\mu \overline{\alpha}_\nu$.

FUNCTIONS OF A REAL VARIABLE

On the extended real line, the symbols $[a,b]$ and (a,b) denote **closed and open intervals**, respectively, i.e., the sets of real numbers x such that $a \leq x \leq b$ and $a < x < b$. A point is an **interior point** of a set E of real numbers if there exists an open interval, contained entirely within E, which contains this point. A set is **open** if all of its points are interior points. A set is **closed** if its complement (in the real numbers) is open. Two sets which have no points in common are **disjoint**. A **bounded** set, of which a finite interval is an example, is one all of whose points are less, in absolute value, than some finite real number. Every nonvoid *bounded* open set can be *uniquely* represented as the countable union of pairwise disjoint open intervals, called **component intervals**, the end points of which do not belong to the given set.

The length of a nonvoid open interval (a,b), that is, $b - a$, is the **measure** of the interval. Generalizing this concept, the **measure** of a nonvoid bounded open set is defined as the sum of the lengths of all of its component intervals. The **measure** of a bounded closed set is the greatest lower bound of the measures of all possible bounded open sets containing the given closed set. For an arbitrary bounded set E, the greatest lower bound of the measures of all bounded open sets containing E is the **outer measure** of E. In analogous fashion, the **inner measure** of E is the least upper bound of the measures of all closed sets contained within E. The set E is **measurable** if its outer and inner measures are equal, the common value being taken as the measure $m(E)$ of the set. More generally, an arbitrary (perhaps unbounded) set E is **measurable** if for each integer N the subset E_N, which consists of those points of E that are also contained in the interval $[-N, N]$, is measurable. If E is measurable, $m(E) = \lim_{N \to \infty} m(E_N)$.

Occasionally, a particular property will be valid for all the points of a set E, *except* for those which also belong to a certain measurable subset E_0 of E. If the measure of E_0 is zero, the property is said to hold **almost everywhere** (a.e.) on E.

The **real-(complex-) valued function** f of the real variable x is defined on a set E of real numbers if to every x in E there is associated a real (complex) number $f(x)$. $\mathrm{Re}(f)$ and $\mathrm{Im}(f)$ designate the **real and imaginary parts** of such a function, while $\overline{f}$ and $|f|$ denote the **complex conjugate** and **absolute value**, respectively. If E is measurable and for every real a so also are the subsets of E consisting of those x for which $\mathrm{Re}(f) > a$ and $\mathrm{Im}(f) > a$, the function f is **measurable.** Every function defined on a measurable set but equal to zero almost everywhere is measurable and is called a **zero (trivial, null) function.** Two measurable functions that differ by a *zero function*, i.e., that are equal almost everywhere, are **equivalent.** No distinction is generally made between equivalent functions. In other words, a particular function is said to represent the *unique* solution of a given equation, for example, if every other solution within the class under consideration is equivalent to it.

A real-valued measurable function f defined on a measurable set E that takes on only nonnegative values is a **nonnegative** function. The Lebesgue integral of such a function can be defined as follows: The interval $[0, \infty]$ is subdivided by means of the points $0 = W_0 < W_1 < \cdots < W_n = \infty$ and $E_\nu (\nu = 0, 1, \ldots, n-1)$ is used to designate those subsets of E for which the inequalities $W_\nu \leq f < W_{\nu+1} (\nu = 0, 1, \ldots, n-2)$ and $W_{n-1} \leq f \leq W_n$ hold; the least upper bound of the lower Lebesque sums

$$s = \sum_{\nu=0}^{n-1} W_\nu \, m(E_\nu)$$

taken over all possible subdivisions of $[0, \infty]$, is the **Lebesgue integral** of the function f on the set E and is designated by the symbol

$$\int_E f(x)\,dx$$

When E is the interval $[a,b]$, we write simply

$$\int_a^b f(x)\,dx$$

For a more general real-valued measurable function f defined on a measurable set E, the auxiliary nonnegative measurable functions f^+ and f^- where

$$f^+(x) = \begin{cases} f(x) & \text{if } f(x) \ge 0 \\ 0 & \text{otherwise} \end{cases}$$

$$f^-(x) = \begin{cases} -f(x) & \text{if } f(x) < 0 \\ 0 & \text{otherwise} \end{cases}$$

are introduced. If the Lebesgue integral over E of *at least one* of the functions f^+, f^- is *finite*, the difference

$$\int_E f^+(x)\,dx - \int_E f^-(x)\,dx$$

is the **Lebesgue integral** of f over the set E and is denoted by the same symbol as above. If the Lebesgue integrals of *both* f^+ and f^- are *finite*, the given function f is said to be **Lebesgue integrable** or **summable** on the set E. Any complex-valued measurable function is summable if both its real and imaginary parts are summable. Bounded measurable functions defined on bounded measurable sets are necessarily summable.

A summable function f whose **norm**

$$\|f\| \equiv \left[\int_E f(x)\,\overline{f}(x)\,dx \right]^{\frac{1}{2}}$$

$$= \left[\int_E |f(x)|^2\,dx \right]^{\frac{1}{2}}$$

is finite is a **square-integrable** or $\mathfrak{L}^2$ **function**. Two $\mathfrak{L}^2$ functions f and g

defined on the same set E have a complex **inner product** (f,g) given by

$$(f,g) = \int_E f(x)\overline{g}(x) \, dx$$

This inner product satisfies

$$(\alpha_1 f_1 + \alpha_2 f_2, g) = \alpha_1(f_1, g) + \alpha_2(f_2, g)$$
$$(f, \beta_1 g_1 + \beta_2 g_2) = \overline{\beta}_1(f, g_1) + \overline{\beta}_2(f, g_2)$$
$$(g,f) = \overline{(f,g)}$$

and is related to the norm functional by the identity

$$(f,f) = \|f\|^2$$

Two $\mathfrak{L}^2$ functions f and g are **orthogonal** to one another if $(f,g) = 0$. Clearly an $\mathfrak{L}^2$ function is orthogonal to itself (has vanishing norm) *if and only if* it is a zero function.

For $\mathfrak{L}^2$ functions the classical inequalities of **Schwarz** and **Minkowski** are valid. In terms of the above symbols, these take the form

$$|(f,g)| \leq \|f\| \cdot \|g\| \qquad \text{(Schwarz)}$$
$$\|f + g\| \leq \|f\| + \|g\| \qquad \text{(Minkowski)}$$

The collection of all square-integrable functions defined on a given measurable set E constitutes a linear vector space with an inner product, commonly called $\mathfrak{L}^2(E)$ or simply $\mathfrak{L}^2$ when no confusion will arise. The norm functional on this space permits measurement of the distance $\|f - g\|$ between two functions f and g in the space and thereby the consideration, in simplified fashion, of concepts such as sequential convergence.

An element f of a linear space V with a norm functional (**metric**) such as the one above is a **limit point** of a subset U of V if it is possible to select a sequence of elements $f_1, f_2, \ldots,$ from U with the property that for every positive real number ε there exists a natural number N such that

$$\|f - f_\nu\| < \varepsilon \qquad \text{for all } \nu > N$$

This notion is often expressed alternatively by saying that the sequence $\{f_\nu\}$ **converges** to the element f. In $\mathfrak{L}^2$ we refer to this as **mean convergence** and to f as the **limit in the mean**. As an important consequence

$$\int_E f(x) \, dx = \lim_{\nu \to \infty} \int_E f_\nu(x) \, dx$$

whenever E is *bounded*.

If all of the limit points of U are contained within U itself, U is **closed**. If, in addition, U is a linear manifold, then U is a **subspace** of V. A set contained within a subspace of V is **dense** in that subspace if every element of the subspace is a limit point of the set. The subspace is **separable** if it contains a *countable* dense set. This means that separable normed vector spaces have countable bases. The functions $\{x^\nu\}$ form such a basis for the separable space $\mathfrak{L}^2$.

Not every infinite sequence $\{f_\nu\}$ of elements of a normed linear space V need be convergent. A subset U of V is (conditionally sequentially) **compact** if every infinite sequence $\{f_\nu\}$ of elements of U *at least* contains a *convergent subsequence*. More generally, a sequence $\{f_\nu\}$ is a **Cauchy sequence** if to every $\varepsilon > 0$ there corresponds a natural number N with the property that

$$\|f_\mu - f_\nu\| < \varepsilon \qquad \text{for all } \mu, \nu > N$$

Every convergent sequence is necessarily a Cauchy sequence. When the converse is also true, the given space is **complete**. It is a fundamental result of Riesz and Fischer that $\mathfrak{L}^2$ is complete. In an abstract setting any complete linear space with a norm functional analogous to the $\mathfrak{L}^2$ norm discussed above is a **Banach space**. If, in particular, the norm is generated by an inner product, the space becomes a **Hilbert space**. A linear transformation on such spaces which is characterized by the fact that it preserves distances is an **isometry**.

Returning now to consideration of real-valued measurable functions f defined on measurable subsets E of the real line, we recall that such a function is **continuous** at a point x_0 in E if to every positive number ε there is associated a positive number δ (which usually depends upon both x_0 and ε) with the property that $|f(x) - f(x_0)| < \varepsilon$ for every x in E such that $|x - x_0| < \delta$. The function is a **continuous function** on E if it is continuous at each point of E. Moreover, the function is **uniformly continuous** on E if, given $\varepsilon > 0$, the *same* value of δ above serves equally well for *every* point x_0 in E. If E is closed and bounded, every continuous function is automatically uniformly continuous. These various properties carry over to complex-valued functions f if they are valid for both the real and imaginary parts of f.

A sequence $\{f_\nu\}$ of continuous functions on a measurable set E is **equibounded** if there exists a finite real number A with the property that $|f_\nu| < A$ *for all* ν. The sequence is termed **equiuniformly continuous** if each f_ν is uniformly continuous and, given $\varepsilon > 0$, the *same* value of δ suffices for *all* the functions. A famous result of Arzela and Ascoli asserts that every equibounded, equiuniformly continuous sequence of functions has a *subsequence* which *converges* to a *continuous* limit function f.

Moreover, the convergence is **uniform** in that for every positive number ε there exists a natural number N (depending only on ε) such that

$$|f(x) - f_\mu(x)| < \varepsilon$$

for all $\mu > N$ and all x in E.

Essentially, by imitating the definitions and methods that have been used above for functions of a single variable, the various concepts can be carried over to functions of two (or more) real variables. In essence, rectangles and squares and their areas will now play the role of intervals and their lengths, care being taken to recognize that a nonvoid interval, while of positive linear measure, has zero planar measure. One essential difficulty encountered in this extension concerns successive integrations. In this regard, however, we have the classical result of Fubini that *the Lebesgue integral of a summable function of two variables can be calculated by successive integrations.* The norms of square-integrable functions $K(x,y)$ of two real variables,

$$\|K\| \equiv \left[\int_E \int_F |K(x,y)|^2 \, dx \, dy \right]^{\frac{1}{2}}$$

therefore, can be unambiguously defined and Minkowski's inequality generalized. Although other norms, in particular, the operator norms often associated with linear transformations on a Hilbert space, are typically not used in this book, the following normlike quantities are advantageously employed on occasion:

$$k_1(x) \equiv \|K\|_y \equiv \left[\int_F |K(x,y)|^2 \, dy \right]^{\frac{1}{2}}$$

$$k_2(y) \equiv \|K\|_x \equiv \left[\int_E |K(x,y)|^2 \, dx \right]^{\frac{1}{2}}$$

FUNCTIONS OF A COMPLEX VARIABLE

If f is a complex-valued function defined on an open set E of the *real* line, the **derivative** of f at the point z_0 in E is given by the number

$$\lim_{z \to z_0} \left[\frac{f(z) - f(z_0)}{z - z_0} \right]$$

whenever this limit exists. For a complex-valued function defined on a connected open set (**region**) of the *complex plane*, the derivative is similarly defined, but now, of course, for the limit to exist it must be independent of the particular direction of approach of z to z_0. A function that has a derivative at every point of a region of the complex plane is said to be **differentiable in the region**. Such a function is **regular analytic** and the region is a **region of regularity** of the function.

A regular analytic function f is infinitely differentiable. Moreover, within any region of regularity D it can be uniquely represented by convergent *power series* of the form

$$\sum_{\nu=0}^{\infty} a_\nu (z - z_0)^\nu$$

where $|z - z_0| < R$ with z_0 in D and $a_\nu = f^{(\nu)}(z_0)/\nu!$. For each z_0, the *largest* value of R for which the series converges determines **the circle of convergence** about that z_0 as center. Often these circles of convergence extend beyond D, and the series can thus be used to **analytically continue** f into other areas of the complex plane. The totality of points of regularity of f thus obtained is the **domain of regularity** of f. If it is impossible to include some point in the interior of a circle of convergence of some power series representing the (analytically continued) function f, this point is a **singular point** of the function.

A function that is regular analytic throughout the whole finite complex plane is an **entire** or **integral** function. More typically, a complex-valued function has singularities distributed here and there in the complex plane. For a single-valued function f, a singular point z_0 is an **isolated singularity** if there is some circular region (**neighborhood**) containing z_0 throughout which f is regular analytic except at the point z_0 itself. In such a situation f has a (uniquely determined) *Laurent expansion*

$$f(z) = \sum_{\nu=-\infty}^{\infty} a_\nu (z - z_0)^\nu$$

which converges for all z for which $0 < |z - z_0| < R$, where R is the distance from z_0 to the nearest other singular point. The isolated singularity z_0 is then classified into one of three types as follows:

1. If $a_\nu = 0$ for $\nu < 0$, z_0 is a **removable singularity** and can be removed by redefining f at z_0 to have the value a_0.
2. If $a_{-n} \neq 0$ for some positive n but $a_\nu = 0$ for all $\nu < -n$, z_0 is a **pole** of order n. A pole of order 1 is a **simple** pole.
3. If $a_\nu \neq 0$ for *infinitely* many negative values of ν, z_0 is an **essential singularity**.

The term **analytic** function refers to a single-valued function that is regular analytic in a region of the complex plane except for a number of (isolated) singular points. An analytic function with no singularities other than poles in the whole finite complex plane is **meromorphic.** In every *bounded* region such a function has at most a *finite* number of poles. Moreover, every such function can be expressed as the *quotient* of two *entire* functions having no zeros in common.

A **simple closed contour** is a closed curve that may be cut by parallels to the coordinate axes (in the complex plane) in no more than two points. For a function f regular analytic in a neighborhood N of a point z_0, a famous theorem of Cauchy asserts

$$\oint_C f(z)\ dz = 0$$

where C is a simple closed contour lying entirely within N and encircling z_0 in the positive (counterclockwise) sense. If z_0 is an isolated singularity of f, which is otherwise single valued and regular in the neighborhood of z_0, then

$$\frac{1}{2\pi i}\oint_C f(z)\ dz = a_{-1}(z_0)$$

The value of this integral, namely, the coefficient with index -1 in the Laurent expansion of f about z_0, is the **residue** of f at z_0. More generally, if f is analytic in a region D of the complex plane and C is a simple closed contour lying within D and encircling (but not passing through) a number of isolated singularities of f, say $z_1, z_2, \ldots, z_n$, then

$$\frac{1}{2\pi i}\oint_C f(z)\ dz = \sum_{\nu=1}^{n} a_{-1}(z_\nu)$$

where the $a_{-1}(z_\nu)$ are the residues associated with the singular points $z_\nu (\nu = 1, 2, \ldots, n)$.

References

Aalto, S. K. (1966a): Reduction of Fredholm Integral Equations with Green's Function Kernels to Volterra Equations, *Tech. Rept.* 26, Dept. of Math., Oregon State University, Corvallis.

_____(1966b): Solution of Systems of Fredholm Equations with Green's Matrix Kernels, *MRC Tech. Summary Rept.* 714, Math. Res. Center, University of Wisconsin, Madison.

Abel, N. H. (1823): Solution de quelques problèmes à l'aide d'intégrales définies, *Mag. Naturvid.* (*Christiania*) **2**; see also (1881), "Oeuvres Complètes," nouvelle éd., vol. I, pp. 11-27, Grondahl & Son, Christiania.

_____(1826a): Auflösung einer mechanischen Aufgabe, *J. Reine Angew. Math.*, **1**:153–157; French translation in "Oeuvres Complètes," *op. cit.*, pp. 97–101.

_____(1826a): Untersuchungen über die Reihe: $1 + m/1x + m(m - 1)/1.2\ x^2 + m(m - 1)(m - 2)/1.2.3\ x^3 + \cdots$ u.s.w., *J. Reine Angew. Math.*, **1**:311–339; French translation in "Oeuvres Complètes," *op. cit.*, pp. 219–250.

Ahlfors, L. V. (1953): "Complex Analysis," McGraw-Hill, New York.

Aitken, A. C. (1958): "Determinants and Matrices," 9th ed., Oliver & Boyd, London.

Ambarzumian, V. A. (1943): Diffuse Reflection of Light by a Foggy Medium, *C. R. (Dokl.) Acad. Sci. URSS (N.S.)*, **38**:229–232.

Anselone, P. M., ed., (1964): "Nonlinear Integral Equations," Wisconsin, Madison, Wisconsin.

Aronszajn, N. (1948): Rayleigh-Ritz and A. Weinstein Methods for Approximation of Eigenvalues, I, II, *Proc. Nat. Acad. Sci. U.S.A.*, **34**:474–480. 594–601.

_____(1951): Approximation Methods for Eigenvalues of Completely Continuous Symmetric Operators; in "Proc. Symp. Spectral Theory Differen. Probs.," Math. Dept., Oklahoma A & M, Stillwater, Oklahoma.

Arzela, C. (1889): Funzioni di linee, *Atti Accad. Naz. Lincei Rend.*, (4) **5**₁: 342–348.

_____ (1895): Sulle funzioni di linee, *Mem. Accad. Sci. Ist Bologna*, (5) **5**: 55–74.

Ascoli, G. (1883–1884): Le curve limiti di una varietà data di curve, *Atti Accad. Naz. Lincei Mem.*, (3) **18**:521–586.

Atkinson, K. E. (1967): The Numerical Solution of the Eigenvalue Problem for Compact Integral Operators, *Trans. Amer. Math. Soc.*, **129**:458–465.

Bary, N. K. (1964): "A Treatise on Trigonometric Series," vols. I and II, (translated from the 1961 Russian eds. by M. F. Mullins), Pergamon, New York.

Bazley, N. W. (1961): Lower Bounds for Eigenvalues, *J. Math. Mech.*, **10**: 289–308.

_____ and D. W. Fox (1961): Truncations in the Method of Intermediate Problems for Lower Bounds to Eigenvalues, *J. Res. Nat. Bur. Stand. Sect. B*, **65**:105–111.

_____ and _____ (1962a): A Procedure for Estimating Eigenvalues, *J. Math. Phys.*, **3**:469–471.

_____ and _____ (1962b): Error Bounds for Eigenvectors of Self-Adjoint Operators, *J. Res. Nat. Bur. Stand. Sect. B*, **66**:1–4.

_____ and _____ (1966): Comparison Operators for Lower Bounds to Eigenvalues, *J. Reine Angew. Math.*, **223**:142–149.

Beckenbach, E. F., and R. Bellman (1965): "Inequalities," Springer-Verlag, New York.

Beesack, P. R. (1969a): Comparison Theorems and Integral Inequalities for Volterra Integral Equations, *Proc. Amer. Math. Soc.*, **20**:61–66.

_____(1969b): Inequalities Involving Iterated Kernels and Convolutions, *Publ. Elek. Fak. Univ. Beograd (Mat. Fiz. Ser.)*, (276):11–16.

Bellman, R. (1950): A Note on the Summability of Formal Solutions of Linear Integral Equations, *Duke Math. J.*, **17**:53–55.

_____, R. E. Kalaba, and M. C. Prestrud (1963): "Invariant Imbedding and Radiative Transfer in Slabs of Finite Thickness," American Elsevier, New York.

_____and R. Latter (1952): On the Integral Equation $\lambda f(x) = \int_0^a K(x-y) f(y)$ dy, *Proc. Amer. Math Soc.*, **3**:884–891.

Belward, J. A. (1969): The Solution of an Integral Equation of the First Kind on a Finite Interval, *Quart. Appl. Math.*, **27**:313–321.

Bernau, S. J. (1967): Extreme Eigenvectors of a Normal Operator, *Proc. Amer. Math. Soc.*, **18**:127–128.

Bernstein, S. N. (1914a): Sur la convergence absolue des séries trigonométriques, *C. R. Acad. Sci. Paris*, **158**:1661–1663.

_____(1914b): On the Absolute Convergence of Trigonometric Series, *Soobshch. Khar'kov. Mat. Obshch.*, (2) **14**:139–144 (in Russian); (1915), *ibid.*, 200–201; see also (1952), "Collected Works," vol. I, pp. 217–223 (in Russian).

_____(1934): Sur la convergence absolue des séries trigonométriques, *C. R. Acad. Sci. Paris*, **199**:397–400.

Bertram, G. (1957): Fehlerabschätzung für das Ritz-Galerkinsche Verfahren bei Eigenwertproblemen, *Z. Angew. Math. Mech.*, **37**:191–201; (1959), *ibid.*, **39**: 236–246.

Bessel, F. W. (1828): Ueber die Bestimmung des Gesetzes einer periodischen Erscheinung, *Astron. Nachr.*, **6**:333–348; see also (1876), "Abhandlungen," vol. II, pp. 364–372, Leipzig; reprinted (1971) by Chelsea, New York.

Birkhoff, G. (1957): Extensions of Jentzsch's Theorem, *Trans. Amer. Math. Soc.*, **85**:219–227.

Boas, R. P. (1954): "Entire Functions," chap. 2, Academic, New York.

Bôcher, M. (1909): "An Introduction to the Study of Integral Equations," Cambridge, London, p. 17.

Boggio, T. (1907): Un théorème sur les équations intégrales, *C. R. Acad. Sci. Paris*, **145**:619–622.

Boland, W. R. (1969a): Convexity Theorems for the Eigenvalues of Certain Fredholm Operators, *J. Math. Anal. Appl.*, **25**:162–181.

_____(1969b): Convexity Conditions for the Kernels and Eigenvalues of a Class of Fredholm Integral Equations, *J. Math. Anal. Appl.*, **28**:609–618.

Bosanquet, L. S. (1930): On Abel's Integral Equation and Fractional Integrals, *Proc. London Math. Soc.*, (2) **31**:134–143.

Brodskii, M. S., and M. S. Livsic (1958): Spectral Analysis of Non-Selfadjoint Operators and Intermediate Systems, *Usp. Mat. Nauk (N.S.)*, **13** (1) (79): 3–85; English translation in (1960), *Amer. Math. Soc. Transl.*, (2) **13**: 265–346.

Bückner, H. (1949): Ein unbeschränkt anwendbares Iterationsverfahren für Fredholmsche Integralgleichungen, *Math. Nachr.*, **2**:304–313; see also (1952), "Die praktische Behandlung von Integral-Gleichungen," chap. 3, Springer-Verlag, Berlin.

_____(1948): A Special Method of Successive Approximations for Fredholm Integral Equations, *Duke Math. J.*, **15**:197–206.

Burgmeier, J. W. (1970): On the Reduction of a Class of Fredholm Integral Equations to Equivalent Matrix Problems, *J. Math. Anal. Appl.*, **31**:529–544.

Burkill, J. C. (1951): "The Lebesgue Integral," Cambridge, London.

Busbridge, I. W. (1960): "The Mathematics of Radiative Transfer," Cambridge, London.

Cairo, L., and T. Kahan (1965): "Variational Techniques in Electromagnetism," (translated from the 1962 French ed. by G. D. Sims), Gordon and Breach, New York.

Carleman, T. (1916): "Über das Neumann-Poincarésche Problem für ein Gebeit mit Ekken," doctoral dissertation, Uppsala, Sweden.

―――― (1917): Sur le genre du dénominateur $D(\lambda)$ de Fredholm, *Ark. Mat., Astron., Fys.*, **12** (15):5 pp.; see also (1960), "Edition Complète des Articles," 1–5, Malmö.

―――― (1918): Über die Fourierkoeffizienten einer stetigen Funktion, *Acta Math.*, **41**:377–384; see also (1960), "Édition Complète des Articles," pp. 15–22, Malmö.

―――― (1921): Zur Theorie der linearen Integralgleichungen, *Math. Z.*, **9**: 196–217; see also (1960), "Édition Complète des Articles," pp. 79–100, Malmö.

―――― (1922): Über die Abélsche Integralgleichung mit konstanten Integrationsgrenzen, *Math. Z.*, **15**:111–120; see also (1960), "Édition Complète des Articles," pp. 161–170, Malmö.

Carrier, G. F. (1948): On the Determination of the Eigenfunctions of Fredholm Equations, *J. Math. and Phys.*, **27**:82–83.

―――― (1949): A Generalization of the Wiener-Hopf Technique, *Quart. Appl. Math.*, **7**:105–109.

Case, K. M. (1957): On Wiener-Hopf Equations, *Ann. Phys. (New York)*, **2**: 384–405.

Cauchy, A. (1829): Sur l'équation à l'aide de laquelle on détermine les inégalités séculaires des mouvements des planètes; in (1891), "Oeuvres Complètes," ser. II, vol. IX, pp. 174–195, Gauthier-Villars, Paris.

―――― (1844): Mémoires sur les fonctions complémentaires, *C. R. Acad. Sci. Paris*, **19**:1377–1378; see also (1893), "Oeuvres Complètes," ser. I, vol. VIII, *op. cit.*, pp. 378–385.

Chandrasekhar, S. (1950): "Radiative Transfer," Oxford, London; reprinted (1960) by Dover, New York.

Chang, S. H. (1947): A Generalization of a Theorem of Lalesco, *J. London Math. Soc.*, **22**:185–189.

―――― (1949): On the Distribution of the Characteristic Values and Singular Values of Linear Integral Equations, *Trans. Amer. Math. Soc.*, **67**: 351–367.

―――― (1952): A Generalization of a Theorem of Hille and Tamarkin with Applications, *Proc. London Math. Soc.*, (3) **2**:22–29.

———— (1954): A Relation Between Characteristic Values and Singular Values of Linear Integral Equations, *Sci. Sinica*, **3**:237–245.

Chu, S. C., and F. T. Metcalf (1967): On Gronwall's Inequality, *Proc. Amer. Math. Soc.*, **18**:439–440.

Clarkson, J. A., and C. R. Adams (1933): On Definitions of Bounded Variation for Functions of Two Variables, *Trans. Amer. Math. Soc.*, **35**:824–854.

———— and ———— (1934): Properties of Functions $f(x, y)$ of Bounded Variation, *Trans. Amer. Math. Soc.*, **36**:711–730.

Cochran, J. A. (1962): "Problems in Singular Perturbation Theory," doctoral dissertation, Stanford University, Stanford, Calif., chap. 2.

———— (1965): The Existence of Eigenvalues for the Integral Equations of Laser Theory, *Bell Syst. Tech. J.*, **44**:77–88.

———— (1968): On the Uniqueness of Solutions of Linear Differential Equations, *J. Math. Anal. Appl.*, **2**:418–426.

————and E. W. Hinds (1972): Improved Error Bounds for the Eigenvalues of Certain Normal Operators, *SIAM J. Num. Anal.*, **9**.

Coddington, E. A., and N. Levinson (1955): "Theory of Ordinary Differential Equations," chaps. 7, 11, and 12, McGraw-Hill, New York.

Cole, R. H. (1968): "Theory of Ordinary Differential Equations," Appleton-Century-Crofts, New York.

Collatz, L. (1939): Genäherte Berechnung von Eigenwerten, *Z. Angew. Math. Mech.*, **19**:224–249, 297–318.

———— (1940): Schrittweise Näherungen bei Integralgleichungen und Eigenwertschranken, *Math. Z.*, **46**:692–708.

———— (1941): Einschliessungssatz für die Eigenwerte von Integralgleichungen, *Math. Z.*, **47**:395–398.

———— (1960): "The Numerical Treatment of Differential Equations," 3d ed., (translated from a supplemented version of the 2d German ed. by P. G. Williams), Springer-Verlag, Berlin.

Copson, E. T. (1935): "Theory of Functions of a Complex Variable," Clarendon, Oxford.

Courant, R. (1920): Über die Eigenwerte bei den Differentialgleichungen der mathematischen Physik, *Math. Z.*, **7**:1–57.

———— (1923): Zur Theorie der linearen Integralgleichungen, *Math. Ann.*, **89**: 161–178.

———— (1943): Variational Methods for the Solution of Problems of Equilibrium and Vibrations, *Bull. Amer. Math. Soc.*, **49**:1–23.

————and D. Hilbert (1953): "Methods of Mathematical Physics," vol. **I**, chap. 3, Interscience, New York.

Cryer, C. W. (1967): On the Calculation of the Largest Eigenvalue of an Integral Equation, *Numer. Math.*, **10**:165–176.

Davis, H. T. (1930): "The Theory of the Volterra Integral Equation of Second Kind," Indiana University Studies, vol. 17, Bloomington, Ind.

———— (1960): "Introduction to Nonlinear Differential and Integral Equations," U.S. Atomic Energy Commission, Washington, D.C., chap. 13; reprinted (1962) by Dover, New York.

DePree, J. D. (1969): Reduction of Linear Integral Equations with Difference Kernels to Nonlinear Integral Equations, *J. Math. Anal. Appl.*, **26**: 539–544.

Diaz, J. B. (1958): Upper and Lower Bounds for Eigenvalues; in "Calculus of Variations and its Applications," *Proc. 8th Symp. Appl. Math.*, McGraw-Hill, New York.

————and F. T. Metcalf (1968): A Functional Equation for the Rayleigh Quotient for Eigenvalues, and Some Applications, *J. Math. Mech.*, **17**: 623–630.

Dini, U. (1892): "Grundlagen für eine Theorie der Functionen einer veränderlichen reellen Grösse," (translated from the 1878 Italian ed.), pp. 148–149, B. G. Teubner, Leipzig.

Dirichlet, P. G. L. (1829): Sur la convergence des séries trigonométriques qui servent à représenter une fonction arbitraire entre des limites données, *J. Reine Angew. Math.*, **4**:157–169; see also (1889), "Werke," vol. I, pp. 117–132, G. Reimer, Berlin; reprinted (1969) by Chelsea, New York.

Doetsch, G. (1923): Die Integrodifferentialgleichungen von Faltungstypus, *Math. Ann.*, **89**:192–207.

———— (1925): Bemerkung zu der Arbeit von V. Fock, *Math. Z.*, **24**:785–791.

Dolph, C. L. (1956): The Mathematician Grapples with Linear Problems Associated with the Radiation Condition, *IRE Trans. Antennas Propagat.*, **AP-4**:302–311.

Dunford, N., and J. T. Schwartz (1958): "Linear Operators, Part I: General Theory," Interscience, New York.

———— and ———— (1963): "Linear Operators, Part II: Spectral Theory, Self-Adjoint Operators in Hilbert Space," Interscience, New York.

Enskog, D. (1917): "Kinetische Theorie der Vorgänge in mässig verdünnten Gasen," dissertation, Uppsala, Sweden.

Erdélyi, A. (1964): The Integral Equations of Asymptotic Theory; in C. H. Wilcox (ed.), "Asymptotic Solutions of Differential Equations and Their Applications," Wiley, New York; also see (1960), Singular Volterra Integral Equations and Their Use in Asymptotic Expansions, *MRC Tech. Summary Rept.* 194, Math. Res. Center, University of Wisconsin, Madison.

————(1962): A Result on Non-linear Volterra Integral Equations; in G. Szegö et al. (eds.), "Studies in Mathematical Analysis and Related Topics," Essays in Honor of George Pólya, Stanford, Stanford, Calif.

————(1968): Some Dual Integral Equations, *SIAM J. Appl. Math.*, **16**:1338–1340.

———— , ed. (1954): "Tables of Integral Transforms," vol. I, chap. 5, Bateman Manuscript Project, McGraw-Hill, New York.

———— and I. N. Sneddon (1962): Fractional Integration and Dual Integral Equations, *Can. J. Math.*, **14**:685–693.

Fan, K. (1949): On a Theorem of Weyl Concerning Eigenvalues of Linear Transformations, I, *Proc. Nat. Acad. Sci. U.S.A.*, **35**:652–655; II (1950), *ibid.*, **36**:31–35.

———— (1951): Maximum Properties and Inequalities for the Eigenvalues of Completely Continuous Operators, *Proc. Nat. Acad. Sci. U.S.A*, **37**:760–766.

Fichera, G. (1966): Approximation and Estimates for Eigenvalues; in J. H. Bramble (ed.), "Proc. Symp. Numer. Solution Partial Differen. Eqs.," Academic, New York; see also (1965), "Linear Elliptic Differential Systems and Eigenvalue Problems," Lecture Notes in Mathematics 8, Springer-Verlag, Berlin.

Fischer, E. (1905): Über quadratische Formen mit reellen Koeffizienten, *Monatsh. Math. Phys.*, **16**:234–249.

———— (1907): Sur la convergence en moyenne, *C. R. Acad. Sci. Paris*, **144**:1022–1024, 1148–1151.

Fock, V. A. (1924): Über eine Klasse von Integralgleichungen, *Math. Z.*, **21**:161–173.

————(1942): Sur certaines équations intégrales de physique mathematique, *C. R. (Dokl.) Acad. Sci. URSS (N. S.)*, **36**:133–136; (1944) *Mat. Sb. (N.S.)*, **14**:3–50 (in Russian).

Fourier, J. B. J. (1888): "Oeuvres," vol. I, (Théorie analytique de la chaleur), chaps. 5–9, pp. 391–410, Gauthier-Villars, Paris.

Fox, A. G., and T. Li (1961): Resonant Modes in a Maser Interferometer, *Bell Syst. Tech. J.*, **40**:453–488.

———— and ———— (1963): Modes in a Maser Interferometer with Curved and Tilted Mirrors, *Proc. IEEE*, **51**:80–89.

Fox, C. (1963): Integral Transforms based upon Fractional Integration, *Proc. Cambridge Phil. Soc.*, **59**:63–71.

Fox, D. W., and W. C. Rheinboldt (1966): Computational Methods for Determining Lower Bounds for Eigenvalues of Operators in Hilbert Space, *Tech. Memo.* TG-810, Appl. Phys. Lab., The Johns Hopkins University, Baltimore, Md.; also in *SIAM R.*, **8**:427–462.

Fox, L., and E. T. Goodwin (1953): The Numerical Solution of Non-Singular Linear Integral Equations, *Phil. Trans. Roy. Soc. London, Ser. A*, **245**:501–534.

Frank, P., and R. von Mises (1937): "Die Differential und Integralgleichungen der Mechanik und Physik," I. M. Rosenberg, New York.

Fredholm, I. (1900): Sur une nouvelle méthode pour la résolution de problème de Dirichlet, *Kongr.Vetensk.-Akad. Stockholm*, **57**:39–46; see also (1955), "Oeuvres Complètes," pp. 61–68, Malmö.

———— (1902a): Sur une classe de transformations rationnelles, *C. R. Acad. Sci.*

Paris, **134**:219–222; see also (1955), "Oeuvres Complètes," pp. 73–76, Malmö.

———— (1902b): Sur une class d'équations fonctionnelles," *C. R. Acad. Sci. Paris*, **134**:1561–1564; see also (1955),"Oeuvres Complètes," pp. 77–79, Malmö.

———— (1903): Sur une classe d'équations fonctionnelles, *Acta Math.*, **27**: 365–390; see also (1955),"Oeuvres Complètes," pp. 81–106, Malmö.

Friedman, A. (1969): Monotonicity of Solutions of Volterra Integral Equations in Banach Space, *Trans. Amer. Math. Soc.*, **138**:129–148.

————and M. Shinbrot (1967): Volterra Integral Equations in Banach Space, *Trans. Amer. Math. Soc.*, **126**:131–179.

Friedman, B. (1956): "Principles and Techniques of Applied Mathematics," chap. 3, Wiley, New York.

Friedrichs, K. O. (1965): "Perturbation of Spectra in Hilbert Space," Lectures in Applied Mathematics, vol. III, Amer. Math. Soc., Providence, R. I.

Frobenius, G. (1908): Über Matrizen aus positiven Elementen, I, *Sitzungsber. Deut. Akad. Wiss. Berlin*, 471–476; II (1909), *ibid.*, 514–518; see also (1968), "Gesammelte Abhandlungen," vol. III, pp. 404–409, 410–414, Springer-Verlag, Berlin.

———— (1912): Über Matrizen aus nichtnegativen Elementen, *Sitzungsber. Deut. Akad. Wiss. Berlin*, 456–477; see also (1968), "Gesammelte Abhandlungen," *op. cit.*, pp. 546–567.

Fubini, G. (1910): Equazioni integrali e valori eccezionali, *Ann. Mat. Pura Appl.*, (3) **17**:111–140; see also (1958), "Opere Scelte," vol. II, pp|. 318–346, Cremonese, Rome.

Gantmacher, F. R. (1959): "Applications of the Theory of Matrices," (rev. translation by J. L. Brenner of part 2 of 1954 Russian ed.), Interscience, New York.

————and M. G. Krein (1961): "Oscillation Matrices and Kernels and Small Vibrations of Mechanical Systems," (translated from the 1950 2d Russian ed.), AEC-tr-4481, U.S. Atomic Energy Commission, Oak Ridge, Tenn.

Garabedian, P. R. (1964): "Partial Differential Equations," chaps. 4, 7, and 9–11, Wiley, New York.

Garbe, E. (1915): Zur Theorie der Integralgleichung dritter Art, *Math. Ann.*, **76**: 527–547.

Gel'fand, I. M., and N. Ya. Vilenkin (1964): "Generalized Functions," vol. 4, Applications of Harmonic Analysis (translated by A. Feinstein from a revision of the 1961 Russian ed.), chap. 1, Academic, New York.

Gheorghiu, S. A. (1928): "Sur l'équation de Fredholm," thesis, Paris.

Gohberg, I. C., and I. A. Fel'dman (1968): On Wiener-Hopf Integral-Difference Equations, *Dokl. Akad. Nauk SSSR*, **183**:25–28; translated in *Soviet Math. Dokl.*, **9**:1312–1316.

———— and M. G. Krein (1957): The Basic Propositions on Defect Numbers, Root Numbers and Indices of Linear Operators, *Usp. Mat. Nauk (N.S.)*,

12 (2) (74):43–118; translated in (1960), *Amer. Math. Soc. Transl.*, (2) **13**:185–264.

_____ and _____ (1969): "Introduction to the Theory of Linear Nonselfadjoint Operators," Translations of Math. Mono., vol 18, Amer. Math. Soc., Providence, R. I.

Goldfain, I. A. (1946): On a Class of Linear Integral Equations, *Uch. Zap. Mosk. Gos. Univ. 100, Mat.*, **1**:104–112; translated in (1958), *Amer. Math. Soc. Transl.*, (2) **10**:283–290.

Gould, S. H. (1966): "Variational Methods for Eigenvalue Problems," 2d ed., Toronto, Canada.

Goursat, E. (1907): Sur quelques propriétés des équations intégrales, *C. R. Acad. Sci. Paris*, **145**:752–754.

_____ (1908): Sur un théorême de la théorie des équations intégrales, *ibid.*, **146**: 327–329.

_____ (1909): Sur quelques points de la théorie des équations intégrales, *Bull. Soc. Math. France*, **37**:197–204.

_____(1933): Détermination de la résolvante d'une classe d'équations integrales, *Bull. Sci. Math.*, (2) **57**:144–150.

Gram, J. P. (1883): Ueber die Entwickelung reeller Functionen in Reihen mittelst der Methode der Kleinsten Quadraten, *J. Reine Angew. Math.*, **94**:41–73.

Gronwall, T. H. (1919): Note on the Derivatives with Respect to a Parameter of the Solutions of a System of Differential Equations, *Ann. Math.*, (2) **20**: 292–296.

Grothendieck, A. (1955): Produits tensoriels topologiques et espaces nucléaires, *Mem. Amer. Math. Soc.*, **16**:1–140.

Hadamard, J. (1893): Résolution d'une question relative aux déterminants, *Bull. Sci. Math.*, (2) **17**:240–246; see also (1968), "Oeuvres," vol. I, pp. 239–245, Centre Nat. Rech. Sci., Paris.

Halberg, C. J. A, and V. A. Kramer (1960): A Generalization of the Trace Concept, *Duke Math. J.*, **27**:607–617.

Halmos, P. R. (1958): "Finite-Dimensional Vector Spaces," 2d ed., Van Nostrand, Princeton, N. J.

Hamburger, H. L., and M. E. Grimshaw (1951): "Linear Transformations in *n*-Dimensional Vector Space," Cambridge, London.

Hamel, G. (1949): "Integralgleichungen," 2d ed., Springer-Verlag, Berlin.

Hammerstein, A. (1923): Über die Entwickelung des Kernes linearer Integralgleichungen nach Eigenfunktionen, *Sitzungsber. Deut. Akad. Wiss. Berlin Phys. Math. Kl.*, 181–184.

_____ (1928): Über Entwickelung gegebener Funktionen nach Eigenfunktionen von Randwertaufgaben, *Math. Z.*, **27**:269–311.

Hardy, G. H., and J. E. Littlewood (1928a): Some Properties of Fractional Integrals. I, *Math. Z.*, **27**:565–606; see also (1969), "Collected Papers of G. H. Hardy," vol. III, pp. 564–607, Clarendon, Oxford.

_____ and _____ (1928b): A Convergence Criterion for Fourier Series, *Math.*

Z., **28**:612–634; see also (1969),"Collected Papers of G. H. Hardy," *op. cit.*, pp. 28–51.

———— and ———— (1928c): Notes on the Theory of Series (IX): On the Absolute Convergence of Fourier Series, *J. London Math. Soc.*, **3**: 250–253; see also (1969),"Collected Papers of G. H. Hardy," *op. cit.*, pp. 52–56.

————, ————, and G. Pólya (1952): "Inequalities," 2d ed., Cambridge, London.

Harrington, R. F. (1968): "Field Computation by Moment Methods," Macmillan, Ne n York.

Hecke, E. (1922): Über die Integralgleichungen der kinetischen Gastheorie, *Math. Z.*, **12**:274–286; see also (1959), "Mathematische Werke," pp. 361–373, Vandenhoeck and Ruprecht, Göttingen.

Heins, A. E. (1956): The Scope and Limitations of the Method of Wiener and Hopf, *Comm. Pure Appl. Math.*, **9**:447–466; reprinted in N. Aronszajn et al. (eds.), "Trans. Symp. Partial Differen. Eqs.," Interscience, New York.

———— and N. Wiener (1946): A Generalization of the Wiener-Hopf Integral Equation, *Proc. Nat. Acad. Sci. U.S.A.*, **32**:98–101.

Heinz, E. (1951): Beiträge zur Störungstheorie der Spectralzerlegung, *Math. Ann.*, **123**:415–438.

Hellinger, E., and O. Toeplitz (1927): Integralgleichungen und Gleichungen mit unendlichvielen Unbekannten, "Encyklopädie der mathematischen Wissenschaften," vol. II C 13, pp. 1335–1601, Leipzig; reprinted (1953) by Chelsea, New York.

Helstrom, C. W. (1960): "Statistical Theory of Signal Detection," chap. 4, Pergamon, New York.

———— (1965): Solution of the Detection Integral Equation for Stationary Filtered White Noise, *IEEE Trans. Info. Theory*, **IT-11**:335–339.

Henrici, P. (1962): "Discrete Variable Methods in Ordinary Differential Equations," Wiley, New York.

Hersch, J. (1961): Caractérisation variationnelle d'une somme de valeurs propres consécutives; généralisation d'inégalités de Pólya-Schiffer et de Weyl, *C. R. Acad. Sci. Paris*, **252**:1714–1716.

Heywood, B. (1907): Sur quelques points de la théorie des fonctions fondamentales relatives à certaines équations intégrales, *C. R. Acad. Sci. Paris*, **145**: 908–910.

Higgins, T. P. (1964): A Hypergeometric Function Transform, *SIAM J. Appl. Math.*, **12**:601–612.

Hilbert, D. (1904): Grundzüge einer allgemeinen Theorie der linearen Integralgleichungen, *Nachr. Akad. Wiss. Göttingen*, 49–91; also (1905), *ibid.*, 213–259 and 307–338; (1906), *ibid.*, 157–227 and 439–480; (1910), *ibid.*, 355–417. Reprinted together under the same title (1912), B. G. Teubner, Leipzig and Berlin; also reprinted (1953) by Chelsea, New York.

———— (1912): Begründung der elementaren Strahlungstheorie, *Nachr. Akad. Wiss. Göttingen*, 773–789; *Phys. Z.*, **13**:1056–1064; see also (1935),

"Gesammelte Abhandlungen," vol. III, pp. 217–230, Julius Springer, Berlin; reprinted (1966) by Chelsea, New York.

———(1935): "Gesammelte Abhandlungen," *op. cit.*, pp. 94–145.

Hille, E., and J. D. Tamarkin (1930): On the Theory of Linear Integral Equations, I, *Ann. Math.*, (2) **31**:479–528; (1934), II, *ibid.*, **35**:443–455.

——— and ——— (1931): On the Characteristic Values of Linear Integral Equations, *Acta Math.*, **57**:1–76.

Hobson, E. W. (1957): "The Theory of Functions of a Real Variable and the Theory of Fourier Series," vols. I and II, Dover, New York.

Hochstadt, H. (1967): The Mean Convergence of Fourier-Bessel Series, *SIAM R.*, **9**:211–218.

———(1970): On Eigenfunction Expansions Associated with the Classical Orthogonal Polynomials, *J. Diff. Eq.*, **8**:542–553.

Hölder, E. (1937): Über die Vielfachheiten gestörter Eigenwerte, *Math. Ann.*, **113**:620–628.

Hopf, E. (1928): Über lineare Integralgleichungen mit positivem Kern, *Sitzungsber. Deut. Akad. Wiss. Berlin, Phys. Math. Kl.*, 233–245.

——— (1934): "Mathematical Problems of Radiative Equilibrium," Cambridge, London.

Horn, A. (1950): On the Singular Values of a Product of Completely Continuous Operators, *Proc. Nat. Acad. Sci. U.S.A.*, **36**:374–375.

Hutson, V. (1964-1965): Asymptotic Solutions of Integral Equations with Convolution Kernels, *Proc. Edinburgh Math. Soc.*, (2) **14**:5–19.

——— (1965): On a Generalization of the Wiener-Hopf Integral Equations, *J. Math. Mech.*, **14**:807–819.

Iglisch, R. (1941): Bemerkungen zu einigen von Herrn Collatz angegebenen Eigenwertabschätzungen bei linearen Integralgleichungen, *Math. Ann.*, **118**: 263–275.

Ince, E. L. (1956): "Ordinary Differential Equations," chaps. 9 and 11, Dover, New York.

Isaacson, E., and H. B. Keller (1966): "Analysis of Numerical Methods," Wiley, New York.

Jacobi, C. G. J. (1841): De formatione et proprietatibus Determinantium, *J. Reine Angew. Math.*, **22**:285–318; see also (1884), "Gesammelte Werke," vol. 3, pp. 355–392, Berlin; reprinted (1968) by Chelsea, New York.

Jentzsch, R. (1912): Über Integralgleichungen mit positivem Kern, *J. Reine Angew. Math.*, **141**:235–244.

Jones, D. S. (1956): A Critique of the Variational Method in Scattering Problems, *IRE Trans. Antennas and Propagat.*, **AP-4**:297–301.

Kac, M., W. L. Murdock, and G. Szegö (1953): On the Eigenvalues of Certain Hermitian Forms, *J. Rat. Mech. Anal.*, **2**:767–800.

Kadota, T. T. (1967): Term-by-Term Differentiability of Mercer's Expansion, *Proc. Amer. Math. Soc.*, **18**:69–72.

Kagiwada, H. H., and R. E. Kalaba (1968): An Initial Value Method for Fredholm Integral Equations of Convolution Type, *Int. J. Comput. Math.*, **2**:143–155.

———, ———, and A. Schumitzky (1969): A Representation for the Solution of Fredholm Integral Equations, *Proc. Amer. Math. Soc.*, **23**:37–40.

———, ———, and B. J. Vereeke (1970): Invariant Imbedding and Fredholm Integral Equations with Displacement Kernels on an Infinite Interval, *Int. J. Comput. Math.*, **2**:221–229.

Kantorovich, L., and V. Krylov (1958): "Approximate Methods of Higher Analysis," (translated by C. D. Benster), Interscience, New York.

Kanwal, R. P. (1970): An Integral Equation Perturbation Technique in Applied Mathematics, *J. Math. Mech.*, **19**:625–656.

Kaplan, S., and S. P. Morgan (1964): The Use of the Rayleigh-Ritz Method in Nonselfadjoint Problems, *IEEE Trans. Microwave Theory Tech.*, **MTT-12**: 254–255.

Karlin, S. (1959): Positive Operators, *J. Math. Mech.*, **8**:907–937.

——— (1964): The Existence of Eigenvalues for Integral Operators, *Trans. Amer. Math. Soc.*, **113**:1–17.

——— (1968): "Total Positivity," vol. I, Stanford, Stanford, Calif.

Kato, T. (1949a): On the Upper and Lower Bounds of Eigenvalues, *J. Phys. Soc. Japan*, **4**:334–339.

——— (1949b): On the Convergence of the Perturbation Method, I, *Progr. Theor. Phys.*, **4**:514–523; (1950), II, *ibid.*, **5**:95–101, 207–212.

——— (1951): On the Convergence of the Perturbation Method, *J. Fac. Sci. Univ. Tokyo Sect. I*, **6**:145–226.

——— (1966): "Perturbation Theory for Linear Operators," Springer-Verlag, New York.

Keller, H. B. (1968): "Numerical Methods for Two-Point Boundary-Value Problems," chap. 4, Blaisdell, Waltham, Mass.

Kellogg, O. D. (1922): On the Existence and Closure of Sets of Characteristic Functions, *Math. Ann.*, **86**:14–17.

Kneser, A. (1906): Ein Beitrag zur Theorie der Integralgleichungen, *Rend. Circ. Mat. Palermo*, **22**:233–240; see also (1922), "Die Integralgleichungen," sect. 63, Friedr. Vieweg & Sohn, Braunschweig.

———(1907): Die Theorie der Integralgleichungen und die Darstellung willkürlicher Funktionen in der mathematischen Physik, *Math. Ann.*, **63**:477–524.

Koehler, F. (1953): Estimates for the Errors in the Rayleigh-Ritz Method, *Pacific J. Math.*, **3**:153–164.

Kohn, W. (1947): A Note on Weinstein's Variational Method, *Phys. Rev.*, **71**: 902–904.

Krein, M. G. (1937): On the Characteristic Values of Differentiable Symmetric Kernels, *Mat. Sb. (N.S.)*, **2**:725–730 (in Russian).

_____ (1958): Integral Equations on a Half-Line with Kernel Depending upon the Difference of the Arguments, *Usp. Mat. Nauk (N.S.)*, **13** (5) (83): 3–120; translated in (1962), *Amer. Math. Soc. Transl.*, (2) **22**:163–288.

_____ (1959): "Criteria for Completeness of the System of Root Vectors of a Dissipative Operator, *Usp. Mat. Nauk (N.S.)*, **14** (3) (87):145–152; translated in (1963), *Amer. Math. Soc. Transl.*, (2) **26**:221–229.

_____ and M. A. Rutman (1948): Linear Operators Leaving Invariant a Cone in a Banach Space, *Usp. Mat. Nauk (N.S.)*, **3** (1) (23):3–95; translated as *Amer. Math. Soc. Transl. 26*; reprinted in *Amer. Math. Soc. Transl.*, (1) **10**:199–325.

Krylov, N. (1931): Les méthodes de solution approchée des problèmes de la physique mathématique, *Mem. Sci. Math.*, **49**:1–68.

_____ and N. Bogoliubov (1929): Sur le calcul des racines de la transcendante de Fredholm les plus voisines d'un nombre donné par les méthodes des moindres carrés et de l'algorithme variationnel, *Izv. Akad. Nauk SSSR, O.P.M.*, **ser. VII**:471–488.

Kwan, C-C. (1965): Sur l'éxistence des valeurs propres et non-nulles des équations intégrales dans la théorie de 'lasers', *Sci. Sinica*, **14**:1077–1078.

Lalesco, T. (1907a): Sur l'ordre de la fonction entière $D(\lambda)$ de Fredholm, *C. R. Acad. Sci. Paris*, **145**:906–907.

_____ (1907b): Sur la fonction $D(\lambda)$ de Fredholm, *ibid.*, **145**:1136–1137.

_____ (1912): "Introduction à la théorie des équations intégrales," pt. III, Hermann and Fils, Paris.

_____ (1914–1915): Un théorème sur les noyaux composés, *Bull. Sect. Sci. Acad. Roumaine*, **3**:271–272.

_____ (1918): Sur un point de la théorie des noyaux symétrisables, *C. R. Acad. Sci. Paris*, **166**:410–411.

Lanczos, C. (1950): An Iteration Method for the Solution of the Eigenvalue Problem of Linear Differential and Integral Operators, *J. Res. Nat. Bur. Stand.*, **45**:255–282.

Latta, G. E. (1956): The Solution of a Class of Integral Equations, *J. Rat. Mech. Anal.*, **5**:821–834.

Latter, R. (1958–1959): Approximate Solutions for a Class of Integral Equations, *Quart. Appl. Math.*, **16**:21–31.

Lax, P. D. (1962): A Procedure for Obtaining Upper Bounds for the Eigenvalues of a Hermitian Symmetric Operator, in G. Szegö et al. (eds.), "Studies in Mathematical Analysis and Related Topics," Essays in Honor of George Pólya, Stanford, Stanford, Calif.

Lehmann, N. J. (1949): Beiträge zur numerischen Lösung linearer Eigenwertprobleme, I, *Z. Angew. Math. Mech.*, **29**:341–356; (1950), II, *ibid.*, **30**: 1–16.

Leonard, A. and T. W. Mullikin (1965a): Integral Equations with Difference Kernels on Finite Intervals, *Trans. Amer. Math. Soc.*, **116**:465–473.

_____ and _____ (1965b): The Resolvent Kernel for a Class of Integral Operators with Difference Kernels on a Finite Interval, *J. Math. and Phys.*, **44**:327–340.

Lewis, D. C. (1950): Comments on the Classical Theory of Integral Equations, *J. Wash. Acad. Sci.*, **40**:65–71.

Li, T. (1963): Mode Selection in an Aperture-Limited Concentric Maser Interferometer, *Bell Syst. Tech. J.*, **42**:2609–2620.

Libin, A. S. (1968): Trace Formulas for Selfadjoint Operators, *Dokl. Akad. Nauk SSSR*, **181**:1327–1330; translated in *Soviet Math. Dokl.*, **9**:1037–1040.

Lidskii, V. B. (1959a): Non-Selfadjoint Operators with Trace, *Dokl. Akad. Nauk SSSR*, **125**:485–487; translated in (1965), *Amer. Math. Soc. Transl.*, (2) **47**:43–46.

_____ (1959b): Conditions for Completeness of a System of Root Subspaces for Non-Selfadjoint Operators with Discrete Spectra, *Tr. Mosk. Mat. Obsc.*, **8**:83–120; translated in (1963), *Amer. Math. Soc. Transl.*, (2) **34**:241–282.

_____ (1962): Summability of Series in the Principal Vectors of Non-Selfadjoint Operators, *Tr. Mosk. Mat. Obsc.*, **11**:3–35; translated in (1964), *Amer. Math. Soc. Transl.*, (2) **40**:193–228.

Liouville, J. (1832): Sur quelques questions de géométrie et de mécanique, et sur un nouveau genre de calcul pour résoudre ces questions, *J. École Polytech.*, sect. 21, **13**:1–69.

_____ (1837): Sur le développement des fonctions. . .II, *J. Math. Pures Appl.*, (1) **2**:16–35.

_____ (1838): Sur la théorie des équations différentielles linéaires et sur le développement des fonctions en séries, *ibid.*, **3**:561–614.

_____ (1839): Note sur quelques intégrales définies, *ibid.*, **4**:225–235.

_____ (1880): Leçons sur les fonctions doublement périodiques faites en 1847, *J. Reine Angew. Math.*, **88**:277–310.

Lipschitz, R. (1880): "Lehrbuch der Analysis," vol. 2, Max Cohen & Sohn, Bonn.

Livsic, M. S. (1954): On Spectral Decomposition of Linear Non-Selfadjoint Operators, *Mat. Sb. (N.S.)*, **34** (76):145–199; translated in (1957), *Amer. Math. Soc. Transl.*, (2) **5**:67–114.

Lonseth, A. T. (1954): Approximate Solutions of Fredholm-Type Integral Equations, *Bull. Amer. Math. Soc.*, **60**:415–430.

Lovitt, W. V. (1950): "Linear Integral Equations," Dover, New York.

Maehly, H. J. (1952): Ein neues Variationsverfahren zur genäherten Berechnung der Eigenwerte hermitescher Operatoren, *Helv. Phys. Acta*, **25**:547–568.

Marek, I. (1962): Iterations of Linear Bounded Operators in Non Self-Adjoint Eigenvalue Problems and Kellogg's Iteration Process, *Czech. Math. J.*, **12**:536–554.

Marshall, A. W., and I. Olkin (1964): Inclusion Theorems for Eigenvalues from Probability Inequalities, *Numer. Math.*, **6**:98–102.

Marty, J. (1910): Sur une équation intégrale, *C. R. Acad. Sci. Paris*, **150**: 515–518; Développement suivant certaines solutions singulières, *ibid.*, 603–606; Éxistence de solutions singulières pour certaines équations de Fredholm, *ibid.*, 1031–1033; Valeurs singulières d'une équation de Fredholm, *ibid.*, 1499–1502.

Mazurkiewicz, S. (1915): On the Fredholm Determinant. I. Kernels Satisfying Hölder Conditions, *Towar. Nauk. Warszaw., Spraw. Pos., Wydz. III*, **8**: 656–662; II. Differentiable Kernels, *ibid.*, 805–810 (in Polish).

McCarthy, C. A. (1967): "c_p," *Israel J. Math.*, **5**:249–271.

Mercer, J. (1909): Functions of Positive and Negative Type, and Their Connection with the Theory of Integral Equations, *Phil. Trans. Roy. Soc. London, Ser. A*, **209**:415–446.

—— (1920): Symmetrizable Functions and Their Expansion in Terms of Biorthogonal Functions, *Proc. Roy. Soc. London, Ser. A*, **97**:401–413.

Meyer zur Capellen, W. (1933): Kleine Änderungen des Kerns einer symmetrischen, homogenen linearen Integralgleichung, *Z. Angew. Math. Mech.*, **13**:323–324.

Mikhlin, S. G. (1944): On the Convergence of Fredholm Series, *Dokl. Akad. Nauk SSSR (N.S.)*, **42**:373–376.

—— (1964): "Integral Equations and Their Applications to Certain Problems in Mechanics, Mathematical Physics and Technology," 2d ed. (translated by A. H. Armstrong), Pergamon, New York.

—— and K. L. Smolitskiy (1967): "Approximate Methods for Solution of Differential and Integral Equations," (corrected translation by Scripta Technica from the 1965 Russian ed.), American Elsevier, New York.

Moiseiwitsch, B. L. (1966): "Variational Principles," Interscience, New York.

Mollerup, J. (1923): Sur l'itération d'une fonction par un noyau donné, *Rend. Circ. Mat. Palermo*, **47**:375–395.

Moore, E. H. (1910): Introduction to a Form of General Analysis; in "New Haven Mathematical Colloquium," pp. 1–150, Yale, New Haven, Conn.

Morgan, S. P. (1963): On the Integral Equations of Laser Theory, *IEEE Trans. Microwave Theory Tech.*, **MTT-11**:191–193.

—— (1964): Note on Complex Symmetric Kernels and the Rayleigh-Ritz Procedure, *SIAM R.*, **6**:265–268.

Morse, P. M., and H. Feshbach (1953): "Methods of Theoretical Physics, Part I," McGraw-Hill, New York.

Muir, T. (1906): "The Theory of Determinants in the Historical Order of Development," 2d ed., vol. 1; reprinted (1960) by Dover, New York.

Muskhelishvili, N. I. (1953): "Singular Integral Equations," (revised translation from the 1946 Russian ed. by J. R. M. Radok), P. Noordhoff N. V., Groningen, Holland.

Mysovskih, I. P. (1959): On Error Bounds for Eigenvalues Calculated by Replacing the Kernel by an Approximating Kernel, *Mat. Sb. (N.S.)*, **49** (91): 331–340; translated in (1964), *Amer. Math. Soc. Transl.*, (2) **35**:251–262.

Naimark, M. A. (1956): Spectral Analysis of Non-Selfadjoint Operators, *Usp. Mat. Nauk (N.S.)*, **11** (6) (72):183–202; translated in (1962), *Amer. Math. Soc. Transl.*, (2) **20**:55–75.

Natanson, I. P. (1955): "Theory of Functions of a Real Variable,"(translated by L. F. Boron), Ungar, New York.

Neumann, C. (1877): "Untersuchungen über das logarithmische und Newtonsche Potential," B. G. Teubner, Leipzig.

Noble, B. (1958): "Methods Based on the Wiener-Hopf Technique," Pergamon, New York.

——— (1964): The Numerical Solution of Nonlinear Integral Equations and Related Topics; in P. M. Anselone (ed.), "Nonlinear Integral Equations," Wisconsin, Madison, Wisc.

Nohel, J. A. (1964): Problems in Qualitative Behavior of Solutions of Nonlinear Volterra Equations; in P. M. Anselone (ed.), "Nonlinear Integral Equations," Wisconsin, Madison, Wisc.

Northover, F. H. (1970): An Embedding Technique for the Solution of Linear Integral Equations, *J. Math. Anal. Appl.*, **29**:305–350.

Olagunju, P. A., and T. T. West (1964): The Spectra of Fredholm Operators in Locally Convex Spaces, *Proc. Cambridge Phil. Soc.*, **60**:801–806.

O'Malley, R. E., Jr. (1965): "Two-Parameter Singular Perturbation Problems," doctoral dissertation, Stanford, Stanford, Calif.

——— (1968): Topics in Singular Pertubations; in R. W. McKelvey (ed.), "Advances in Mathematics," vol. 2, Academic, New York.

Pal'cev, B. V. (1970): Asymptotic Behavior of the Eigenvalues . . . , *Dokl. Akad. Nauk SSSR*, **194**:774–777; translated in *Soviet Math. Dokl.*, **11**:1299–1302.

Paley, R. E. A. C., and N. Wiener (1933): Notes on the Theory and Application of Fourier Transforms, III, IV, V, VI, VII, *Trans. Amer. Math. Soc.*, **35**: 761–791.

——— and ——— (1934): "Fourier Transforms in the Complex Domain," vol. XIX, Amer. Math. Soc. Colloq. Publ., Providence, R.I.

Pao, Y-P. (1967): Fourier Transforms, Generalized Functions and Two Classes of Integral Equations, pt. 1, *Arch. Rat. Mech. Anal.*, **27**:120–132; pt. 2, ibid., **27**:133–152.

Parodi, M. (1965): Sur la convergence de la série de Neumann relative à une équation de Fredholm, *C. R. Acad. Sci. Paris*, **261**:3516–3517.

Parseval, M. A. (1805): Intégration générale et complète de deux équations importantes dans la mécaniques des fluids, *Mem. Inst. Sci. (Mem. Acad. Sci. Paris)*, **1**:524–545.

Pearson, C. E. (1957–1958): On the Finite Strip Problem, *Quart. Appl. Math.*, **15**:203–208.

Pell, A. J. (1911a): Biorthogonal Systems of Functions, *Trans. Amer. Math. Soc.*, **12**:135–164.

_____ (1911b): Applications of Biorthogonal Systems of Functions to the Theory of Integral Equations, *Trans. Amer. Math. Soc.*, **12**:165–180.

Perron, O. (1907): Zur Theorie der Matrizen, *Math. Ann.*, **64**:248–263.

Peters, A. S. (1963): A Note on the Integral Equation of the First Kind with a Cauchy Kernel, *Comm. Pure Appl. Math.*, **16**:57–61.

_____ (1968): Abel's Equation and the Cauchy Integral Equation of the Second Kind, *ibid.*, **21**:51–65.

_____ (1969): Some Integral Equations Related to Abel's Equation and the Hilbert Transform, *ibid.*, **22**:539–560.

Petrovsky, I. (1957): "Lectures on the Theory of Integral Equations," (translated by H. Kamel and H. Komm), Graylock, Rochester, N.Y.

P. I. C. C. (Provisional International Computation Center) (1960): "Symposium on the Numerical Treatment of Ordinary Differential Equations, Integral and Integro-Differential Equations," Birkhäuser Verlag, Basel–Stuttgart.

Pincus, J. D. (1966): The Spectral Theory of Self-Adjoint Wiener-Hopf Operators, *Bull. Amer. Math. Soc.*, **72**:882–887.

Pipes, L. A. (1963): "Matrix Methods for Engineering," chap. 1, Prentice-Hall, Englewood Cliffs, N.J.

Plemelj, J. (1904): Zur Theorie der Fredholmschen Funktionalgleichungen, *Monatsh. Math. Phys.*, **15**:93–128.

Pogorzelski, W. (1966): "Integral Equations and Their Applications," vol. I, (translated from the 1953–1960 Polish eds.), Pergamon, New York.

Poincaré, H. (1890): Sur les équations aux dérivées partielles de la physique mathématique, *Amer. J. Math.*, **12**:211–294; see also (1954), "Oeuvres," vol. IX, pp. 28–113, Gauthier-Villars, Paris.

Pólya, G. (1954): Estimates for Eigenvalues; in "Studies in Mathematics and Mechanics" (presented to Richard von Mises), Academic, New York.

_____ and M. Schiffer (1953–1954): Convexity of Functionals by Transplantation, *J. Anal. Math.*, **3**:245–345.

Porath, G. (1968): Störungsrechnung für lineare Volterrasche Integralgleichungen, *Math. Nachr.*, **37**:83–98.

Pringsheim, A. (1897): Über zwei Abel'sche Sätze, die Stetigkeit von Reihensummen betreffend, *Sitzungsber. Akad. Wiss. München Math. Phys. Kl.*, **27**: 343–356.

Radon, J. (1919): Über lineare Funktionaltransformationen und Funktionalgleichungen, *Sitzungsber. Akad. Wiss. Wien Abt. 2a*, **128**:1083–1121.

Rapoport, I. M. (1948): On a Class of Singular Integral Equations, *Dokl. Akad. Nauk SSSR (N.S.)*, **59**:1403–1406 (in Russian).

——— (1949): On Certain 'Twin' Integral and Integro-Differential Equations, *Sb. Tr. Inst. Mat. Akad. Nauk Ukr. RSR*, **12**:102–118 (in Russian).

Rayleigh, Lord (J. W. Strutt), (1870): On the Theory of Resonance, *Phil. Trans. Roy. Soc. London*, **161**:77–118; see also (1964), "Scientific Papers," vol. I, pp. 33–75, Dover, New York.

——— (1894): "The Theory of Sound," 2d ed., vol. 1, sects. 90 and 91, Macmillan, London; reprinted (1945) by Dover, New York.

——— (1896): "The Theory of Sound," 2d ed., vol. 2, *ibid.*.

——— (1899): On the Calculation of the Frequency of Vibration of a System in its Gravest Mode, with an Example from Hydrodynamics, *Phil. Mag.*, **47**: 566–572; see also "Scientific Papers," vol. IV, *op. cit.*, pp. 407–412.

Reid, W. T. (1931): Expansion Problems Associated with a System of Integral Equations, *Trans. Amer. Math. Soc.*, **33**:475–485.

——— (1951): Symmetrizable Completely Continuous Linear Transformations in Hilbert Space, *Duke Math. J.*, **18**:41–56.

Reissner, E. (1941): On a Class of Singular Integral Equations, *J. Math. and Phys.*, **20**:219–223.

Rellich, F. (1937): Störungstheorie der Spektralzerlegung, I, *Math. Ann.*, **113**: 600–619; II, *ibid.*, 677–685; (1939), III, *ibid.*, **116**:555–570; (1940), IV, *ibid.*, **117**:356–382; (1942), V, *ibid.*, **118**:462–484.

——— (1950): Störungstheorie der Spektralzerlegung, *Proc. Int. Congr. Math.* (Cambridge, Mass.), **1**:606–613.

———(1953a): New Results in the Perturbation Theory of Eigenvalue Problems; in L. J. Paige and O. Taussky (eds.), "Simultaneous Linear Equations and the Determination of Eigenvalues," Nat. Bur. Stand. Appl. Math. Ser. 29, Washington, D.C.

———(1953b): "Perturbation Theory of Eigenvalue Problems," Lecture Notes, Inst. of Math. Sci., New York University; reprinted (1969) by Gordon and Breach, New York.

Reudink, D. O. (1967): Convolution Transforms Whose Inversions Have the Same Kernel, *SIAM R.*, **9**:721–725.

Riemann, B. (1847): "Versuch einer allgemeinen Auffassung der Integration und Differentiation," unpublished manuscript; see also (1892), "Gesammelte Mathematische Werke," 2d ed., pp. 353–366, Teubner, Leipzig; reprinted (1953) by Dover, New York.

Riesz, F. (1907a): Sur les systêmes orthogonaux de fonctions, *C. R. Acad. Sci. Paris*, **144**:615–619; see also (1960), "Gesammelte Arbeiten," vol. I, pp. 378–381, Budapest.

——— (1907b): Sur les systêmes orthogonaux de fonctions et l'équation de Fredholm, *C. R. Acad. Sci. Paris*, **144**:734–736; see also (1960), "Gesammelte Arbeiten,"*op. cit.*, pp. 382–385.

——— (1907c): Ueber orthogonale Funktionensysteme, *Nachr. Akad. Wiss. Göttingen*, 116–122; see also (1960), "Gesammelte Arbeiten," *op. cit.*, pp. 389–395.

——— (1910): Untersuchungen über Systeme integrierbarer Funktionen, *Math.*

Ann., **69**:449–497; see also (1960), "Gesammelte Arbeiten," *op. cit.*, pp. 441–489.

―――― (1918): Über lineare Funktionalgleichungen, *Acta Math.*, **41**:71–98; see also (1960),"Gesammelte Arbeiten," *op. cit.*, vol. II, pp. 1053–1080.

Riesz, F., and B. Sz.-Nagy (1955): "Functional Analysis," (translated by L. F. Boron), Ungar, New York (principally chaps. 4–6).

Riesz, M. (1949): L'intégral de Riemann-Liouville et le problême de Cauchy, *Acta Math.*, **81**:1–223; see also (1939), "Réunion internationale des mathématiciens, Paris, 1937," Gauthier-Villars, Paris, 18 pp.

Ritz, W. (1908): Über eine neue Methode zur Lösung gewisser Variationsprobleme der mathematischen Physik, *J. Reine Angew. Math.*, **135**: 1–61; see also (1911),"Gesammelte Werke," pp. 192–250, Paris.

―――― (1909): Theorie der Transversalschwingungen einer quadratischen Platte mit freien Rändern, *Ann. Physik.*,(**4**) **28**:737–786; see also (1911), "Gesammelte Werke," *op. cit.*, pp. 265–316.

Roark, A. L., and L. F. Shampine (1965): On a Paper of Roark and Wing, *Numer. Math.*, **7**:394–395.

―――― and ―――― (1968): On the Eigenproblem for Displacement Integral Equations, *Numer. Math.*, **12**:170–179.

―――― and G. M. Wing (1965): A Method for Computing the Eigenvalues of Certain Integral Equations, *Numer. Math.*, **7**:159–170.

Rosenblatt, M. (1963a): Asymptotic Behavior of Eigenvalues for a Class of Integral Equations with Translation Kernels; in M. Rosenblatt (ed.), "Proceedings of a Symposium on Time Series Analysis," Wiley, New York.

―――― (1963b): Some Results on the Asymptotic Behavior of Eigenvalues for a Class of Integral Equations with Translation Kernels, *J. Math. Mech.*, **12**: 619–628.

Rutman, M. A. (1938): On a Special Class of Completely Continuous Operators, *Dokl. Akad. Nauk SSSR*, **18**:625–627; French translation in *C. R. Acad. Sci. URSS*.

――――(1940): Sur les opérateurs totalement continus linéaires laissant invariant un certain cône, *Rec. Math. (Mat. Sb.)*, *(N.S.)*, **8** (50):77–96.

Sakaljuk, K. D. (1960): Abel's Generalized Integral Equation, *Dokl. Akad. Nauk SSSR*, **131**:748–751; translated in *Soviet Math. Dokl.*, **1**:332–335.

Salem, R. (1954): On a Problem of Smithies, *Ned. Akad. Wetensch. Proc. Ser. A*, **57**:403–407.

Sarymsakov, T. A. (1949): On a Property of the Characteristic Numbers of an Integral Equation with a Nonnegative and Continuous Kernel, *Dokl. Akad. Nauk SSSR (N.S.)*, **67**:973–976 (in Russian).

Schatten, R. (1950): "A Theory of Cross-Spaces," *Ann. Math. Studies No. 26*, Princeton, Princeton, N. J.

―――― (1960): "Norm Ideals of Completely Continuous Operators," Springer-Verlag, Berlin.

―――― and J. Von Neumann (1948): The Cross-Space of Linear Transformations, III, *Ann. Math.*, **49**:557–582; see also J. Von Neumann (1962), "Collected Works," vol. IV, pp. 409–434, Pergamon, New York.

Schmeidler, W. (1950): "Integralgleichungen mit Anwendungen in Physik und Technik," Akad. Verlag, Leipzig.

———— (1965): "Linear Operators in Hilbert Space,"(revised translation of the 1954 German ed.), Academic, New York.

Schmidt, E. (1907): Zur Theorie der linearen und nichtlinearen Integralgleichungen, Erster Teil, *Math. Ann.*, **63**:433–476; Zweite Abhandlung, *ibid.*, **64**:161–174.

Schrödinger, E. (1926): Quantisierung als Eigenwert-problem (Dritte Mitteilung), *Ann. Physik.*, (4) **80**:437–490; see also (1928), "Collected Papers on Wave Mechanics," pp. 62–101, Blackie, Glasgow.

Schur, I. (1909): Über die charakteristischen Wurzeln einer linearen Substitution mit einer Anwendung auf die Theorie der Integralgleichungen, *Math. Ann.*, **66**:488–510.

———— (1945): Ein Satz ueber quadratische Formen mit komplexen Koeffizienten, *Amer. J. Math.*, **67**:472–480.

Schwartz, J. (1962): Subdiagonalization of Operators in Hilbert Space with Compact Imaginary Part, *Comm. Pure Appl. Math.*, **15**:159–172.

Schwarz, H. A. (1885): Ueber ein die Flächen kleinsten Flächeninhalts betreffendes Problem der Variationsrechnung, *Acta Soc. Sci. Fenn.*, **15**:315–362; see also (1890), "Gesammelte Mathematische Abhandlungen," vol. I, pp. 223–269, Julius Springer, Berlin; reprinted (1971) by Chelsea, New York.

Scott, M. R., and J. W. Burgmeier (1969): A Method for Obtaining Bounds on Eigenvalues by Solving Non-homogeneous Integral Equations, *Res. Rep.* SC-RR-69-552, Sandia Labs., Albuquerque, N. M.

Shinbrot, M. (1958–1959): A Generalization of Latta's Method for the Solution of Integral Equations, *Quart. Appl. Math.*, **16**:415–421.

———— (1960): "Difference Kernels," doctoral dissertation, Stanford University, Stanford, Calif.

———— (1964): On Singular Integral Operators, *J. Math. Mech.*, **13**:395–406.

———— (1969): On the Range of General Wiener-Hopf Operators, *J. Math. Mech.*, **18**:587–601.

————(1970): The Solution of Some Integral Equations of Wiener-Hopf Type, *Quart. Appl. Math.*, **28**:15–36.

Silberstein, J. P. O. (1953): On Eigenvalues and Inverse Singular Values of Compact Linear Operators in Hilbert Space, *Proc. Cambridge Phil. Soc.*, **49**: 201–212.

Slepian, D., and H. O. Pollak (1961): Prolate Spheroidal Wave Functions, Fourier Analysis and Uncertainity—I, *Bell Syst. Tech. J.*, **40**:43–63.

Smithies, F. (1937): The Eigen-values and Singular Values of Integral Equations, *Proc. London Math. Soc.*, (2) **43**:255–279.

———— (1939): Singular Integral Equations, *Proc. London Math. Soc.*, (2) **46**: 409–466.

_____ (1941): The Fredholm Theory of Integral Equations, *Duke Math. J.*, **8**: 107–130.

_____ (1962): "Integral Equations," Cambridge, London.

Sneddon, I. N. (1962): "Lectures on Fractional Integration and Dual Integral Equations," Appl. Math. Res. Group, North Carolina State College, Raleigh, N.C.

Sobolev, V. V. (1963): "A Treatise on Radiative Transfer," (translated from the 1957 Russian ed.), Van Nostrand, Princeton, N.J.

Stadter, J. T. (1966): Extension of an Eigenvalue Estimation Method of D. H. Weinstein, *Tech. Memo.* TG-803, Appl. Phys. Lab., The Johns Hopkins University, Baltimore, Md.

Stakgold, I. (1967): "Boundary Value Problems of Mathematical Physics," vol. I, Macmillan, New York; (1968), *ibid.*, vol. II.

_____ (1969): On Weinstein's Intermediate Problems for Integral Equations with Difference Kernels, *J. Math. Mech.*, **19**:301–307.

Stenger, W. (1966): On Poincaré's Bounds for Higher Eigenvalues, *Bull. Amer. Math. Soc.*, **72**:715–718.

_____ (1967): An Inequality for the Eigenvalues of a Class of Self-Adjoint Operators, *Bull. Amer. Math. Soc.*, **73**:487–490.

_____ (1968): On the Variational Principles for Eigenvalues for a Class of Unbounded Operators, *J. Math. Mech.*, **17**:641–648.

_____ (1969): On Perturbations of Finite Rank, *J. Math. Anal. Appl.*, **28**: 625–635.

Stinespring, W. F. (1958): A Sufficient Condition for an Integral Operator to Have a Trace, *J. Reine Angew. Math.*, **200**:200–207.

Swann, D. W. (1960): "Applications and Extensions of the Method of Wiener and Hopf for the Solution of Singular and Non-Singular Integral and Integro-Differential Equations," doctoral dissertation, Stanford University, Stanford, Calif.

_____ (1971a): Kernels with Only a Finite Number of Characteristic Values, *Proc. Cambridge Phil. Soc.*, **70**:257–262.

_____ (1971b): Some New Classes of Kernels Whose Fredholm Determinants Have Order Less than One, *Trans. Amer. Math. Soc.*, **160**:427–435.

Sz.-Nagy, B. (1946–1947): Perturbations des transformations autoadjointes dans l'espace de Hilbert, *Comment. Math. Helv.*, **19**:347–366.

_____ (1951): Perturbations des transformations linéaires fermées, *Acta Sci. Math. (Szeged)*, **14**:125–137.

Szász, O. (1922): Über den Konvergenzexponenten der Fourierschen Reihen gewisser Funktionenklassen, *Sitzungsber. Akad. Wiss. München Math. Phys. Kl.*, 135–150; see also (1955), "Collected Mathematical Papers," pp. 684–699, Dept. of Math., Cincinnati, Cincinnati, Ohio.

_____ (1928): Über die Fourierschen Reihen gewisser Funktionenklassen, *Math. Ann.*, **100**:530–536; see also (1955), "Collected Mathematical Papers," *op. cit.*, pp. 758–764.

Temple, G. (1928a): The Theory of Rayleigh's Principle as Applied to Continuous Systems, *Proc. Roy. Soc. London Ser. A*, **119**:276–293.

_____ (1928b): The Computation of Characteristic Numbers and Characteristic Functions, *Proc. London Math. Soc.*, (2) **29**:257–280. See also Temple, G., and W. G. Bickley (1933): "Rayleigh's Principle and It's Applications to Engineering," Oxford, London; reprinted (1956) by Dover, New York.

Titchmarsh, E. C. (1939): "The Theory of Functions," 2d ed., Oxford, London.

_____ (1948): "Introduction to the Theory of Fourier Integrals," 2d ed., chap. 11, Clarendon, Oxford.

Tonelli, L. (1925): Sulla convergenza assoluta delle serie di Fourier, *Rend. Accad. Naz. Lincei*, (6) **2**:145–149; see also (1963),"Opere Scelte," vol. IV, pp. 11–16, Cremonese, Rome.

_____ (1926): Sulla quadratura delle superficie, *Rend. Accad. Naz. Lincei*, (6) **3**: 357–362; see also (1960),"Opere Scelte," vol. I, *op. cit.*, pp. 432–438.

_____ (1928): Su un problema di Abel, *Math. Ann.*, **99**:183–199; see also (1963),"Opere Scelte," vol. IV, *op. cit.*, pp. 178–197.

Trefftz, E. (1933): Über Fehlerschätzung bei Berechnung von Eigenwerten, *Math. Ann.*, **108**:595–604.

Tricomi, F. (1940): Sul 'principio del ciclo chiuso' del Volterra, *Atti. Accad. Sci. Torino*, **76**:74–82.

_____ (1957): "Integral Equations," Interscience, New York.

Turnbull, H. W. (1952): "Theory of Equations," rev. 5th ed., chap. 5, Oliver & Boyd, Edinburgh.

Turner, R. E. L. (1969): A Note on Eigenvalues of Normal Transformations, *Proc. Amer. Math. Soc.*, **20**:30–34.

Ukai, S. (1971): Asymptotic Distribution of Eigenvalues of the Kernel in the Kirkwood-Riseman Integral Equation, *J. Math. Phys.*, **12**:83–92.

Ursell, F. (1969): Integral Equations with a Rapidly Oscillating Kernel, *J. London Math. Soc.*, **44**:449–459.

Vainikko, G. M. (1964): Asymptotic Evaluations of the Error of Projection Methods in the Problem of Eigenvalues, *Zh. Vychisl. Mat. Mat. Fiz.*, **4**: 405–425; translated in (1964), *USSR Comput. Math. Math. Phys.*, **4** (3): 9–36.

_____ (1965): Evaluation of the Error of the Bubnov-Galerkin Method in an Eigenvalue Problem, *Zh. Vychisl. Mat. Mat. Fiz.*, **5**:587–607; translated in (1965), *USSR Comput. Math. Math. Phys.*, **5** (4):1–31.

Vergerio, A. (1917): Sulle equazioni integrali di prima specie a nucleo non simmetrico, *Rend. Circ. Mat. Palermo*, **42**:285–302.

Visser, C., and A. C. Zaanen (1952): On the Eigenvalues of Compact Linear Transformations, *Ned. Akad. Wetensch. Proc. Ser. A*, **55**:71–78.

Volterra, V. (1896): Sulla inversione degli integrali definiti, I, *Atti Accad. Sci.*

Torino, **31**:311–323; II, *ibid.*, 400–408; III, *ibid.*, 557–567; IV, *ibid.*, 693–708; also see *Rend. Accad. Naz. Lincei*, (5) **5**:177–185. See also (1956), "Opere Mathematiche," vol. II, pp. 216–225, 226–232, 233–241, 242–254, 255–262, Accad. Naz. Lincei, Rome.

———— (1896): Sulla inversione degli integrali multipli, *Atti Accad. Naz. Lincei*, (5) **5**:289–300; see also (1956), "Opere Mathematiche," *op. cit.*, pp. 263–275.

———— (1897): Sopra alcune questioni di inversione di integrali definiti, *Ann. Mat. Pura Appl.*, (2) **25**:139–178; see also (1956), "Opere Mathematiche," *op. cit.*, pp. 279–313.

———— (1913): "Leçons sur les équations intégrales et les équations intégro-différentielles," Gauthier-Villars, Paris.

———— (1959): "Theory of Functionals and of Integral and Integro-Differential Equations," (reprint of rev. translation of 1927 Spanish ed.), Dover, New York.

Von Neumann, J. (1937): Some Matrix-Inequalities and Metrization of Matrix-Space, *Tomsk Univ. R.*, **1**:286–300; see also (1962), "Collected Works," vol. IV, pp. 205–219, Pergamon, New York.

Waraszkiewicz, Z. (1929): Remarque sur un théorème de M. Zygmund, *Pol. Akad. Umiej. Krakow Wydz. Mat. -Przy. A*, 275–279.

Wavre, R. (1943): L'itération directe des opérateurs hermitiens et deux théorèms qui en dépendent, *Comment. Math. Helv.*, **15**:299–317.

———— (1944): L'itération directe des opérateurs hermitiens, *Comment. Math. Helv.*, **16**:65–72.

Weidmann, J. (1965): Ein Satz über nukleare Operatoren im Hilbertraum, *Math. Ann.*, **158**:69–78.

———— (1966): Integraloperatoren der Spurklasse, *Math. Ann.*, **163**:340–345.

Weinberger, H. F. (1952a): An Optimum Problem in the Weinstein Method for Eigenvalues, *Pacific J. Math.*, **2**:413–418.

———— (1952b): Error Estimation in the Weinstein Method for Eigenvalues, *Proc. Amer. Math. Soc.*, **3**:643–646.

———— (1959): A Theory of Lower Bounds for Eigenvalues, *Tech. Note* BN-183, Inst. Fluid Dynam. Appl. Math., University of Maryland, College Park.

———— (1960): Error Bounds in the Rayleigh-Ritz Approximation of Eigenvectors, *J. Res. Nat. Bur. Stand., Sect. B*, **64**:217–225.

———— (1962): "Variational Methods for Eigenvalue Problems," Lecture Notes by G. P. Schwartz, Dept. of Math., University of Minnesota, Minneapolis.

Weinstein, A. (1935): Sur la stabilité des plaques encastrées, *C. R. Acad. Sci. Paris*, **200**:107–109.

———— (1937): Étude des spectres des équations aux dérivées partielles de la théorie des plaques élastiques, *Mem. Sci. Math.*, **88**:1–62.

———— (1953): Variational Methods for the Approximation and Exact Computation of Eigenvalues; in L. J. Paige and O. Taussky (eds.), "Simultaneous Linear Equations and the Determination of Eigenvalues," Nat. Bur. Stand. Appl. Math. Ser. 29, Washington, D.C.

_____(1966): Some Numerical Results in Intermediate Problems for Eigenvalues; in J. H. Bramble (ed.), "Proceedings of the Symposium on the Numerical Solutions of Partial Differential Equations," Academic, New York.

Weinstein, D. H. (1934): Modified Ritz Method, *Proc. Nat. Acad. Sci. U.S.A*, **20**:529–532.

Weyl, H. (1909): Über beschränkte quadratische Formen, deren Differenz vollstetig ist, *Rend. Circ. Mat. Palermo*, **27**:373–392; see also (1968),"Gesammelte Abhandlungen," vol. I, pp. 175–194, Springer-Verlag, Berlin.

_____ (1911): Ueber die asymptotische Verteilung der Eigenwerte, *Nachr. Akad. Wiss. Göttingen*, 110–117; see also (1968), "Gesammelte Abhandlungen," *op. cit.*, pp. 368–375.

_____ (1912a): Das asymptotische Verteilungsgesetz der Eigenwerte linearer partieller Differentialgleichungen, *Math. Ann.*,**71**:441–479; see also (1968), "Gesammelte Abhandlungen," *op. cit.*, pp. 393–430.

_____ (1912b): Über die Abhängigkeit der Eigenschwingungen einer Membran von deren Begrenzung, *J. Reine Angew. Math.*, **141**:1–11; Über das Spektrum der Hohlraumstrahlung, *ibid.*, 163–181. See also (1968),"Gesammelte Abhandlungen," *op. cit.*, 431–441, 442–460.

_____ (1915): Das asymptotische Verteilungsgesetz der Eigenschwingungen einer beliebig gestalteten elastischen Körpers, *Rend. Circ. Mat. Palermo*, **39**:1–50; see also (1968), "Gesammelte Abhandlungen," *op. cit.*, pp. 511–562.

_____ (1917): Bemerkungen zum Begriff des Differentialquotienten gebrochener Ordnung, *Vierteljahrsschr. Naturforsch. Ges. Zurich*, **62**: 296–302; see also (1968), "Gesammelte Abhandlungen," *op. cit.*, pp. 663–669.

_____ (1949): Inequalities Between the Two Kinds of Eigenvalues of a Linear Transformation, *Proc. Nat. Acad. Sci. U.S.A*, **35**:408–411; see also (1968), "Gesammelte Abhandlungen," *op. cit.*, vol. IV, pp. 390–393.

Whittaker, E. T., and G. N. Watson (1958): "A Course of Modern Analysis," 4th ed., chap. 11, Cambridge, London.

Wiarda, G. (1930): "Integralgleichungen unter besonderer Berücksichtigung der Anwendungen," Leipzig, p. 126.

Widder, D. V. (1941): "The Laplace Transform," chap. 2, Princeton, Princeton, N.J.

_____(1963): Two Convolution Transforms Which Are Inverted by Convolutions, *Proc. Amer. Math. Soc.*, **14**:812–817.

Widom, H. (1958): On the Eigenvalues of Certain Hermitian Operators, *Trans. Amer. Math. Soc.*, **88**:491–522.

_____ (1961): Extreme Eigenvalues of Translation Kernels, *Trans. Amer. Math. Soc.*, **100**:252–262.

_____ (1963): Asymptotic Behavior of the Eigenvalues of Certain Integral Equations, I, *Trans. Amer. Math. Soc.*, **109**:278–295; (1964), II, *Arch. Rat. Mech. Anal.*, **17**:215–229.

Wielandt, H. (1954): Einschliessung von Eigenwerten Hermitescher Matrizen nach dem Abschnittsverfahren, *Arch. Math.*, **5**:108–114.

_____ (1944): Das Iterationsverfahren bei nicht selbstadjungierten linearen Eigenwertaufgaben, *Math. Z.*, **50**:93–143.

_____ (1956): Error Bounds for Eigenvalues of Symmetric Integral Equations; in J. H. Curtiss (ed.), "Proc. 6th Symp. Appl. Math. Amer. Math. Soc.," Mc-Graw-Hill, New York.

Wiener, N. (1932): Tauberian Theorems, *Ann. Math.*, (2) **33**:1–100; see also (1964), "Selected Papers," pp. 261–360, SIAM and M.I.T., Cambridge, Mass.

_____ (1933): "The Fourier Integral and Certain of Its Applications," Cambridge, London.

_____ and E. Hopf (1931): Über eine Klasse singulärer Integralgleichungen, *Sitzungsber. Deut. Akad. Wiss. Berlin Phys. Math. Kl.*, 696–706; see also "Selected Papers of Norbert Wiener," *op. cit.*, pp. 361–371.

Wilf, H. S. (1970): "Finite Sections of Some Classical Inequalities," chap. 2, Springer-Verlag, New York.

Wilkins, J. E., Jr. (1944): Definitely Self-Conjugate Adjoint Integral Equations, *Duke Math. J.*, **11**: 155–166.

Willett, D. W. (1964): Nonlinear Vector Integral Equations as Contraction Mappings, *Arch. Rat. Mech. Anal.*, **15**:79–86.

Williams, W. E. (1963): A Class of Integral Equations, *Proc. Cambridge Phil. Soc.*, **59**:589–597.

Wilson, E. B., Jr. (1965): Lower Bounds for Eigenvalues, *J. Chem. Phys.*, **43**: S172–S174.

Wing, G. M. (1965): On a Method for Obtaining Bounds on the Eigenvalues of Certain Integral Equations, *J. Math. Anal. Appl.*, **11**:160–175.

_____ (1967a): Some Convexity Theorems for Eigenvalues of Fredholm Integral Equations, *J. Math. Anal. Appl.*, **19**:330–338.

_____ (1967b): On Certain Fredholm Integral Equations Reducible to Initial Value Problems, *SIAM R.*, **9**:655–670.

_____ (1968): On a Generalization of a Method of Bellman and Latter for Obtaining Eigenvalue Bounds for Integral Operators, *J. Math. Anal. Appl.*, **23**: 384–396.

Wouk, A. (1964): Direct Iteration, Existence, and Uniqueness; in P. M. Anselone (ed.), "Nonlinear Integral Equations," Wisconsin, Madison, Wisc.

Yamamoto, T. (1968): On the Eigenvalues of Compact Operators in a Hilbert Space, *Numer. Math.*, **11**: 211–219.

Zaanen, A. C. (1943): Ueber vollstetige symmetrische und symmetrisierbare Operatoren, *Nieuw Arch. Wisk.*, (2) **22**:57–80.

_____ (1946): On the Theory of Linear Integral Equations, I, II, III, IV, V, VI, *Ned. Akad. Wetensch. Indag. Math.*, **8**:91–101, 102–109, 161–170, 264–278, 352–366, 367–380.

_____ (1947): On the Theory of Linear Integral Equations, VII, VIII, VIIIa, *Ned. Akad. Wetensch. Indag. Math.*, **9**:215–226, 271–279, 320–325.

_____ (1950): Normalizable Transformations in Hilbert Space and Systems of Linear Integral Equations, *Acta Math.*, **83**:197–248.

_____ (1953): "Linear Analysis," chaps. 9–17, Interscience, New York.

Zimmerberg, H. J. (1948): Definite Integral Systems, *Duke Math. J.*, **15**: 371–388.

Zygmund, A. (1928): Remarque sur la convergence absolue des séries de Fourier, *J. London Math. Soc.*, **3**:194–196.

Name Index

Subject Index

Index of Symbols